# Evolutionary Biochemistry of Proteins

# Evolutionary Biochemistry of Proteins
*Homologous and Analogous Proteins from Avian Egg Whites, Blood Sera, Milk, and Other Substances*

### Robert Earl Feeney & Richard Gall Allison
College of Agricultural and Environmental Sciences,
University of California, Davis

**WILEY-INTERSCIENCE** A DIVISION OF JOHN WILEY & SONS,
NEW YORK • LONDON • SYDNEY • TORONTO

Copyright © 1969 by John Wiley & Sons, Inc.

All rights reserved. No part of this book may be reproduced by any means, nor transmitted, nor translated into a machine language without the written permission of the publisher.

*Library of Congress Catalog Card Number: 69-19099*
SBN 471 25685 4

Printed in the United States of America

# Preface

In undertaking the preparation of this book we were confronted with a choice of preparing a very large volume or volumes or selecting a few proteins that typify the subjects of evolutionary biochemistry. The second choice was the only feasible one in such a rapidly advancing field.

Blood plasma, milk, and egg white were selected as protein sources because of their biosynthetic and compositional relationships. Three kinds of protein, transferrins, lysozymes, and inhibitors of proteolytic enzymes were selected because each exists in all three different sources and because of their different biochemical properties and their different evolutionary relationships. The transferrins consist of quite similar proteins, all of which appear to be homologous. Their biochemical activity is the formation of a very specific complex with the metal iron. The lysozymes consist of some proteins sufficiently similar to be considered homologous and other proteins sufficiently different to be considered analogous. The lysozymes have the biochemical activity of enzymatically hydrolyzing certain glycosidic bonds. The inhibitors of proteolytic enzymes have very different structures from one another and very different evolutionary origins. The inhibitors have in common the capacity to combine with and inhibit certain proteolytic enzymes. These three different kinds of proteins appear to represent proteins which have evolved by different routes and which vary in their conservatisms insofar as replacement of amino acids during evolution.

No attempt has been made to review thoroughly the proteins of blood plasma, milk, and egg whites. An attempt has rather been made to discuss their evolutionary and comparative relationships. Obvious omissions are in the details of many of the blood plasma and milk constituents; for example, the caseins of milks are only casually described. It is hoped that the omissions, both large and small, will be regarded in the light of the primary purpose of the book, which is to show the great utility of the comparative biochemical approach to modern molecular biology.

An additional hope is that the molecular interrelationships described in this book will help to show the interrelationships between different scientific disciplines and between different organizational units of the scientific community. Perhaps we can help those in Medical, Agricultural, and Liberal Arts Schools to see the common interests and goals in their respective scientific departments!

We are deeply grateful to the many individuals who read portions of the manuscript and offered valuable comments. Particular appreciation is due to P. R. Azari, R. E. Canfield, W. Gaffield, J. Goodman, F. C. Greene, R. Haynes, R. Jenness, B. Kassell, S. K. Komatsu, Woo-Hoe Liu, J. McIntire, G. E. Means, P. Melnychyn, D. T. Osuga, J. A. Rupley, C. A. Ryan, Y. Tomimatsu, F. Utter, and J. R. Vandenheede. Special thanks are due to E. Babas, C. Brumley, J. Miller and J. Tweedie for assistance in preparing and proofreading of the manuscript.

*Davis, California*
*December 1968*

ROBERT E. FEENEY
RICHARD G. ALLISON

# Contents

1 **The Intimacies of Genetics, Evolution, and Modern Protein Biochemistry**     1

   Biochemical Information, Evolutionary Relationships, The Use of Comparative Biochemistry for Studying Molecular Function, Detection and Quantitation of Homologous and Analogous Proteins, Teleological Interrelationships of Proteins of Blood Serum, Milk, and Egg White.

2 **Chicken Egg-White Proteins**     24

   History of Chicken Egg-White Proteins, General Composition, Chemical Fractionation and Purification of Egg-White Proteins, Properties of Egg-White Proteins, Genetics of Chicken Egg-White Proteins, Biological Aspects, Food Uses, Chemical and Physical Changes in Eggs Occurring On Incubation.

3 **Egg-White Protein of Different Avian Species**     58

   General Properties of Eggs of Different Avian Species, Amounts of Proteins in Egg Whites, Immunological Comparisons, Comparative Electrophoretic Studies, Properties of Proteins of Different Species, Taxonomic Relationships.

4 **Milk Proteins**     92

   Composition, Properties of Proteins, Biosynthesis, Comparative Biochemistry, Bio-Utilization.

5  Blood Plasma Proteins                                              117

   Composition and Properties of Constituents, Comparative
   Biochemistry of Genetics, Relationship of Milk, Blood,
   and Egg Proteins.

6  The Transferrins                                                   144

   Preparation, Chemical Composition, Physical Properties,
   Comparative Properties of the Metal-Free and Metal
   Complexes of Transferrins, Biosynthetic Aspects, Biological
   Functions.

7  Lysozyme                                                           172

   The Development of our Present Knowledge of Lysozyme,
   General Physical and Chemical Properties, Enzymatic
   Activity and Active Site Studies, Evolution and Genetics.

8  Inhibitors of Proteolytic Enzymes                                  199

   General Properties, Comparative Biochemistry, Mechanisms
   of Inhibitory Activity, Possible Functions and Roles of
   Protein Inhibitors of Proteolytic Enzymes.

   References                                                         245
   Author Index                                                       273
   Subject Index                                                      285

# Evolutionary Biochemistry of Proteins

# 1
# The Intimacies of Genetics, Evolution, and Modern Protein Biochemistry

Proteins are presently considered the primary products of the genetic process. Those proteins that are enzymes synthesize all the other molecules of the species not derived directly from the environment. As in most scientific fields, there are overlapping areas in genetics, evolution, and protein biochemistry, but these areas have become so mutually involved and so beneficial to one another that sometimes only the fundamental objectives of a research program may serve to distinguish the fields. New fields have arisen and each field is being used as an important tool in the others.

The comparative biochemistry, biochemical genetics, and evolution of proteins are receiving wide attention today. The most obvious genetic variations at the molecular level, such as those of plant pigments, have been studied for many years by biological scientists. It has, however, been primarily during the last two decades that the great advances in the field of molecular evolution have been attained. At the beginning of this period stand the superb studies of Ingram and Pauling and co-workers on human hemoglobin [1-3]. The term, biochemical lesion, which these workers coined for the replacement of the single amino acid valine by glutamic acid in a variant of human hemoglobin resulting in sickle-cell anemia, has become a part of the elementary vocabulary of biology. Scientists are now working with many thousands of different genetic variants of proteins, both naturally occurring and man made. Only a little over two decades ago there was almost no activity in this field.

The current interplays between these various fields are the fascinating subjects of biology. The evidences for these interplays are easily seen from the titles of recent books: "Evolving Genes and Proteins" edited by

Vernon Bryson and Henry J. Vogel [4]; "A Molecular Approach to Phylogeny" by Marcel Florkin [5]; "Molecular Biology of the Gene" by James D. Watson [6]; "The Genetic Code: The Molecular Basis for Genetic Expression" by Carl R. Woese [7]; and "Atlas of Protein Sequence and Structure" by Richard V. Eck and Margaret O. Dayhoff [8]. Several journals are now devoted to these areas and separate departments in universities and research institutes have been formed to specialize in them.

In this book an attempt is made to show these interrelationships in the study of proteins found in three related animal fluids: vertebrate blood sera, mammalian milks, and avian egg whites. During the evolutionary development of the proteins of these fluids, several of the proteins appear to have evolved from common origins.

## BIOCHEMICAL INFORMATION

### THE DNA, RNA TO PROTEIN PROCESS

The genetic message is carried through generations encoded in the deoxyribonucleic acid, DNA. For protein synthesis, this information is transferred to messenger ribonucleic acid (m-RNA) through a template mechanism. The information is finally given to the protein synthesizing sites on ribosomes by m-RNA acting as a template for the assembly of the polypeptide chains in the proper sequence. Throughout all of these processes, the order or sequence of the message is maintained, although expressed in different ways. The final product, the polypeptide, has the sequence of amino acids dictated by the sequence of the original DNA. The biochemist usually calls the entire process the storage and transfer of biochemical information. There is, however, a further type of information built into the protein by virtue of its sequence of amino acids. It is here that the protein biochemist takes over. This will be discussed in detail in the next section. The protein biochemist is vitally interested in all the preceding events as well. A primary interest is in the relative ease of changes (from mutations) in the different purines and pyrimidines in the DNA, because these affect the relative ease of changes in amino acids in the proteins and the length of the polypeptide chain. Nevertheless, the preprotein phases of the genetic process are the property of the nucleic acid chemist, or the biochemical geneticist. No attempt is made here to cover this highly important and interesting subject. Anfinsen's superb book of a decade ago [9] describes the historical aspects and developments at that time. There are now many reviews on specialized areas [10] and more general treatments from the biological viewpoints have been done by Stebbins [11], Watson [6], and Woese [7]. A summary of the "Genetic Code" is included for reference purposes (Table 1-1).

## The Information in the Protein

The polypeptide chain, as it leaves the RNA template, is currently believed to contain all the information necessary for the subsequent formation of the biologically significant form of the protein and for its ultimate function [8–10]. The dogma is that "The primary structure dictates the secondary, tertiary and quaternary structure (conformation) in any given environment." Stated another way, taking into consideration the contributing effects of pH, buffer-binding, allosteric interactions, etc., the final conformation is entirely predicted on the sequence of amino acids in the polypeptide chains.

TABLE 1-1.

|  |  | First Letter A | | C | | G | | U | |
|---|---|---|---|---|---|---|---|---|---|
| Second Letter | A | AAA<br>AAG | Lys | CAA<br>CAG | Gln | GAA<br>GAG | Glu | UAA<br>UAG | Terminate |
| | | AAC<br>AAU | Asn | CAC<br>CAU | His | GAC<br>GAU | Asp | UAC<br>UAU | Tyr |
| | C | ACA<br>ACG<br>ACC<br>ACU | Thr | CCA<br>CCG<br>CCC<br>CCU | Pro | GCA<br>GCG<br>GCC<br>GCU | Ala | UCA<br>UCG<br>UCC<br>UCU | Ser |
| | G | AGA<br>AGG | Arg | CGA<br>CGG | Arg | GGA<br>GGG | Gly | UGA<br>UGG | Term.<br>Trp |
| | | AGC<br>AGU | Ser | CGC<br>CGU | | GGC<br>GGU | | UGC<br>UGU | Cys |
| | U | AUA | Ilu | CUA | Leu | GUA | Val | UUA<br>UUG | Leu |
| | | AUG | Met | CUG | | GUG | | | |
| | | AUC<br>AUU | Ilu | CUC<br>CUU | | GUC<br>GUU | | UUC<br>UUU | Phe |

The genetic code. The m-RNA codons are shown with their corresponding amino acids. Most of these allocations are now quite certain for *E. coli*. The "nodoc" of the t-RNA is complementary to the codon and reads in the opposite direction. The format is taken from Eck's early prediction of the nature of the degeneracy in the code. The synonym codons have a regular pattern which is highly unlikely to have arisen by chance. Most probably the degeneracy is a surviving relic of an early nonspecific function of the third nucleotide. By the time the organism had a sophisticated enough protein enzyme system to discriminate differences in this third position, it was unable to change the significance of the triplets. Any change in the translation of a widely used codon would have disrupted the structure of most of the proteins of the cell, a highly deleterious event for the organism. Reproduced by permission of the *Atlas of Protein Sequence and Structure 1967-68*, Margaret O. Dayhoff and Richard V. Eck, National Biomedical Research Foundation, Silver Spring, Maryland, (1968).

Before assumption of all functions of the protein, however, there are a series of chemical events that must occur. These all appear to happen subsequently to the removal of the polypeptide chain from the template, although absolute proof of the sequence of events appears lacking. Probably the first of these events is the addition of nonamino side groups, such as carbohydrate residues [12]. Another is the formation of cross-links between the polypeptide chains, formed by uncommon amino acids such as desmosine and isodesmosine (from lysines) in elastin [13]. Doubtlessly of most general significance is the formation of cross-links by the disulfide bonds of cystine. Primarily through the pioneering researches in C. B. Anfinsen's laboratory [14] it is now widely accepted that the correct pairing of cysteines (oxidation of cysteines to cystine) to form disulfide bonds in a protein is based upon the sequence of amino acids in a protein. An abbreviated summary of the current concept is the following: (a) the release of the polypeptide chain from the template; (b) the enzymatic formation of covalent linkages to form side chains of additional groups like carbohydrates; (c) the oxidation and correct pairing of cysteines to form cystines, and thus disulfide cross-links. Simultaneous with this process or nearly so is the assumption of the "native" conformation of the protein—the correct amount of $\alpha$-helix and foldings characteristic of the particular protein. A sketch of this process is given in Fig. 1-1.

The ultimate information carried in the protein is its biochemical function. This function may be catalytic, as in an enzyme, or some other equally important action, such as found in the various structural proteins [15]. The function of the protein is obviously the final purpose of the entire genetic process and this function is not merely a "test tube" property but rather a large number of properties that must be important in the environment of the cell. Although the sequence of amino acids in a protein is useful for taxonomy and certain studies in biochemical genetics, sequence alone is presently a completely sterile piece of knowledge. The important information is function! Many examples are possible here; one of the best known comes from studies in Yanofsky's laboratory [10] on the microbial enzyme tryptophan synthetase A. Here inactive enzymes (proteins derived from the enzyme but lacking catalytic activity themselves), from mutants are obtained and then active enzymes from "revertant" mutants are obtained (Fig. 1-2). The important aspect is that a revertant active enzyme may be obtained by a mutation giving a substitution of an amino acid at a different place in the protein than where the first substitution occurred. Thus, changing an amino acid at one position in the polypeptide chain causes the enzyme to be inactive, but an active enzyme can be obtained again from the inactive enzyme by changing another amino acid far away sequentially from the first one. There are many properties that the

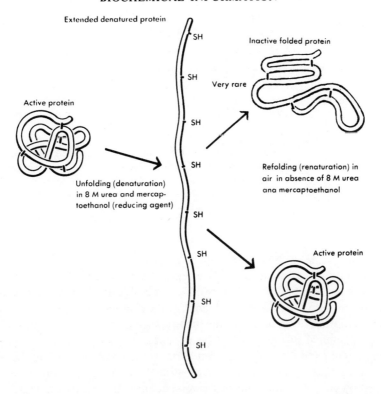

Fig. 1-1. Schematic illustration of the fate of S-S bonds during protein denaturation and renaturation. When the denaturing agents are removed, most of the polypeptide chains resume the native configuration with the original S-S bonds. Only a few polypeptide chains fold up in an inactive form characterized by a different set of S-S bonds than those found in the native molecules. Reprinted from [6]. (The reversible denaturation of a native protein is used to support the hypothesis that proteins are synthesized in an opened and denatured form and subsequently assume their native states.)

present knowledge of proteins does not allow us to deduce from only the sequence of amino acids in a protein. Proteins can have very different primary structures and yet have very similar characteristics insofar as what would appear to be a specific property. In addition proteins can have very similar primary structures and yet have very different properties. This is why a change in one amino acid, not even a part of the "catalytic site," can result in an inactive enzyme, and a further change in another amino acid remote from the first can give an active revertant.

The biochemical and physical properties of a protein are thus a critical part of the information carried in a protein. Protein chemists may eventu-

# GENETICS, EVOLUTION, PROTEIN BIOCHEMISTRY

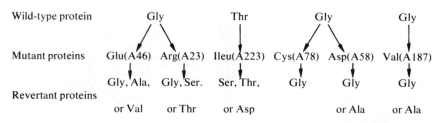

Fig. 1-2. Amino acid changes in revertant tryptophan synthetase A proteins. Each series represents changes at a different position in the protein. If two or more amino acids are listed as replacing a single amino acid, it means that different proteins are produced which are identical except for the amino acid at that position. Reprinted from [10], Academic Press, 1966.

ally be able to take the data of a primary sequence of a protein and compute its conformation in a given environment and its physical and biochemical properties. They will probably be able to do the reciprocal of this: to compute the primary sequence of the amino acids from the properties of the protein! Even when this occurs, the careful study of the physical and biochemical properties will be necessary in most studies related to biochemical genetics and evolution. Today the physical and biochemical studies offer a way of studying more sophisticated changes in a structure caused by minor changes in primary sequence as well as the significant characteristics of the functional protein.

There are a dozen or so proteins about whose primary structure and conformation a lot is known. The research to obtain this information is a combination of the determinations of primary sequence, X-ray diffraction studies, enzyme "active site" mapping by chemical and enzymatic means, and a variety of other different approaches. The following are six proteins that have received considerable attention and whose primary sequence and three dimensional structure are apparently known: hemoglobin, myoglobin, ribonuclease, α-chymotrypsin (and chymotrypsinogen), cytochrome c, and muramidase (lysozyme).

## EVOLUTIONARY RELATIONSHIPS

The two main interacting processes in evolution are mutation and natural selection. Mutations at the gene level allow for many different sequences in proteins. Selection acts upon the functioning protein whose properties are considered dependent on its primary structure. Most mutations are probably poor, or even lethal for the individual, and we do not find them expressed in living organisms; only those expressed are available for study.

## THE EVOLUTIONARY STORY IN PROTEIN STRUCTURE

The properties of proteins show the ultimate biochemical relationships of organisms. They are fundamental stones in unravelling evolutionary pathways and taxonomic relationships. The big problem is the supply of sufficient knowledge about enough proteins to solve the problems. It is desirable to make thorough examinations of a series of an homologous protein from many species, as it is being done for hemoglobin by many laboratories and for cytochrome c primarily in the laboratories of Emil Smith [16] and Emmanuel Margoliash [17]. Indeed, as recently suggested by Sibley [18] such examinations of a series of only one or two homologous proteins may provide extensive information for biochemical taxonomy. This approach, however, can give significant results with only more distantly related species or with certain groups of species. The reason for this is related to the necessity for consideration of the total function of the protein, including its interrelationships (interactions) with other proteins. The reason is also related to the amount of variation in a particular protein, i.e., the amount of conservativeness. For example, most avian egg-white proteins evidence extensive variation and are useful for studying close relationships, whereas the cytochromes are highly conservative and nearly worthless for studying close relationships, but excellent for studying distant evolutionary events.

Because of the particular biochemical function of hemoglobin in the human erythrocyte, there are many examples of the effects of variation in its sequence at a clinical level. Table 1-2 lists several symptoms or diseases found to be a direct result of the variation of amino acids in hemoglobin. As is discussed later, this is only part of the picture with the hemoglobins. A rapidly moving field is now the sophisticated study of the molecular functions and interactions of the subunits.

Cytochrome c is probably the most universal enzyme directly in the oxidation-reduction path in cells. From the work of Margoliash [17, 19] and Smith [16] and others it has been possible to map broad evolutionary pathways from plants up through mammals. Figure 1-3 is such a map. Figure 1-4 is a sequence for one of the most recent cytochrome c studies, the sequence of wheat germ cytochrome c as done by F. C. Stevens, A. N. Glazer, and E. L. Smith [16]. Figure 1-5 illustrates the constant residues presently known for all the mammalian-type cytochromes.

TABLE 1-2. HEMOGLOBIN ABNORMALITIES ARISING FROM POINT MUTATIONS

| Hemoglobin | Position + | From | To |
|---|---|---|---|
| | | Alpha Chain | |
| J Toronto | 5 | Ala | Asp |
| I Texas | 16 | Lys | Glu |
| Sealy | 47 | Asp | His |
| G Philadelphia | 68 | Asn | Lys |
| $X^b$ | 68 | Asn | Lys |
| Stanleyville 2 | 78 | Asn | Lys |
| M Kankakee | 87 | His | Tyr |
| J Cape Town | 92 | Arg | Gln |
| O Indonesia | 116 | Glu | Lys |
| | | Beta Chain | |
| Tokuchi | 2 | His | Tyr |
| $S^a$ | 6 | Glu | Val |
| $X^b$ | 6 | Glu | Lys |
| C | 6 | Glu | Lys |
| C Harlem[c] | 6 | Glu | Val |
| Genova | 28 | Leu | Pro |
| Hammersmith | 42 | Phe | Ser |
| M Saskatoon | 63 | His | Tyr |
| Zurich | 63 | His | Arg |
| C Harlem | 73 | Asp | Asn |
| Agenogi | 90 | Glu | Lys |
| Hopkins 1 | 95 | Lys | Glu |
| Koln | 98 | Val | Met |
| Kansas | 102 | Asn | Thr |
| New York | 113 | Val | Glu |
| D Los Angeles | 121 | Glu | Gln |
| O Arabia | 121 | Glu | Lys |
| K Cameroon | 129 | Ala | Glu or Asp |
| Kenwood | 143 | His | Asp |

[a] Changes associated with sickling of red cells.
[b] Abnormalities in alpha and beta chains.
[c] Two abnormalities in one chain.

## Divergence and Convergence — Homology and Analogy

The differentiation between divergence and convergence has probably been one of the most troublesome problems in studies on evolution. Divergent evolution results in adaptive changes in a structure to different structures, whereas convergent evolution results in adapted changes of different (and unrelated) structures to structures resembling one another. In this sense homologous proteins would have similar genetic origins and arise from divergent evolution. Homologous proteins probably arise from gene duplication. The duplicate genes would initially produce the same

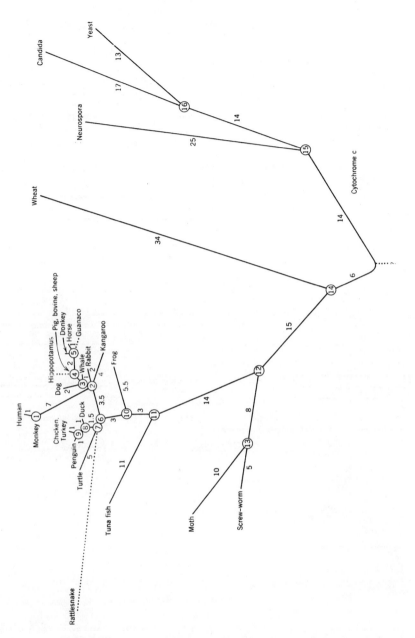

Fig. 1-3. Phylogenetic tree of cytochrome c. The topology has been inferred from the sequences as explained in the text. The numbers of inferred amino acid changes per 100 links are shown on the tree. The point of earliest time cannot be determined directly from the sequences; we have placed it by assuming that on the average, species change at the same rate. Reproduced by permission of the *Atlas of Protein Sequence and Structure 1967-68*, Margaret O. Dayhoff and Richard V. Eck, National Biomedical Research Foundation, Silver Spring, Maryland, (1968).

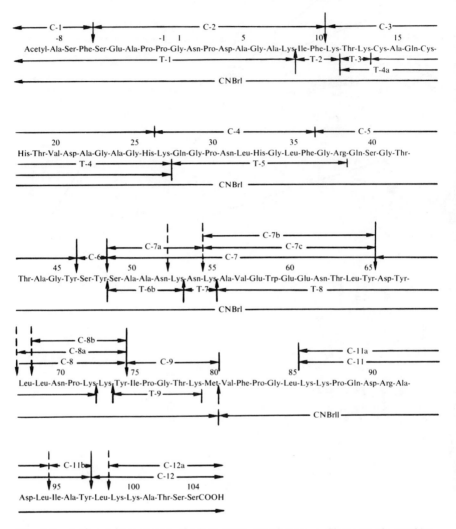

Fig. 1-4. Amino acid sequence of wheat germ cytochrome c. Chymotryptic peptides, tryptic peptides, and cyanogen bromide peptides are indicated. Reprinted from [16].

protein. A mutation in either of the genes would be the first step toward the development of homologous proteins. Analogous proteins would have different genetic origins and arise from convergent evolution. The semantics therefore are ones of origin.

Florkin [5] has recently discussed these definitions as well as isologous, a term useful in taxonomical matters (Fig. 1-6). Isologous compounds have certain similarities in chemical structure, "chemical kinship," regardless of the genetic origins or sources. Thus a particular lipid from an animal and a microorganism would be isologous, although the lipid might be synthesized by different enzymatic routes in the different organisms.

In the case of some of the proteins discussed in this book it is sometimes difficult to differentiate between analogous and homologous proteins. A few seem undoubtedly to be analogous, such as the proteolytic enzyme inhibitors from the soybean, avian egg whites, and blood serum. In other cases, some proteins such as the transferrins in blood serum, milk, and egg white (Table 1-3) appear to be homologous although their evolutionary origins remain as yet unknown. As a general practice, proteins might be considered homologous when they have similar functional properties and many physical-chemical properties in common and similar genetic origins appear probable. This can obviously lead to erroneous conclusions. Rigorous proof of homology or analogy requires knowledge of amino acid sequence, function, source of material, and other information of a biological nature. The biochemist needs the assistance and cooperation of the biologist in such decisions.

One of the more interesting aspects of these analogous and homologous proteins is their occurrence in such widely dispersed biological systems. Thus one animal can have a variety of homologous and analogous examples of a biologically active protein. For example, lysozyme is found in

```
-8                      1           5              10
- - - - - - - - Gly - - - - Gly - - - Phe - - -

  15              20          25            30
Cys - - Cys.His - - - - - - - - Lys - Gly.Pro - Leu - Gly
 └─Heme─┘

35              40          45            50             55
- - - Arg - - Gly - - - Gly - - Tyr - - Ala.Asn - - - -

        60          65          70            75
- - Trp - - - - - - - - Tyr.Leu - Asn.Pro.Lys.Lys.Tyr.Ile.Pro.Gly.

        80          85          90            95
Thr.Lys.Met - Phe - Gly - - Lys - - - Arg - - - - - - - -

100         104
- - - - -
```

Figure 1-5. Constant residues presently known from all the species of mammalian-type cytochrome c which have been investigated. (16). *J. Biol. Chem.* 242, 2764 (1967).

## BASIC CONCEPTS

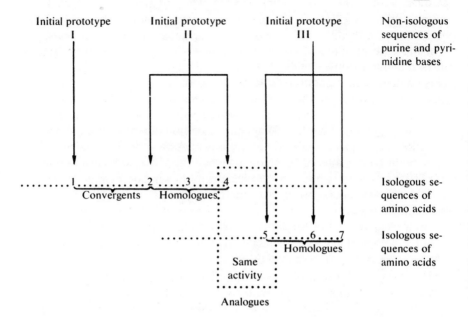

Fig. 1-6. Isology, homology, analogy and convergence. The roman figures designate sequences of purine and pyrimidine bases. The arabic figures designate sequences of amino acids. Reprinted from [5]. "A Molecular Approach to Phylogeny," Elsevier, New York (1966).

TABLE 1-3. DISTRIBUTION OF LYSOZYMES, "TRYPSIN" INHIBITORS, AND TRANSFERRINS

| Source | Protein | | |
| --- | --- | --- | --- |
| | Lysozyme | "Trypsin" inhibitor | "Transferrin" |
| Avian egg white | + | + | + |
| Blood serum | + | + | + |
| Milk | + | + | + |
| Pancreatic secretion | + | + | − |
| Human tears | + | − | − |
| Seminal fluid | − | + | − |
| Bacteriophage | + | − | − |
| Some plants | + | + | − |

human tears, blood serum, milk, pancrease and, apparently, at low levels in many body tissues. These lysozymes are probably homologous; in turn, they exist in similar organs, tissues, and secretions in many other animals, and also exist in widely different places as plants, the tail of some bacteriophages, and in avian egg white. Some of these lysozymes from widely different sources may well be analogous; they will be discussed in Chapter 7. Examples of obviously analogous proteins in the same species are some of the inhibitors of proteolytic enzymes. They occur in blood serum, milk, egg white, pancrease, and other organs or fluids of animals and birds. There are examples of obviously homologous proteins with important differences in functional activity but produced in the same organ; the pancreatic proteolytic enzymes trypsin and $\alpha$-chymotrypsin are such proteins [20]. Then there are examples of homologous proteins as polypeptide chains that are associated with one another physically and functionally and yet the homology is not obvious. Among the best known of these are hemoglobin molecules consisting of four noncovalently associated subunits, two $\alpha$ and two $\beta$ chains (Table 1-2). The $\alpha$ and $\beta$ chains are homologous and the tetramer is thus composed of two sets of homologs. Another example of this type is the mammalian gamma globulin molecule that is a tetramer, but it consists of four *covalently* linked subunits (and also two sets of homologs) to form a molecule with a molecular weight of approximately 150,000 g. These homologs are the heavy and light chains with molecular weights approximating 50,000 g and 25,000 g, respectively. There is still, however, a further type of homology, an intrachain homology. The homologous heavy and light chains have homologous characteristics but, in addition, there are homologous sequences repeated in the individual peptide chains of each. In fact the ancestral unit is a small protein with a molecular weight of approximately 12,000 g.

An intrachain homology also appears to exist in the transferrins [21], which are one of the protein groups discussed in detail in subsequent chapters.

## Multiple Molecular Forms of Proteins

The existence of multiple forms of proteins has interested many biochemists and biologists during the past decade [22–25]. Table 1-4 lists the main types of multiple forms of proteins recognized today. Inciting factors have been the findings of multiple forms of enzymes in the same tissues and the subsequent coining of the term isozyme by Markert. Isozyme is now a common term in biochemical vernacular and it is well recognized that many multiple forms of proteins exist. The existence of several forms of several proteins has, however, been recognized for many years. Some of the earliest observations were on those forms caused by the presence or absence of prosthetic groups carrying a charge. A classical example was the early observations that ovalbumin from chicken egg white existed in three forms, $A_1$, $A_2$, and $A_3$, which differed by the number of phosphates per molecule—2, 1, or no phosphates [26, 27]. During the past fifteen years attention has been given to the existence of multiple forms due to small differences in primary amino acid structure rather than to different amounts of prosthetic groups. Colvin, Smith, and Cook [28] focused attention on the subject by noting that many proteins show heterogeneity at low ionic strength and introduced the term of microheterogeneity. Although it is not generally believed that heterogeneity is a general property of proteins, it is well recognized that many multiple forms occur commonly.

TABLE 1-4.  MAIN TYPES OF MULTIPLE MOLECULAR FORMS OF PROTEINS

| |
|---|
| (1) Difference in primary structure (isozymes) |
|     (a) Difference in prosthetic groups |
|     (b) Difference in A.A. of monomer |
|     (c) Polymer of different units |
| (2) Polymers of subunits |
| (3) Conformational forms |

In addition to the multiple forms due to differences in primary structure of the fundamental protein unit, there are multiple forms due to other reasons. One type of multiple form is caused by the formation of polymers from either the same fundamental protein unit or from two (or possibly more) different protein units [23]. If the difference between the monomer and polymer is due to the formation of polymers of the same subunit, the fundamental differences observed will be those due to a difference in size, although there may be differences due to a conformational change. If, however, the polymer is formed from different subunits there will be differences in the size and in physical-chemical properties, depending on the properties and proportions of the subunits in the polymer.

Another type of multiple molecular form is due to molecules of proteins with the same primary structure existing in several physical-chemical forms as influenced by the environment. These are termed "conformational forms" [29]. The work of Foster and co-workers [30] on serum albumin is now one of the classical examples. These workers found that serum albumin existed in several conformational forms and that there was a progressive transition of one form to the other over a narrow pH range at approximately pH 4. Sogami and Foster [31] have suggested that serum albumin exists in several multiple forms due to differences in primary structure. Each of these forms is capable of undergoing this transition at low pH to give a large number of transitional forms. There are also many other such examples reported. Both chicken egg-white ovomucoid and chicken egg-white ovotransferrins, as will be discussed later, undergo extensive transition in conformation below pH 4.

Recently there has been much attention given to biological control mechanisms and the relationship of multiple molecular forms of enzymes [32], [33]. The multiple forms of enzymes are controlled to different degrees by the other interacting substances such as the end products of the reaction. One of the interesting examples of this phenomenon is the recent observation of Hathaway and Criddle [34] on the association and dissociation of lactic dehydrogenase into subunits with enzymatic properties differing from that of the tetramer. These dissociations are controlled by concentrations of substrate and products. These observations provide an interesting example of a teleological function of multiple molecular forms. Nevertheless, it is still hard to understand why many multiple molecular forms persist. For example, the cassowary has five or six multiple forms of its egg-white ovotransferrin. No function has yet been proven for the ovotransferrin itself.

BIOSYNTHETIC ASPECTS

In any consideration of homologous proteins in an individual organism, the biosynthetic origins of the different homologs must be considered. The same applies to multiple molecular forms of a protein. In other words, how are these different forms of the protein synthesized and what are their genetic origins?

There would appear to be a number of means by which homologous proteins may develop in different organs or fluids of one individual:

1. The first and most obvious involves a common synthetic site and a transport of the proteins to the different parts of the organism; for example, one may be the "leakage" of blood proteins across membranes of the mammary gland into colostral milk. Such proteins in blood and milk would not be homologs because they are structurally identical.

2. In the second type the proteins are passed by leakage or secretion and are chemically modified. Such a modification might be the addition or subtraction of a prosthetic group. These proteins would be homologous. No documented examples appear to be available, however.

3. In the third type the apparently homologous forms are coded by the same gene but the biosynthesis is directed independently in different or-

gans. It has been suggested that this is true of the serum transferrin and the ovotransferrin in chicken egg white [35].

4. In a fourth type the synthetic processes in the different organs are unrelated and are coded by different but related genes.

The biosynthesis of multiple molecular forms due to differences in primary structure or prosthetic groups is also poorly understood at this time. One good possibility, however, is that such multiple forms are controlled more or less independently by different genes originating from duplicate genes. In such a situation a mutation in one of the duplicate genes might cause the formation of a multiple form. Such multiple forms would obviously also be homologous. Additional mutations could result in sufficient changes in properties that the two proteins would no longer resemble each other sufficiently to be classified as multiple forms. Another concept is that multiple forms due to differences in primary amino acid structure may result from errors in the biosynthetic mechanism. According to this idea, under certain conditions one amino acid might be substituted for another at the time of the biosynthesis of the protein at the template, or perhaps there might be errors involving the RNA. The percentage of the different forms would then be a statistical matter related to the probability of the occurrence of the error [36]. This concept is not popular at this time.

## THE USE OF COMPARATIVE BIOCHEMISTRY FOR STUDYING MOLECULAR FUNCTION

The protein biochemist is not only interested in collaborating with the geneticists for studies in genetics and evolution, but he is also interested in inherited differences in the structures of proteins for the sake of studying molecular function. His objective is to relate differences in molecular structure of proteins to differences in their biochemical or physical properties. He thus uses the differences as a tool for understanding molecular function. This approach is now frequently complemented by other commonly used procedures. These include studying the effects of variations in the structures of enzyme substrates and coenzymes, as well as the effects of physical treatments and chemical modifications of the enzymes. As all protein chemists know, serious drawbacks to physical and chemical modifications are that they are frequently nonspecific and cause unintended side reactions. A principal advantage in using the differences provided by nature is that they are "built in" modifications, rather than made by the chemical or physical manipulations of the investigator. A serious drawback to the use of genetic variants is that the modification provided by nature is usually unknown, at least initially, and may require considerable research to determine the structural differences. The two problems can go hand in hand — providing evolutionary information and providing information about the protein.

One of the earlier applications of comparative biochemistry to understanding function was the studies of Linus Pauling and co-workers on the variant of human hemoglobin causing sickle-cell anemia. There are hundreds of such applications of comparative or genetic biochemistry, although most of them have not been studied to the depth that hemoglobin has been studied. The hemoglobin studies include investigations of the properties of the tetramers formed from variants of $\alpha$- and $\beta$-chains. Itano and co-workers [37, 38] studied the reversible dissociations of HbI ($\alpha_2^I \beta_2^A$) and HbS ($\alpha_2^A \beta_2^S$) and obtained evidence for hybrids ($\alpha_2^A \beta_2^A$ and $\alpha_2^I \beta_2^S$). Other similar hybrids have since been found [10]. An abnormal human hemoglobin (H) results from the association of four $\beta$-chains to form ($\beta_4$). Recently many interesting studies have been made of the formation of tetramers from $\alpha$-chains of one species and the $\beta$-chains of another. The degrees of "fitting" in the tetramer and the responses to oxygenation are indicative of structural variation (and taxonomic relationship) and are providing very useful information for understanding function.

The microbiologists' mutants have proven a powerful approach to this field [10]; for example, when the thermostabilities of different mutationally altered A proteins of tryptophan synthetase were examined, both more heat stable and less heat stable proteins were found. In two particular cases in which the heat stabilities were changed and amino acid substitutions were determined, the substitution involved a change in the hydrophobic nature of the residue. This is in agreement with the current belief that hydrophobic forces are important for the stabilization of proteins in aqueous solutions.

Kaplan [39] has shown differences between the isozymic forms of lactic dehydrogenase by the use of analogs (derivatives) of the normal nicotinamide coenzyme. The different isozymes have different specific activities with the coenzyme analogs. Table 1-5 lists the ratios of enzymatic activities of the $H_4$ and $M_4$ isozymes of the lactic dehydrogenases of different vertebrates when the normal coenzyme DPN is substituted by the acetyl pyridyl derivative. This means that the "fitting" and other more delicate biochemical relationships may vary considerably with small changes in structures of enzymes and may be difficult to observe by conventional procedures. These studies have suggested that much of the older work comparing enzymes, either isozymes or homologous enzymes of different species, now might be profitably reexamined. Most comparative studies

TABLE 1-5. CATALYTIC PROPERTIES OF VERTEBRATE LACTIC DEHYDROGENASES [39]

| Species | AcPyDPN/DPN | |
|---|---|---|
| | $H_4$ | $M_4$ |
| Chicken | 0.15 | 1.1 |
| Turkey | 0.17 | 2.0 |
| Duck | 0.18 | 0.8 |
| Pigeon | 0.11 | – |
| Caiman | 0.07 | 0.7 |
| Turtle | – | 0.6 |
| Bullfrog | 0.18 | 0.3 |
| Sturgeon | 0.08 | 0.2 |
| Halibut | – | 1.0 |
| Mackerel | 0.12 | 0.7 |
| Dogfish | 0.26 | 0.9 |
| Lamprey | – | 0.2 |
| Cow | 0.06 | 0.3 |
| Man | 0.04 | 0.2 |
| Rabbit | 0.05 | – |

Reprinted from *Evolving Genes and Proteins*, p. 243, Academic Press, New York, 1965.

have been done with the normal coenzymes and these studies should be repeated by the approach that Kaplan and other workers have employed.

During the last decade the "serine" proteolytic enzymes have been intensively studied. A review article on the "Evolution of Structure and Function of Proteases" [20] has recently appeared. In the animal pancreas one of the very interesting examples of homology is found: two homologous proenzymes activated to give such different specificities that their homology was once not considered probable—trypsin and α-chymotrypsin. Not only are the zymogens, trypsinogen, and chymotrypsinogen activated by the same enzyme (trypsin), but the mechanism of activation is similar, the hydrolytic cleavage of a peptide bond. There is little doubt that these enzymes had a common ancestor and that they probably arose through gene duplication. These two homologous enzymes with different specificities have given the unique opportunity for the study of the relationships between their differences in structure and function. Two other enzymes, elastase and α-lytic protease, also show homologies with trypsin and α-chymotrypsin (Fig. 1-7).

Early in the 1950's one of us became interested in the use of comparative and genetic biochemistry to study protein functions. In fact his primary purpose in starting a program at the San Diego Zoological Gardens in 1958 was to find transferrins with different iron-binding characteristics.

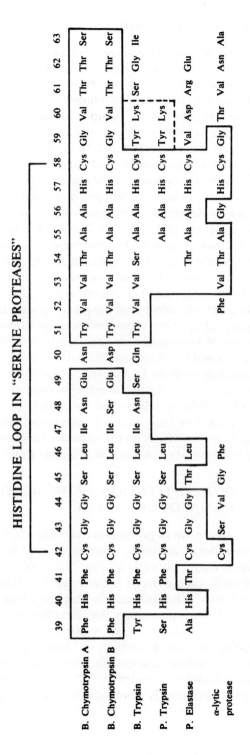

Fig. 1-7. The amino acid sequences about the active histidine residue (No. 57) in a variety of serine proteases. The solid lines enclose regions of general homology; the broken lines enclose a region of limited homology. In all these cases, a disulfide bridge between cysteines 42 and 58 form a loop. The residue numbering is that for bovine chymotrypsin A. The deletions in the last three have been made to maximize the homology. Reprinted from [20].

The ones studied to that time were the chicken ovotransferrin and human, bovine, and porcine serum transferrin. These all had closely similar molecular weights, adsorption spectra, and capacities for binding strongly two atoms of iron. The irons were thought to interact. It was hoped to find transferrins that did not have the "magic number" of two as regards the iron-binding in order to use this difference as a tool for studying the possible interaction of the iron. This did not prove to be the case, however, with the ovotransferrins of a large number of different species examined; i.e., they all had quite similar properties. Very recently, however, it has been found that the irons do not interact [40] and it now appears that the transferrin molecule is a single peptide chain consisting of two nearly similar parts [21]. An unexpected finding in the Zoo work concerned the avian egg-white ovomucoids. This was the observation that the ovomucoids of the egg whites of some avian species had inhibitory activity against $\alpha$-chymotrypsin, in contrast to the well-studied chicken ovomucoid that was an inhibitor of trypsin. This, in retrospect, now seems obvious when one considers the discussions of the homologous relationships of trypsin and $\alpha$-chymotrypsin. One should expect sophisticated differences in the interactions of homologs with homologs! These interesting relationships will be discussed in subsequent chapters.

More recently a research program in Antarctica has made it possible for our laboratory to initiate a study of the proteins from cold adapted Antarctic fish [41]. These fish are adapted to the $-1.8°C$ of McMurdo Sound by the Ross Ice Shelf and are unable to survive at temperatures about 10°C [42]. Studies are underway on the structure-function relationships of blood and muscle proteins as a function of temperature.

## DETECTION AND QUANTITATION OF HOMOLOGOUS AND ANALOGOUS PROTEINS

Because the various properties of proteins are capable of a large number of relatively independent variations, determination of homologous and analogous proteins is frequently very difficult. The amounts of homologous proteins in different materials can some times be estimated from the electrophoretic patterns. The difficulty is that the electrophoretic mobilities of the different constituents can change sufficiently from species to species that different components can overlap one another electrophoretically in one species and not in another, and hence give absolutely incorrect results both qualitatively and quantitatively. As in any type of quantitative analysis a standard or a reference of some type is required. A standard or reference with proteins is more difficult in some analyses because of the large variation in the properties of proteins and a possibility of interference by other proteins of very similar properties or other types of chemical substances. One of the most desirable standards to have is a standard of the pure protein itself. It would be desirable to isolate each of the protein constituents undergoing analysis from each source of each species. Then some specific property of this protein would be used to analyze quantitatively the original material and to refer this analysis to the values obtained with the purified protein. In lieu of this very formidable task, a reference or standard may be employed providing the danger of the assumptions involved are continually realized.

Such an approach in the case of the avian egg-white proteins has been employed in our laboratory. In these analyses, the chicken egg-white proteins are used as standards for the egg-white proteins of other avian species. In most all cases, attempts are also made to check the validity of this procedure by observing a variety of properties. The problems involved are many, for example, if one assumes that the chromogenic capacity of the ovotransferrins with iron is the same for the ovotransferrins of all egg whites, one must also assume that the molecular weights are the same. We would not expect this to be the case as certainly minor differences in molecular weights must undoubtedly exist through the deletion or addition of amino acids. Minor weight differences would be negligibly important in comparisons with such biochemical assays, but larger differences must always be considered possible. Similarly large errors may also be encountered by using adsorption at 280 m$\mu$ or colorimetric tests, because these values may vary greatly with differences in a few amino acids. Probably of greater importance would be differences in biochemical activity or specificity. This could be quite possible with an enzyme such as lysozyme, which might have an altered rate of reaction with the substrate as a result of relatively minor alterations in structure. An assay based on chicken lysozyme as a standard would correspondingly show more or less lysozyme than would actually be present. It is obvious that the only way to solve this problem would be to make an independent determination on the lysozyme activity. This by present-day procedures would undoubtedly require isolation and purification as described above. In fact, this identical approach with the lysozyme of kiwi egg white has shown that kiwi lysozyme assays differently from chicken egg-white lysozyme. As discussed in Chapter 7, there is no good assay available for the determinations of kiwi lysozyme. An even more striking example of the effects of differences in biological activity is the enzyme inhibitory activity of golden pheasant ovomucoid. Chicken ovomucoid is an inhibitor of trypsin, and the assay for ovomucoid based on chicken ovomucoid as a standard is the determination of its trypsin-inhibitory activity. This assay would be totally unsuitable for golden pheasant ovomucoid which is an inhibitor of $\alpha$-chymotrypsin rather than trypsin [43].

## TELEOLOGICAL INTERRELATIONSHIPS OF PROTEINS OF BLOOD SERUM, MILK, AND EGG WHITE

One of the first questions asked about many of the interesting protein constituents is what function do they serve. Unfortunately in many cases this is impossible to state with the knowledge presently available and in many of the remaining cases we can only make broad guesses. This is not unusual to proteins, as any student of comparative and evolutionary biology can certify. In many cases it is also difficult to find the functions of anatomical structures. In the case of proteins it is even more difficult, possibly because we know the properties of proteins primarily as they are isolated and studied in test tubes, rather than *in situ* in an animal or plant tissue. One must also remember, however, that the particular proteins discussed are somewhat unique and perhaps there are some unrecognized evolutionary or developmental interrelationships that influence their distribution.

Most of these proteins are primarily distributed in materials directly associated with nutrition or development. Milk and egg white are obviously in these categories; they serve as nutritional or protective materials for the developing young animal. Even the lysozyme associated with the tail of the bacteriophage might be considered in this category because it is essential for the escape of the phage from the bacterial cell. Blood serum proteins have, of course, a variety of functions, but one of their most important functions is as a transporting and distributing vehicle for the nutrients in the animal body. Several of the proteins to be discussed are found in plants, where they are usually isolated from the seeds. This again may show a relationship to developmental processes. Closely related to the teleological aspects are the genetic and biosynthetic aspects. Why should these homologous proteins find their way to such diverse tissues? Do they have similar functions in different tissues? What are the survival factors that lead to differences in structure or biochemical activities in different tissues? Who would have dared guess a few years ago that the $\alpha$-lactalbumin of cow's milk would now be considered homologous with mammalian and chicken egg-white lysozymes?

Examples of differences in structure and biochemical activity within a homologous series are found in many places. One example is the difference in the pH at which the metal ion dissociates from the transferrins. The metals of the complexes of bovine and human lactotransferrins have been found to dissociate at a much more acid pH than for the serum transferrins and the ovotransferrins [40]. Why should the metals of the milk transferrins dissociate at a lower pH? An obvious teleological interpretation would implicate the acidity of the stomach. This does not appear to be a proper one because the acidity of the stomach is much greater than that at which the metal begins to dissociate. Another example is the rather extensive differences in the isoelectric points of the ovotransferrins in some of the avian species. Many ovotransferrins have an isoelectric point of approximately pH 6, whereas the cassowary ovotransferrin has an isoelectric point of pH 9. A third example of differences in biological function is the difference found in the specificity of the ovomucoids of different avian species. The ovomucoid of the chicken is apparently primarily an inhibitor of bovine trypsin, whereas the ovomucoid of many other birds also inhibits chymotrypsin. In addition, in two cases the ovomucoid apparently inhibits chymotrypsin and trypsin very weakly or not at all. There are many other examples impossible to explain satisfactorily with our present knowledge. This is, indeed, an area in which biochemists will have the opportunity of helping the biologists by unraveling these problems.

# 2
# Chicken Egg-White Proteins

Chicken egg-white proteins have been the subject of numerous fundamental investigations by competent protein chemists. Their primary interest in egg white has been due to the interesting properties and easy availability of these proteins as well as to its importance as embryological material. In contrast to the requirements for obtaining many proteins from animal sources, eggs, milk, or blood may be obtained from an animal without injuring it. Eggs are even easier to obtain than milk or blood because the bird does not need to be handled in order to obtain an egg. In the case of chicken eggs, the situation is excellent for the biochemist because of the genetic control of commercial layers and the availability of large numbers of eggs from the same individual bird. As a consequence several of its proteins have become standard proteins in biochemical research. Most widely known are the ovalbumin, avidin, and the muramidase (lysozyme), but other proteins such as the ovomucoid and ovotransferrins have been receiving attention in many different studies more recently. In most of the studies on all five of these proteins, the main interest has apparently been on the use of the protein to study fundamental problems in protein biochemistry. All five of the proteins have properties that are quite different from one another.

## HISTORY OF CHICKEN EGG-WHITE PROTEINS

The history of the egg-white proteins is closely allied with the history of protein chemistry itself. This is because, in common with milk and blood proteins, egg proteins have been one of the primary sources of materials for study by protein chemists. There was considerable work done before

1900 and several of the principal protein constituents were named and identified in the order of their solubility by Eicholz in 1898 [44]. These were, in the order of decreasing solubility, ovomucoid, ovalbumin, and ovomucin. In 1900 Osborne and Campbell [45] described the presence of ovomucin, crystallizable ovalbumin, noncrystallizable conalbumin, and noncoagulable ovomucoid. Approximately thirty years later, Hektoen and Cole [46], and Sorenson [47], reported on the globulins and described refinements in fractionation with ammonium sulfate.

The first definitive separation and identification of the proteins were given in the classical report of Longsworth, Cannan, and MacInnes [26]. These investigators applying the Tiselius apparatus for free boundary electrophoresis reported eight constituents as compared to five or six by previous investigators. In addition, they identified two ovalbumin constituents, which they named $A_1$ and $A_2$, and reported the presence of three globulins as $G_1$, $G_2$, and $G_3$.

Investigations since then have been primarily concerned with refinements of isolation and analysis, and identification of biological or biochemical activities in the egg white. One of the outstanding contributions was the direct crystallization from egg white of lysozyme (muramidase) in a greater than eighty per cent yield by Alderton and Fevold in 1946 [48]. Lysozyme activity had been reported in egg white, human tears, and other materials in 1922 by Sir Alexander Fleming [49]. It had been provisionally identified as one of the globulins before the work of Alderton and Fevold but had been obtained only in very small yield. It is probable that in 1946 lysozyme was the most easily obtainable crystalline enzyme from an animal source. Lysozyme has since been named officially muramidase.

Another biological activity discovered early was identified eventually as due to the avidin. In 1924 Boas [50] reported that inclusion of dried egg white in the diet of rats caused poor growth of the rats. Other symptoms observed by Boas and others included evidence for nervous disorders and eventually death. The phenomenon was called egg-white injury, and in some circles at least, the material in the egg white was reputedly called Chinese dried-egg-white injury factor, because the source of the egg white was from China. It was later discussed by Parsons and Kelly [51] that the symptoms could be also caused by fresh egg white and that there was a material in egg yolk which would counteract or prevent the disease when fed along with the egg white. The identification of the nature of the reaction and the compounds involved were reported in 1941 by Eakin and co-workers [52]. As is now well known, the reaction was caused by the trace protein avidin complexing with the B vitamin biotin in the diet and causing a deficiency thereof. The protein was named avidin because it avidly formed a complex with biotin. Of the various names that have been given to biologically active proteins avidin appears to have been accepted and worn as well as any. Another biochemical activity in egg white that was known for many years was a capacity to inhibit the digestive enzyme, bovine trypsin. Although various workers had suggested that it might be ovomucoid which had this activity, it was not until the more definitive and extensive researches of Lineweaver and Murray in 1947 [53] established that the ovomucoid is the protein responsible for the trypsin-inhibitory capacity.

Conalbumin (ovotransferrin) had been known for approximately a decade before any biochemical activity was attributed to it. This biochemical activity was the capacity to bind iron and form a rather intense salmon-pink coloration in egg white. Hindsight here makes it very difficult to see how such a simple reaction as this could have been missed. On the other hand we must realize that protein chemists were accustomed to discolorations when metal ions were added to crude protein mixtures and any such reaction with iron in egg white might have been discounted. The discovery of the iron-binding activity came in a round-about way again with the discovery of a biological activity in 1944 by Schade and Caroline [54]. In studying the antibacterial activity of egg white, they found an inhibition that was apparently not due to avidin because the inhibition of certain organisms was not prevented by the addition of biotin but could be prevented by the addition of iron to the egg white. Two years later, part of a team at the Western Regional Research Laboratory of the U.S. Department of Agriculture reported the identification of the bacteria-inhibiting iron-binding protein of egg white as the conalbumin (ovotransferrin) [55]. This was apparently the first example of a specific iron-binding protein which had bacterial inhibitory capacities. As described in Chapter 6 the ovotransferrin has since been found nearly identical to the blood serum protein, transferrin, and we are now using the term ovotransferrin in place of the older term conalbumin.

## Recent Findings

During the last decade several other biochemically active proteins have been identified in chicken egg white. The first of these proteins separated and identified was named ovoinhibitor in 1958 by Matsushima [56]. This protein was an inhibitor of proteolytic enzymes and thus similar in this respect to ovomucoid. It was different from ovomucoid in that it inhibited fungal and bacterial enzymes as well as trypsin. In 1959 [57] our laboratory reported a riboflavin-binding protein. Like avidin the discovery of such a protein was apparently another first for "eggology." It was the first identified protein that bound riboflavin more tightly than it bound either the riboflavin monophosphate or riboflavin dinucleotide. Its identification had been occluded by previous reports that the riboflavin was associated with the ovotransferrin as some type of metalloflavoprotein. This proved not to be the case. A third protein ovomacroglobulin was noted because of its easy identification in starch-gel electrophoretic patterns [58, 59] and its comparatively high immunochemical reactivity [60, 61]. Trace amounts of other constituents with biological activity have been detected by assays directly on the egg white and, in one case, by purification and partial characterization. Enzymatic activities detected at low levels include a peptidase [62], $\beta$-N-acetylglucosaminadase, and $\alpha$-mannosidase [63]. Fossum and Whitaker [64] have partially purified another inhibitor of proteolytic enzymes, an inhibitor of papain and ficin. A number of other constituents have been detected electrophoretically or chromatographically and several of these have been partially purified. One of these partially characterized was named ovoglycoprotein by Ketterer [65] and another was merely termed "genetic globulins" for convenience sake in genetic studies in our laboratories in 1963 [59]. It would appear likely that there are other proteins with biological activities in egg white, but egg white is relatively devoid of most of the enzymes commonly found in biological tissue [66].

## GENERAL COMPOSITION

### Physical Structure in the Egg

Chicken egg white is composed of several physically different forms. The primary difference is that it exists as layers of thick egg white and thin egg white. The thick egg white composes over half of the total egg white in a fresh egg and is anchored by the chalazae to the surface of the yolk membrane. The chalazae is frequently referred to as the hammock which holds the yolk in the center of the egg. The chalazae and at least part of thick egg white are closely related physically and probably chemically. In fact there is no known way of removing the chalazae other than to tear it from the egg white. It appears that the fibers of chalazae in the thick egg white become finer and finer until they admix and "disappear" into the thick egg white. Figure 2-1 is a simplified sketch of component parts of the egg. There are further differentiations of the physical structures than given in this sketch. There is definitely a multilayered structure at the surface of the yolk. Part of this structure is apparently thick egg white or chalazae, imbedded or contiguous to the vitelline or "true yolk membrane".

Separation of the thick and thin egg white is possible only by crude and approximate means. The most obvious method is to cut the thin and thick

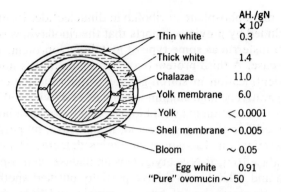

Fig. 2-1. Antihemagglutination (AH) activities of the various components of the egg. Reprinted from [68].

white away from one another in the broken-out egg. Another and more common method is to separate the egg whites by means of passing the egg white through sieves with large diameter holes [67]. There are at least three other differences between the thick and thin egg white [68, 69].

1. Upon dilution and acidification of egg white, approximately four times the amount of precipitate is obtained from thick egg white as from thin egg white. This precipitate is called crude ovomucin.

2. Thick egg white contains approximately four times the amount of inhibitory activity against the hemagglutination of erythrocytes by viruses.

3. Thick egg white contains approximately 30–40% more N-acetylneuraminic acid (sialic acid) than does thin egg white. Comparisons of the contents of sialic acid and antihemagglutination activity are given in Table 2-1.

It is evident from Table 2-1 that there are certain similarities between preparations of chalazae, extracts of crude yolk membranes, and ovomucin. They are relatively high in sialic acid and in inhibitory activity for viral hemagglutination. These comparative compositions support the hypothesis that the fibers in thick egg white, the chalazae, and the contiguous layer of thick egg white which is intimately associated with the true yolk membrane are closely related if not identical substances. These observations agree with the hypothesis of Conrad and Phillips [70] that the chalazae is formed from the ovomucin fibers in the thick egg white as the egg passes down the hen's oviduct. According to this hypothesis, the egg is slowly rotated during its oviductal passage but the yolk resists this rotation. As a consequence, the ovomucin fibers are slowly turned and twisted into what eventually becomes the chalazae. There are, however, no direct observations to substantiate this attractive mechanical theory for the formation of chalazae.

PROTEIN COMPOSITION

Avian egg white is the type of material that the biochemist prefers to select when he desires to separate, fractionate, and prepare in quantity purified proteins. The reason for this is simply that egg white is primarily a solution of proteins with a relatively small amount of sugar and salts. This makes the job of protein separation and purification much easier than with many other biological materials. Indeed, egg white has been the source of several of the standard proteins for biochemists. Chicken ovalbumin and lysozyme have been two of the less expensive and highest purity proteins available for many years. The composition of a representative chicken egg white is given in Table 2-2; it is representative because there are genetic variations in the composition. Ovotransferrin, lysozyme, ovomucoid, and ovoinhibitor are described in more detail in subsequent chapters; they are discussed briefly in this section.

TABLE 2-1. SIALIC ACID CONTENT OF EGG CONSTITUENTS

| Constituent | Sialic Acid | Virus antihemagglutination Activity per g of nitrogen |
|---|---|---|
| | % | |
| Ovalbumin (crystalline) | <0.1 | <0.01 |
| Conalbumin (crystalline) | <0.1 | <0.01 |
| Lysozyme (crystalline) | <0.1 | <0.01 |
| Flavoprotein | 0.5 | ? |
| Ovomucoid | $0.7^a$ | 0.01 |
| Ovomucin | $2.6^a$ | 50 |
| Egg white | | |
| Thin | $0.25^b$ | 0.3 |
| Thick | $0.39^b$ | 1.4 |
| Whole | $0.29^b$ | 0.9 |
| Yolk membrane | $2.2^a$ | 6.0 |
| Chalazae | $2.3^a$ | 11.0 |
| Shell membrane | <0.05 | 0.005 |

Reproduced from [69].
[a] These materials vary in sialic acid content depending upon method of preparation. The value for ovomucoid is an average for the total ovomucoid.
[b] Values for sialic acid are the averages of multiple determinations (>6) on several preparations.

TABLE 2-2. COMPOSITION OF EGG WHITE AND PROPERTIES OF EGG-WHITE PROTEINS

| Protein | % | pI | Mol. Wt. (g) | $S_{20,w}$ (Svedberg units) | $D_{20,w} \times 10^7$ (cm²sec⁻¹) | $\bar{V}$ (cm³/g) | $E_\lambda^{1\%}$ |
|---|---|---|---|---|---|---|---|
| Ovalbumin | 54[a] (71) | 4.5 (72) | 46,000[b] (73) | 3.27 (74)[f] | 7.67 (74) | 0.750 (75) | $E^{1\%}_{280}$ = 7.50 (76) |
| Ovotransferrin | 12 (77) | 6.05 (78) | 76,600[c] (79) | 5.05 (80) | 5.72 (Fe) (80), 5.30 (Fe-free) | 0.732 (80) | $E^{1\%}_{280}$ = 11.6 (72) |
| Ovomucoid | 11 (81) | 4.1 (72) | 28,000[d] (82)(83) | 2.62 (83) | 7.7 (81) | 0.685 (83) | $E^{1\%}_{280}$ = 4.55 (84) |
| Lysozyme | 3.4 (77) | 10.7 (85) | 14,300[e] (86) | 1.91 (87) | 11.2 (85) | 0.703 (87) | $E^{1\%}_{280}$ = 26.35 (87) |
| Ovomucin | 2.9 (82) | nd | nd | nd | nd | nd | nd |
| Ovoinhibitor | 1.5 (82) | 5.1 (81) | 49,000[f] (88), 44,000[g] | nd | nd | 0.693 (88) | $E^{1\%}_{278}$ = 7.4 (88) |
| Ovomacroglobulin | 0.5 (60) | 4.5 (61) | 900,000[f] (61), 760,000[h] | 15.1 (61) | 1.98 (61) | 0.745 (61) | nd |
| Ovoglycoprotein | 1.0 (65) | 3.9 (65) | 24,400[i] (65) | 2.47(65)[l] | nd | nd | $E^{1\%}_{280}$ = 3.8 (65) |
| Flavoprotein + Apoprotein | 0.8 (82) | 4.0 (57) | 32,000[j] (57), 36,000[h] | 2.76(57)[l] | 6.4 (57) | 0.70 (57) | nd |
| Avidin | 0.05 (89) | 10 (72) | 68,300[k] (90) | 4.55(90)[l] | 5.98 (90) | nd | $E^{1\%}_{282}$ = 15.7 (91) |

[a] Estimated from sulfhydryl content of whole egg white.
[b] Osmotic pressure.
[c] Iron binding.
[d] Average value from osmotic pressure and sedimentation velocity measurements. The sedimentation of ovomucoid is strongly dependent on concentration and the molecular weight would be closer to 29,000 if previous results from this laboratory (81) were extrapolated to zero concentration.
[e] Amino acid content.
[f] Light scattering.
[g] Approach to equilibrium.
[h] Sedimentation velocity.
[i] Minimum molecular weight from tryptophan content.
[j] Flavin binding.
[k] Sedimentation equilibrium.
[l] Not reported as extrapolated to zero concentration.

## FRACTIONATION AND PURIFICATION

Ovalbumin is the primary protein of chicken egg white present at four to five times the concentration of the secondary constituents, ovotransferrin, and ovomucoid. The properties of egg white are primarily those of the ovalbumin with the other constituents contributing mainly to the biological properties of the egg white. An obvious exception to this is the ovomucin which appears responsible for the high viscosity of thick egg white.

Chicken egg white has been examined by many different electrophoretic techniques. After the early work of Longworth's group [26] free boundary electrophoresis was the only method for many years until paper and then gel techniques became available. Evans and Bandemer [92] described a general method for paper electrophoresis.

Gel electrophoretic patterns of chicken egg white done under different conditions are given in Figs. 2-2, 2-3, and 2-4. Figures 2-2 and 2-3 are reproductions of starch-gel patterns of Lush [58] and Feeney and co-workers [59], respectively. In Fig. 2-3 components 3, 4, and 6 are considered to be ovalbumin. Components 8, 9, 10, and 11 were in the area of ovomucoid and unidentified proteins, probably mostly globulins. The ovomucoid did not stain under these conditions. Components 12 and 14 were investigated further [59] and were described as globulins $A_1$ and $A_2$. Components 16 and 17 were identified as ovotransferrin. Component 18 was stated to be a high molecular weight substance and later named ovomacroglobulin [61]. Component 19 was definitely identified as lysozyme.

## CHEMICAL FRACTIONATION AND PURIFICATION OF EGG-WHITE PROTEINS

The many different procedures for fractionating chicken egg white for the preparation of the individual proteins have arisen from the techniques

---

TABLE 2-2 (Continued)

References to original report are given in ( ) and should be consulted for methods and interpretation of results. % is the per cent of the dry egg white represented by a given compound.
pI is the approximate isoelectric point determined from electrophoretic mobility or estimated from isolation procedures with ion-exchange celluloses.
$S_{20,w}$ is the sedimentation constant reduced to water at 20° and extrapolated to infinite dilution if reported.
$D_{20,w}$ is the diffusion constant reduced to water at 20°.
V is the partial specific volume of the protein.
$E_\lambda^{1\%}$ is the extinction of a 1% protein solution at the given wavelength in a cell of 1 cm pathlength.
nd is the value not determined or not reported.

# CHICKEN EGG-WHITE PROTEINS

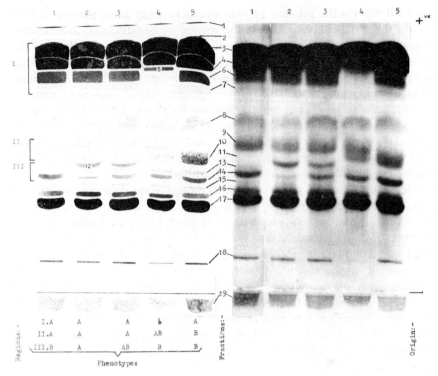

Fig. 2-2. Starch-gel electrophoretic patterns of five different chicken egg whites showing "elements of variation" in three regions. Reprinted from [58].

and methods that were available at the time when a particular protein was first investigated. No one has taken the trouble to develop a general method for the preparation of all, or nearly all of egg-white proteins with newer techniques available during the last four or five years. A close approximation to this was a procedure first developed by our laboratory in 1958 [72] for the fractionation of chicken egg-white proteins by adsorption and elution from carboxymethyl cellulose using a series of step-wise additions of elution buffers (Fig. 2-5). This, however, was primarily a demonstration of the efficacy of the procedures and was not intended as the method for routine separations of the individual proteins. Although no attempt will be made to give a general method for fractionation of all the proteins, it is intended that the ensuing discussion will help to make possible the development of satisfactory methods for obtaining several of the individual proteins that might be desired in a particular laboratory. Methods for the preparation of ovoinhibitor, ovomucoid, lysozyme, and transferrin are discussed in subsequent chapters.

## SALT FRACTIONATION TECHNIQUES

Some of the older techniques that utilize ammonium sulfate for fractional precipitation and crystallization are still among the best procedures for obtaining egg white proteins in quantity. The use of ammonium sulfate or other salts for one or more purification steps is still highly desirable in most cases as a preliminary or adjunctive procedure. However, it has been the experience in our laboratory that all proteins of egg-white produced by salt fractionation, whether crystallized or not, should, if possible, have a further purification with other fractionation techniques. The choice at the moment would appear to be fractionation on one of the ion exchangers such as CM-cellulose or DEAE-cellulose and molecular filtration through a cross-linked dextran such as a Sephadex.

The older procedures of Hektoen and Cole [46] and Sorenson [47] utilized pH adjustment and changes in concentration of ammonium sulfate. These techniques are still excellent for obtaining initial crude preparations or as adjunctive procedures to other methods. One procedure is as follows:

1. The egg white is carefully blended in a Waring Blender [68] and is adjusted to pH 5.5 to 6.0. After standing for 30 to 60 minutes the precipi-

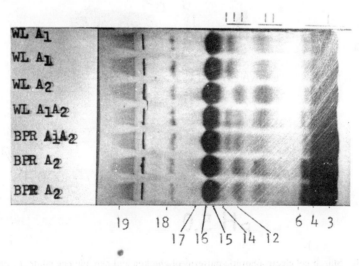

Fig. 2-3. Starch-gel electrophoretic patterns of chicken egg whites showing globulins $A_1$ and $A_2$. The Roman numerals on the top correspond to the genetic loci of Lush [58]; the Arabic numerals on the bottom correspond to the phenotypes of Lush [58]. WL and BPR refer to white leghorn and barred Plymouth Rock, respectively. $A_1$, $A_1A_2$, and $A_2$ refer to phenotypic designation of egg white. Reprinted from [59].

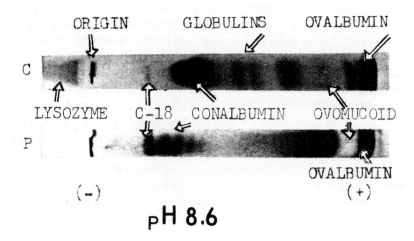

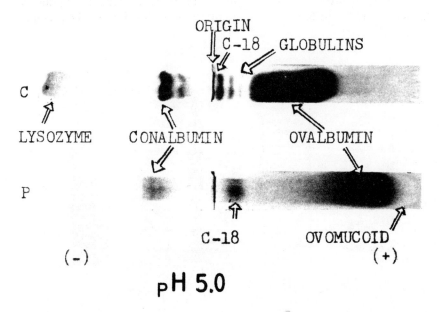

Fig. 2-4. Starch-gel electrophoretic patterns of chicken and penguin egg white. C, chicken egg white; P, penguin egg white. Reprinted from Feeney et al., Pergamon Press, 1966, p. 134.

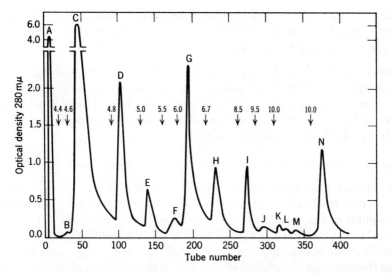

Fig. 2-5. Complete analysis and fractionation of egg white on carboxymethyl-cellulose. The numbers and vertical arrows show the pH and point of change of the eluting buffers. 30 ml of dialyzed egg white were chromatographed on a 2.2 × 14.0 cm column with a flow rate of 2.5 ml per minute. The fraction size was 15 ml. Total recovery of applied nitrogen was 88%. A stepwise elution procedure was employed as described in the text. Eluting buffer was 0.1 M ammonium acetate with the further addition of 0.025 M $Na_2CO_3$ and 0.2 M $Na_2CO_3$ to the buffer at tube No. 310 and No. 359, respectively. Reprinted from [72].

tate, which is primarily crude ovomucin plus globulins, is removed by centrifugation.

2. The egg white is brought to approximately pH 4.6, and 225 grams of solid ammonium sulfate are added per liter. After standing for 30 to 60 minutes the voluminous precipitate is removed by filtration. The precipitate is primarily globulins as well as various other egg-white proteins.

3. Ammonium sulfate is added to the filtrate to give 50% saturation. At this point the solution is seeded with crystals of ovalbumin and the solution is allowed to stand for 4 to 6 days at room temperature to complete crystallization. Crystals are frequently seen in a matter of hours and nearly complete crystallization occurs within 48 hours.

4. The crystallized ovalbumin is removed by filtration or centrifugation. Eight grams of solid ammonium sulfate per 100 ml of filtrate are now added and the solution allowed to stand overnight. A precipitate which is primarily ovotransferrin should form at this stage.

5. The precipitated ovotransferrin is removed by centrifugation. The

filtrate is now made to saturation with ammonium sulfate and allowed to stand overnight. A precipitate which is primarily ovomucoid should form. This precipitate can be removed by centrifugation or filtration.

All of these preparations are, of course, grossly impure with the exception of the ovalbumin. Even the ovalbumin is impure and after as many as five recrystallizations may give a strong immunological test for ovotransferrin and show small amounts of ovotransferrin by physical procedures. Each protein should be further purified.

### FRACTIONATION WITH CELLULOSE ION-EXCHANGE AGENTS

As already stated, the inclusive procedure of Rhodes, Azari, and Feeney [72] for fractionation of egg white on CM-cellulose was presented as an example of the excellence of the method for fractionation and not as a general recommended procedure. All of the fractions of Fig. 2-5 were collected in order to calculate the recovery of the original egg white. Data on these fractions are shown in Table 2-3. The designations of proteins were for the main constituents and small amounts of other proteins were not listed. Further purification of all fractions was necessary. By sequential fractionation it was possible to separate ovalbumin into two main fractions with approximately two and one phosphorous atoms per molecule, respectively, corresponding to ovalbumins $A_1$ and $A_2$. Other investigators have since used various modifications of this procedure by employing both stepwise and gradient elutions [57, 59, 61, 78, 91, 93–96].

### MOLECULAR FILTRATION

Molecular filtration (exclusion) has also been employed in our laboratories for the purification of most of the egg-white proteins. In most all instances a Sephadex has been used, the type depending on the particular protein desired and the impurities present. It has been used for purifying various transferrins and chemical derivatives of transferrins [97], ovomucoids [98, 99], and ovomacroglobulin [61].

### OTHER METHODS

Other procedures that have been employed include most of the techniques common to protein chemistry. Forsythe and Foster [100, 101] employed the low dielectric-solvent precipitation and crystallization method developed originally for blood serum proteins. This method, however, has apparently not been employed by many other investigators; it has been used for the crystallization of ovotransferrin by Warner and Weber [79].

As with other proteins the particular characteristics of the individual

TABLE 2-3. ANALYSIS OF ISOLATED PROTEINS FROM EGG WHITE

| Protein in Fraction | pI | pH of Eluting Buffer | pH of Eluate at Peak | Distribution of Nitrogen in Fractions | | Amount of Protein | | Recoveries by Specific Assay | $\frac{1}{E} \times 10^a$ |
|---|---|---|---|---|---|---|---|---|---|
| | | | | | | From N Content | By Specific Assay | | |
| | | | | mg. | per cent | mg. | mg. | per cent | |
| Ovomucoid + flavoprotein | 3.9–4.3 | 4.3 | 4.3 | 42.2 | 12.3 | 304 | 303 ⎫ 87 | | 1.64 |
| Ovomucoid | 3.9–4.3 | 4.4 | 4.4 | 0.9 | 0.3 | 6 | 3 ⎭ | | 1.63 |
| Ovalbumin A$_1$ | 4.58 | 4.55 | 4.5 | 173.8 | 50.8 | 1106 | 1245 ⎫ | | 1.16 |
| Ovalbumin A$_2$ | 4.65 | 4.75 | 4.65 | 37.2 | 10.8 | 237 | 267 ⎬ 89 | | 0.99 |
| Ovalbumin A$_3$ | 4.75 | 5.0 | 4.85 | 11.0 | 3.2 | 70 | 64 ⎭ | | 0.99 |
| "Globulin" | | 5.5 | 5.2 | 3.6 | 1.5 | 22 | | | 0.63 |
| Conalbumin | 6.5–6.8 | 6.0 | 5.8 | 42.5 | 12.4 | 256 | 233 ⎫ 87 | | 0.86 |
| Conalbumin | 6.5–6.8 | 6.7 | 6.1 | 16.1 | 4.7 | 97 | 103 ⎭ | | 0.88 |
| "Globulin" | | 8.5 | 8.0 | 7.5 | 2.2 | 46 | | | 0.59 |
| "Globulin" | | 9.5 | 9.3 | 2.4 | 0.7 | 15 | | | 0.51 |
| "Globulin" | | 10.0 | 9.4 | 1.1 | 0.3 | 7 | | | 0.50 |
| Avidin | >10 | 10.0 | 9.5 | 0.9 | 0.3 | 6 | | 39[b] | 0.42 |
| "Globulin" | | 10.0 | 9.9 | 1.2 | 0.4 | 8 | | | 0.34 |
| Lysozyme | 10.7–11.3 | 10.0 | 10.0 | 2.1 | 0.6[c] | 11[c] | 70[c] | 92[c] | 0.44 |

The starting material was dialyzed and centrifuged egg white. Assays for specific proteins were carried out by methods described in the text. The isoelectric points listed are ranges quoted from Warner [107] and observations of this laboratory. Reproduced from [72].
Fractionation was on carboxymethyl-cellulose as shown in Fig. 2-5. Proteins are crude mixtures at this stage.
[a]This figure is the reciprocal of the extinction coefficient measured in a 1 cm cell at 280 m$\mu$ at the isoelectric pH multiplied by 10. When it is multiplied by experimentally observed optical densities, the amount of a particular protein is obtained in mg. per ml. The values are for unpurified fractions.
[b]Total recovery of avidin equaled 50 per cent distributed between Fractions K, L, and M.
[c]The assays for lysozyme were conducted on the fractions before dialysis. The nitrogen content was determined after dialysis, and low recovery on an N basis was due to extensive loss on dialysis in this particular experiment.

egg-white proteins may be employed for fractionation purposes. In the case of ovotransferrin the stability of the ion complex may be employed in order to denature by heat contaminating proteins. In the case of the flavoprotein it is very advantageous to fractionate the flavoprotein and then to remove the riboflavin and refractionate the riboflavin-free apoprotein. The apoprotein has an isoelectric point approximately 0.5 pH unit more alkaline than the flavoprotein and can be separated on ion exchangers quite readily from impurities that were originally eluted with the flavoprotein [57].

## PROPERTIES OF EGG-WHITE PROTEINS

The proteins discussed briefly in this chapter include: ovalbumin, ovomucin, ovoflavoprotein, avidin, and ovomacroglobulin. The general properties and chemical compositions of the characterized proteins of egg white are given in Tables 2-2 and 2-4. Reviews of the older literature of the egg-white proteins have been made by Fevold [106] and Warner [107].

*Notes:*
The first column for each protein gives the grams of anhydrous amino acid or carbohydrate/100 g protein. The second column gives the apparent number of residues/mole, assuming a given molecular weight.
A molecular weight has not been assigned to ovomucin, therefore, the second and third column for this protein are given in residues/10,000 g protein as reported from two different laboratories.
The amino acid sequence of lysozyme is known and the amino acid content is presented in residues/mole as calculated from the amino acid content.
The reported molecular weight of ovomacroglobulin is too large to permit a meaningful analysis of residues/mole. The second column under this protein is presented as residues/10,000 g protein.
*nd* is the value not determined or not reported.

TABLE 2-4. COMPOSITION OF EGG-WHITE PROTEINS

|  | Ovalbumin | | Ovotransferrin | | Ovomucoid | | Ovomucin | | |
|---|---|---|---|---|---|---|---|---|---|
|  | g/100g$^a$ | Res/ 46,000g$^a$ | g/100g | Res/ 76,000g$^b$ | g/100g | Res/ 28,000g$^b$ | g/100g$^c$ | Res/ 10,000g$^b$ | Res/ 10,000g$^c$ |
| Alanine | 5.85 | 37.8 | 5.09 | 54.5 | 2.96 | 11.7 | 3.42 | 4.18 | 4.81 |
| Arginine | 5.32 | 15.7 | 6.07 | 29.6 | 3.53 | 6.3 | 3.95 | 2.80 | 2.53 |
| Aspartate | 8.18 | 32.7 | 11.50 | 76.0 | 13.10 | 31.9 | 8.13 | 7.49 | 7.06 |
| Cystine/2$^j$ | 1.34(0.90)$^k$ | 6.0(4.0)$^k$ | 3.34 | 24.9 | 6.38 | 17.5 | 3.27 | 4.72 | 3.20 |
| Glutamate | 14.25 | 50.8 | 11.73 | 69.1 | 6.85 | 14.9 | 10.64 | 8.60 | 8.24 |
| Glycine | 2.42 | 19.6 | 3.91 | 52.1 | 3.28 | 16.1 | 2.56 | 4.70 | 4.48 |
| Histidine | 2.30 | 7.7 | 2.14 | 11.9 | 2.10 | 4.3 | 2.00 | 1.63 | 1.46 |
| Isoleucine | 6.22 | 25.3 | 3.85 | 25.9 | 1.30 | 3.2 | 4.35 | 3.58 | 3.84 |
| Leucine | 8.28 | 33.7 | 7.42 | 49.9 | 4.92 | 12.2 | 6.52 | 5.38 | 5.76 |
| Lysine | 6.02 | 21.6 | 9.73 | 57.8 | 6.22 | 13.6 | 5.17 | 4.68 | 4.03 |
| Methionine | 4.74 | 16.6 | 1.85 | 10.7 | 0.90 | 1.9 | 2.56 | 1.50 | 1.95 |
| Phenylalanine | 6.73 | 21.0 | 4.98 | 25.8 | 2.79 | 5.3 | 4.86 | 3.15 | 3.30 |
| Proline | 3.16 | 15.0 | 3.59 | 28.1 | 2.67 | 7.7 | 3.73 | 4.19 | 3.84 |
| Serine | 5.24 | 27.7 | 5.26 | 46.0 | 3.89 | 12.5 | 5.65 | 6.14 | 6.49 |
| Threonine | 3.03 | 13.8 | 4.86 | 36.6 | 5.27 | 14.6 | 4.79 | 5.41 | 4.74 |
| Tryptophan | 1.35 | 3.8 | 2.86 | 11.7 | 0.00 | 0.0 | 2.05 | nd | 1.10 |
| Tyrosine | 3.64 | 10.26 | 4.25 | 19.8 | 3.91 | 6.7$^p$ | 3.87 | 3.33 | 2.37 |
| Valine | 7.18 | 33.4 | 6.69 | 51.4 | 5.64 | 16.0 | 5.56 | 4.99 | 5.61 |
| Sialic acid | 0.00$^q$ | 0.0$^q$ | 0.00 | 0.0 | 0.32 | 0.3 | 4.0 | 0.61 | 1.4 |
| Hexose | 1.76$^q$ | 5.0$^q$ | 0.92 | 4.3 | 9.64 | 16.7 | 7.4 | 2.84 | 4.1 |
| Glucosamine | 1.33$^r$ | 3.0$^r$ | 0.97 | 4.6 | 12.08 | 21.0 | 5.85 | 2.79 | 3.25 |
| Galactosamine |  |  |  |  |  |  | 1.31 | 0.75 | 0.73 |
| Other |  |  |  |  |  |  | 0.44$^u$ | nd | 1.47$^u$ |

TABLE 2-4 (Continued)

|  | Lysozyme Res/14,307g[d] | Ovoinhibitor g/100g[e] | Ovoinhibitor Res/44,000g[e] | Ovomacroglobulin g/100g | Ovomacroglobulin Res/10,000g[f] | Flavoprotein g/100g | Flavoprotein Res/32,000g[g] | Avidin g/100g[h] | Avidin Res/68,300g[i] |
|---|---|---|---|---|---|---|---|---|---|
| Alanine | 12 | 2.40 | 14.83 | 2.81 | 3.95 | 2.9 | 12.9 | 2.18 | 20.9 |
| Arginine | 11 | 5.19 | 14.61 | 4.06 | 2.60 | 2.5 | 5.2 | 7.53 | 32.9 |
| Aspartate | 8(13)[l] | 8.84 | 33.79 | 7.48 | 6.50 | 6.7 | 18.6 | 10.15 | 60.2 |
| Cystine/2[j] | 8 | 4.22 | 18.17 | 2.35 | 2.30 | 4.8 | 14.9 | 4.04 | 27.0 |
| Glutamate | 2(3)[m] | 8.07 | 27.50 | 10.09 | 7.82 | 15.5 | 33.8 | 8.56 | (45.6)[m] |
| Glycine | 12 | 3.11 | 23.98 | 1.99 | 3.49 | 1.4 | 7.6 | 3.90 | 46.7 |
| Histidine | 1 | 3.16 | 10.12 | 1.71 | 1.25 | 3.6 | 8.5 | 0.86 | 4.3 |
| Isoleucine | 6 | 3.25 | 12.63 | 5.00 | 4.42 | 2.3 | 6.6 | 5.48 | 33.1 |
| Leucine | 8 | 4.20 | 16.32 | 7.11 | 6.28 | 4.9 | 13.7 | 4.96 | 29.9 |
| Lysine | 6 | 5.00 | 17.16 | 5.27 | 4.11 | 6.4 | 16.0 | 7.04 | 37.5 |
| Methionine | 2 | 0.71 | 2.38 | 1.88 | 1.43 | 3.1 | 7.6 | 1.64 | 8.5 |
| Phenylalanine | 3 | 2.35 | 7.04 | 5.08 | 3.45 | 3.0 | 6.5 | 6.24 | 29.0 |
| Proline | 2 | 2.74 | 12.41 | 3.54 | 3.65 | 2.7 | 9.0[n] | 1.33 | 9.4 |
| Serine | 10 | 3.74 | 18.88 | 4.40 | 5.05 | 7.3 | 26.7 | 4.76 | 37.3 |
| Threonine | 7 | 5.26 | 22.88 | 4.60 | 4.55 | 2.3 | 7.2 | 12.33 | 83.3 |
| Tryptophan | 6 | 0.0 | 0.0 | 0.93 | 0.5 | 4.7 | 8.1 | 5.01 | 18.4 |
| Tyrosine | 3 | 3.51 | 9.46 | 4.37 | 2.68 | 4.5 | 8.8[n] | 1.04 | 4.4 |
| Valine | 6 | 3.99 | 18.92 | 5.36 | 5.41 | 1.6 | 5.2 | 4.40 | 30.3 |
| Sialic acid |  |  |  | <0.03 |  |  |  |  |  |
| Hexose |  | 3.5 | 9.50 | 3.5 | 2.16 | 9.7 |  | 5.22 | 22.0 |
| Glucosamine |  | 2.7[s] | 7.37 | 5.2[s] | 3.23 | 5.0 |  | 3.44[s] | 14.6 |
| Galactosamine |  |  |  |  |  |  |  |  |  |
| Other |  |  |  |  |  | 0.7[v] |  |  |  |

[a]Calculated from the results of Habeeb (101a). The results of Lewis et al. (102) as presented by Tristram and Smith (103) were based largely on microbiological analysis and differ slightly from these values.
[b](82), [c](104), [d](86), [e]Calculated from (88), [f](61), [g](105), [h]Calculated from (91), [i]Calculated on basis of mean molecular weight (90).
[j]Represents the sum of cysteine and half-cystine residues.
[k]The value in parenthesis represents cysteine as reported by MacDonnell, Silva, and Feeney (71). The total cystine/2 was calculated assuming one cystine per mole (see text).
[l]The value in parenthesis represents asparagine.
[m]The value in parenthesis represents glutamine.
[n]Reported as the amount released by 72 hr hydrolysis.
[p]A value of 5.5 as reported from our laboratory was corrected but not extrapolated to zero hydrolysis time (95). Spectrophotometric titrations in our laboratory currently indicate 6 tyrosines. The true value may therefore be either 6 or 7 residues.
[q]See text. Mannose calculated assuming 5 moles/46,000 g of ovalbumin.
[r]N-acetylglucosamine calculated assuming 3 moles/46,000 g of ovalbumin (see text).
[s]Hexosamine.
[t]See the text concerning the presence of one or two moles of phosphate/mole of ovalbumin.
[u]Sulfate.
[v]Phosphate (57).

## Ovalbumin

Ovalbumin, the principal protein of chicken egg white, is also the only major constituent for which no unique biochemical activity has been found. This is probably one of the major reasons why this protein has not received more attention during the last two decades. The main property of ovalbumin attracting attention has been its sensitivity to denaturation. Three chemical aspects, however, not directly related to sequence, have also attracted attention: (a) the heterogeneity due to the presence of either one or two phosphates, (b) the number and reactivity of the cysteine sulfhydryl groups, and (c) the structures of the side-chain carbohydrates and linkages of the carbohydrates to the peptide chain.

In 1940 Longsworth, Cannon, and MacInnes [26] first resolved crystalline ovalbumin electrophoretically into two components, which they named ovalbumins $A_1$ and $A_2$. Gertrude Perlmann [108] later showed beautifully that the difference between $A_1$ and $A_2$ was one phosphate group and that $A_1$ and $A_2$ contained two and one phosphates per molecule of protein, respectively. When we recall that these observations of Longsworth and Perlmann were made before the philosophy of isozymes and before the days of ion-exchange chromatography, their findings and conclusions were outstanding. With the advent of ion-exchange chromatography, the separation of the multiple forms became comparatively easy. In 1958 Rhodes, Azari, and Feeney [72] separated $A_1$ and $A_2$ chromatographically and reported a molar ratio of phosphate to protein of 1.99 for $A_1$ and 0.93 for $A_2$. Small amounts of $A_3$ containing no phosphate were also obtained.

Ovalbumin has been termed a classical example of a protein which has masked sulfhydryls in the native state. It was once commonly accepted that there were five "masked" and "unreactive" sulfhydryls and one disulfide in native ovalbumin [106]. In 1951, however, MacDonnell, Silva, and Feeney [71] showed that (a) there were only four sulfhydryls rather than five, (b) the "masked" state was a relative one. All four were unreactive to some reagents in the native protein, but several were reactive to other reagents; for example, three of the four sulfhydryls were reactive to *p*-chloromercuribenzoate in the native protein. (c) Crystalline derivatives or complexes of the native proteins and mercury reagents could be prepared. Boyer [109] confirmed both the total number of sulfhydryl groups as four, and the differential reactivity to the mercury reagent. He also indicated that there might be a third type of reactive species of the four sulfhydryls. Many other investigators have confirmed and extended these studies with different reagents. Table 2-5 is a summary of the sulfhydryls found with four different reagents before and after denaturation [96]. The total number of half cystine residues in the protein may, however, still be in doubt as four sulfhydryls and two disulfides have been reported in the molecule [110].

TABLE 2-5. COMPARISONS OF SULFHYDRYL DETERMINATIONS ON CHICKEN OVALBUMIN

| Reagent | Conditions | Native (moles/mole) | Denatured (moles/mole) |
|---|---|---|---|
| p-Chloromercuribenzoate | pH 4.6 | 2.8 | 4.0$^a$ |
| Iodine | pH 6.5 | 2.9 | 4.0$^a$ |
| N-Ethylmaleimide | pH 7.0 | 0.4–0.6 | 3.8$^{a,b}$ |
| 5,5'-Dithiobis-(2-nitrobenzoic acid) | pH 8.0 | 0.0 | 3.8$^b$ |

Data from [96].
$^a$Denaturant was incubation at pH 12.2 for 10 minutes.
$^b$Denaturant was 0.48% dodecyl sulfate.

A different form of ovalbumin has been reported by Smith [111] to exist in stored eggs. This form was a more stable one and was named "S-ovalbumin." In further studies it was possible to show differences in physical and chemical properties between S-ovalbumin and native ovalbumin [112].

The carbohydrate moiety of ovalbumin has been shown to be connected to the protein through an aspartic acid [113, 114]. Apparently homogeneous preparations of asparaginyl carbohydrate from ovalbumin contain five moles of D-mannose and three of N-acetylglucosamine per aspartic acid. From partial acid hydrolysates of an ovalbumin glycopeptide, Marshall and Neuberger [113] isolated a crystalline compound which was identified as 2-acetamideo-1-(L-$\beta$-aspartamido)-1,2-dideoxy-$\beta$-D-glucose. Montgomery et al. [114] isolated asparaginyl carbohydrates from pronase digests of ovalbumin. Cunningham et al. [115] obtained results indicating heterogeneity in the carbohydrate fraction.

AVIDIN

In 1941 Eakin, Snell, and Williams [52] were able to concentrate avidin by acetone and salt fractionation procedures. They followed the purification by modifying a microbial assay for biotin to an assay for avidin. A unit was defined as the amount of concentrate capable of inactivating 1 $\gamma$ of biotin. Raw egg white varied from 0.8 to 1.2 units/cc. These workers also reported the capacity to inactivate biotin was unaffected by incubation at pH 6.2 and 80° for 2 min. However, 80% of the capacity was lost after treatment at 100° C for 2 min. The resistance of the avidin-biotin complex to digestion by pepsin, trypsin, pancreatin, and papain was shown by György and Rose in 1943 [116] when they failed to obtain release of biotin. Treatment of the complex with 0.45% $H_2O_2$ was found to cause release of biotin active material. Interest in the physiological importance of avidin as a biotin-binding protein prompted other workers to study the secretion of avidin from oviducts and the effect of hormones on this secretion [117].

The first estimation of molecular weight was 44,000 g based on a combining ratio of 1:1 (biotin:avidin). But this was predicted to be too high an equivalent weight due to lack of purity of the preparation [52]. A tentative molecular weight of 66,000 was found by A. Mohammad in 1950 using osmotic pressure measurements [102]. The high isoelectric point of avidin, pH 10, was unaccounted for by initial amino acid analysis and dye-binding techniques and ascribed as possibly due to the presence of nucleic acid present in avidin preparations.

Fraenkel-Conrat and co-workers in 1952 [89] separated three biotin-binding factors: avidin, a complex of avidin and nucleic acid, and a complex of avidin with an uncharacterized phosphate-free acidic component. They suggested that the water-insoluble avidin in previous work was only a fraction of the total biotin-binding activity of the egg white and probably represented an avidin-nucleic acid complex. Free avidin was shown to be an albumin-like protein, although it probably occurred in egg white to varying extents as the above mentioned complexes. Fraenkel-Conrat and co-workers [89] also found two molecules of biotin to be bound by each avidin molecule. These workers investigated the biotin-binding site and proposed a hypothetical structural component of the binding site. Chemically reactive amino, phenolic, imidazole, carboxyl, or disulfide groups were apparently not part of the active site. Other workers had studied the binding site using various analogs of biotin and (a) have established the necessity of the ureido group of biotin for binding, and, (b) the lack of ionic linkage since biotin and oxybiotin were bound even as esters or alcohols [118–120]. Avidin was found to be stable to high and low pH (pH 2.2 and 10.5) for 4 hrs at 23°. The complex of avidin and biotin was found to be completely resistant to these conditions for prolonged periods. Of the chemical modifications used by these workers, only prolonged esterification, treatment with formaldehyde in the presence of an amide, and oxidation with $H_2O_2$ - $Fe^{2+}$ lowered the biotin-binding activity of avidin.

The shift of the absorption spectrum of tryptophan residues to longer wavelengths when avidin reacted with biotin led Green [121] to suggest the involvement of tryptophan in the binding of biotin. Such a shift was interpreted as a transfer of tryptophan into a less polar environment. The study of difference spectra revealed peaks at 233, 288, and 293 m$\mu$ which were due to tryptophan. The change in extinction at 233 m$\mu$ corresponded to a 25% increase in absorbancy. There was almost no change in optical rotation of avidin when it combined with biotin so it was felt that alterations of the absorption band of the peptide chromophore could be eliminated. Spectrophotometric titration of avidin with $10^{-3}M$ N-bromosuccinimide showed that 12 out of 14 tryptophans present were destroyed and a linear relationship was obtained suggesting that tryptophan was specifically attacked. The spectrum of the biotin-avidin complex was unaltered by this treatment with N-bromosuccinimide, suggesting that each molecule of biotin protected 4 residues of tryptophan, and indicated that 3 moles of biotin were found per 64,000 g of avidin. Destruction of an average of two out of four reduced the activity to 10-15% of the original. These observations would explain the sensitivity of avidin to photooxidation [122]. Evidence for an involvement of tryptophan was also provided by fluorescence studies [123].

Methods faster than the microbiological avidin assay have been developed. Most of these depend on the quantitation of excess ($^{14}$C)biotin separated from the ($^{14}$C)biotin-complex. Methods of separating the unbound biotin include: adsorption of the complex onto CM-cellulose followed by centrifugation [124], ultrafiltration [124, 125], and gel filtration [126]. Green used such an assay procedure in kinetic studies to estimate the rate constant for avidin-biotin association to be $7 \times 10^7 M^{-1}$ sec$^{-1}$ and that for dissociation to be $9 \times 10^{-8}$ sec$^{-1}$. This corresponded to a dissociation constant for the complex of about $10^{-15}M$ [124].

Preparation of avidin using CM-cellulose has been reported by several laboratories [72] and the most active preparations bound 13.8 $\mu$g biotin/mg of avidin [91].

Avidin has been dissociated reversibly in 6 $M$ guanidine hydrochloride by Green [127] into subunits of 16-20,000 g. A biotin-induced reassociation was indicated by a spectrophotometric titration at guanidine hydrochloride concentrations below 3 $M$. Careful measurement of the molecular weight of avidin indicated a molecular weight of 66-70,000 g, which suggested a molecule consisting of four subunits [90]. Uncombined avidin is very labile to steaming but the avidin-biotin complex is quite stable in $0.2M$ ammonium carbonate at 100° for a period of 15 minutes, in contrast to earlier reports [128, 129]. The affinity of avidin for biotin was found to be a direct function of ionic strength and to be essentially zero in aqueous solution [129].

McCormick [130] has described the preparation and use of biotin-cellulose for the specific purification of avidin. This procedure takes advantage of the fact that the free carboxyl of biotin is not required for complex formation. The avidin is adsorbed to the column in pH 8.9, 0.2 $M$ ammonium carbonate and eluted by a gradient of decreasing ionic strength.

## FLAVOPROTEIN

Chicken egg-white flavoprotein (or ovoflavoprotein) is one of two egg-white proteins which contains, or can bind, a B-vitamin. The other, avidin, was described in the previous section. Rhodes, Bennett, and Feeney [57] found this protein to exist in two forms; one was the flavoprotein containing riboflavin and the other was the vitamin-free form, the apoprotein. The total concentration of the two forms approximated 0.8% of the egg white (Table 2-2).

As is well known, when the riboflavin content of the diet of a laying hen is low the riboflavin content of the white of the egg is greatly lowered [131]. Table 2-6 shows that the diet of the hen changes the proportion of flavoprotein to apoprotein but does not affect the total amount of the two forms.

TABLE 2-6. INFLUENCE OF HEN'S DIETARY RIBOFLAVIN ON RIBOFLAVIN AND FLAVOPROTEIN-APOPROTEIN CONCENTRATION OF WHITE

| | Egg White Composition | | |
|---|---|---|---|
| Riboflavin in Ration | Riboflavin | Total Riboflavin-binding capacity[a] | Total Calculated Protein[b] |
| mg/lb | μg/ml | μg/ml | mg/ml |
| "None"[c] | 0.25 | 8.8 | 0.75 |
| 1.2[c] | 1.1 | 7.8 | 0.67 |
| 12.0[c] | 2.4 | 9.5 | 0.81 |
| Stock[d] | 3.9 | 9.0 | 0.77 |

Reproduced from [57].
[a]Total binding capacity is determined by amount of riboflavin present and additional amount of riboflavin bound.
[b]Total calculated flavoprotein and apoprotein.
[c]Each figure represents the average value of analyses of egg white of 2 egg from 2 hens each on the respective rations for 16 days. Rations described in text.
[d]Stock ration was a standard egg laying ration.

Rhodes et al. [57] reported on several properties of the flavoprotein. From inspection of the composition given in Table 2-4 it is evident that a possibly unusual characteristic is the presence of phosphate. Removal of this phosphate enzymatically, however, had no influence on the riboflavin binding properties. The binding property was unusually stable to heat. Heating the apoprotein in 0.05 $M$ Tris buffer, pH 7.0, at 95–100°C for 15 minutes did not cause loss of binding capacity. Of all the characteristics of this protein its relative affinities for derivatives or analogs of riboflavin are probably of most general interest. The affinities of enzymes are usually much higher for coenzymes containing riboflavin than they are for riboflavin itself. However, the egg-white protein has been shown to bind riboflavin much more strongly than riboflavin monophosphate (FMN) or riboflavin adenine dinucleotide (FAD). The displacements of several other riboflavin analogs are given in Table 2-7. It is evident that a variety of very different changes in the molecule greatly influence the binding. Because changes in the carbohydrate residue (on position 9 of the isoalloxazine ring) had such a large effect, it has been concluded that riboflavin is either bound through several functional groups or the entire molecule fits into a very specific structure or crevice on the protein [57]. The rigorous requirements of the structures of the flavin for binding seem to make this protein an excellent tool for studying the interaction of small molecules with proteins.

TABLE 2-7. RELATIVE BINDINGS OF ANALOGUES OF RIBOFLAVIN BY APOPROTEIN[a]

| Compound | Apparent Riboflavin Displaced | Riboflavin[b] due to Analogues | Corrected Riboflavin Displaced | Total Riboflavin | | Relative Binding[c] |
|---|---|---|---|---|---|---|
| | | | | Displaced | Bound | |
| | µg/ml | µg/ml | µg/ml | µg | µg | |
| Riboflavin | <0.003 | | | | 10.00 | 1.00 |
| 3-Methylriboflavin | 0.195 | 0.006 | 0.189 | 4.73 | 5.26 | 0.156 |
| Lyxoflavin | 0.114 | 0.008 | 0.106 | 2.65 | 7.35 | 0.060 |
| Galactoflavin | 0.092 | 0.007 | 0.085 | 2.13 | 7.87 | 0.049 |
| Lumiflavin | 0.040 | 0.007 | 0.033 | 1.08 | 8.92 | 0.014 |
| 3-Methyllumiflavin | 0.044 | 0.00 | 0.044 | 1.10 | 8.90 | 0.015 |
| 6,7-Dimethyl-9(formylmethyl)isoalloxazine | 0.050 | 0.008 | 0.042 | 1.05 | 8.95 | 0.015 |
| 6-Chloro-9-sorbitylisoalloxazine | 0.0035 | 0.00 | 0.0035 | 0.09 | 9.91 | 0.002 |

[a] Riboflavin, 10 µg, plus 60µg of analogue was mixed with 1 mg of apoprotein in 0.025 M sodium phosphate buffer, pH 7.0, in a total volume of 3 ml. This was dialyzed against 22 ml of buffer for 2 hours. Assays for riboflavin displacement were run on the dialysate. Reproduced from [57].
[b] These figures represent the riboflavin activity of analogues when traces of riboflavin are present as determined by separate experiments.
[c] Binding relative to riboflavin under the competitive conditions. Figures calculated on basis of following equation:

$$\frac{\mu g \text{ riboflavin displaced}}{\mu g \text{ riboflavin bound}} \times \frac{\text{total } \mu\text{mole riboflavin}}{\text{total } \mu\text{mole analogue}}$$

Similar protein has also been found in chicken blood serum and egg yolk [132]. These are discussed in Chapters 3 and 5.

OVOMACROGLOBULIN

This protein is noted by Lush [58] and described later by Feeney and co-workers [93] as component 18 on starch-gel electrophoresis. It was subsequently found by Miller and Feeney [60, 61] to be the only detectable component of avian egg whites to have a wide spectrum of immunological cross-reactivity. It also appeared to be strongly immunogenic. Deutsch [133] reported that a large part of the capacity of egg white to produce antibodies was probably a minor unrecognized fraction of the egg white. The ovomacroglobulin might be this fraction.

The protein is large with a molecular weight of approximately 800,000 g as determined by sedimentation-velocity, diffusion, and light scattering measurements (Table 2-2). Ovomacroglobulin is the largest recognized egg-white protein, other than ovomucin. There was, however, evidence for its dissociation into smaller units in acid solution.

The chemical composition (Table 2-4) revealed no unusual relationships and the only distinguishing characteristics were the comparatively large size and the immunological relationships. Since the immunological relationships are closely related to the comparative biochemical aspects, this subject is treated in Chapter 3.

OTHER PROTEINS

There are probably one or two dozen minor proteins in chicken egg white. Several of these have been seen in different types of electrophoretic patterns of the egg white itself and probably many more have been seen as minor contaminating substances on purification of the recognized constituents. Twenty years ago Lineweaver and co-workers [66] examined chicken egg white for the presence of enzymatic activities and found it to be essentially devoid of activities with the exception of catalase activity. During the past twenty years there have been many assays developed for numerous different enzymatic activities and the possible presence of other activities in egg white might well now be reexamined.

One of the more obvious minor constituents is the globulin represented as number 12 and 14 in Lush's original electrophoretic pattern (Fig. 2-2). This was partially characterized by Feeney and co-workers [59]. By starting with egg whites from chickens homologous for either variant, two genetic variants were purified and designated globulins $A_1$ and $A_2$. The molecular weight was approximated as $> 36,000$ g and $< 45,000$ g by membrane filtrations and the isoelectric point was estimated as approximately pH 6.

At least five other acidic fractions have been reported to exist in chicken egg white. Varying amounts of an acidic protein containing sulfhydryl groups have been found in the whites of different species [77, 134]. Studies in our laboratory revealed that an acidic fraction of chicken egg white contained high amounts of sialic acid (approximately 4%) and was lower in tyrosine than the main ovomucoid fractions [81]. Ketterer [65] has also reported the presence of an acidic protein containing similar amounts of sialic acid but having no antitryptic activity, which he named ovoglycoprotein. Montreuil and co-workers [135] observed a protein in preparations of ovomucoid resembling that found by Ketterer. Kanamori and Kawabata [136] have recently reported the presence of an acidic substance containing a flavin and having antitryptic activity. Feeney and co-workers [99] reported that materials resembling nucleic acids are in acidic fractions. Donovan et al. [137] have recently identified several sugar nucleotides in egg white. Contaminations with such materials as these might affect electrophoretic patterns or analytical values.

## GENETICS OF CHICKEN EGG-WHITE PROTEINS

Many genetically-controlled variants have been found in the chicken egg white since the advent of the highly discriminatory starch-gel and acrylamide-gel techniques. Earlier studies had shown definite but minor differences in the lysozyme content [138]. Although minor differences in many proteins were reported in different breeds of chickens as determined by filter paper electrophoretic studies, the differences were sufficiently small that no definite conclusions appeared warranted [139].

Lush [58] was apparently the first to prove true polymorphism in the egg-white proteins of the domestic chicken. He noted nine phenotypes by starch-gel electrophoresis (Fig. 2-2) and interpreted the data as suggesting that variations at three regions were due to segregation at three loci with two alleles per locus. Other investigators [59, 140] have confirmed many of Lush's observations. Feeney et al. [59] made a limited study of the genetics and the chemical properties of globulins $A_1$ and $A_2$. Tables 2-8 and 2-9 present part of the data of these genetic studies. These data were considered to confirm the suggestions of Lush that globulins $A_1$ and $A_2$ are controlled by a single gene with two alleles. The ovotransferrins have been the subject of several similar investigations. In addition to the two ovotransferrins exhibited in Figs. 2-2 and 2-3 several other variants have been reported. Since these earlier studies there have been many reports on genetic variations of other constituents. One of the most precise chemical studies was by Lush [141] with strains of chickens producing different ovalbumins. Lush showed that each of the genetic variants also existed with phosphate characteristic of ovalbumins $A_1$ and $A_2$. Removal of the phosphates enzymatically gave changes in electrophoretic patterns (Fig. 2-6). Baker and Manwell [140] have examined variations of several of the minor constituents. Studies on ovomucoid from our laboratory are described in Chapter 8. Lush [142] presents a short perspective on the variations in egg-white proteins in his recent monograph.

TABLE 2-8. DISTRIBUTION OF GLOBULIN PATTERNS IN EGG WHITES OF 40 INBRED LINES OF WHITE LEGHORN CHICKENS IN 1961 GENERATION

| No. of Lines | No. of Hens Tested | Globulin Pattern (No. of Hens) | | |
|---|---|---|---|---|
| | | $A_1$ | $A_1A_2$ | $A_2$ |
| 36 | 112 | 112 | 0 | 0 |
| 1 | 2 | 0 | 2 | 0 |
| 1 | 3 | 1 | 1 | 1 |
| 1 | 2 | 0 | 2 | 0 |
| 1 | 2 | 0 | 1 | 1 |
| Total 40 | 121 | 113 | 6 | 2 |

Reproduced from [59].

TABLE 2-9. DISTRIBUTION OF GLOBULIN PATTERNS 1962 TEST MATINGS[a]

| Dam Phenotype | Sire's Sisters' Phenotypes | No. of Families | Offspring Phenotype (No. of Hens) | | |
|---|---|---|---|---|---|
| | | | $A_1$ | $A_1A_2$ | $A_2$ |
| $A_1$ | $A_1$ | 12 | 27 | 0 | 0 |
| $A_1$ | $A_1$ or $A_1A_2$ | 5 | 4 | 3 | 0 |
| $A_1A_2$ | $A_1A_2$ | 4 | 4 | 7 | 3 |
| $A_2$ | $A_1$ or $A_1A_2$ | 2 | 0 | 2 | 3 |

[a]Parents were selected from survey of inbred line of Table 2-8. Phenotypic designations are based on globulin patterns. Reproduced from [59].

Many investigators have noted minor uncharacterized fractions in egg white by different electrophoretic techniques [143–146]. Some of these are probably still unidentified constituents, but others might have been what we would now identify as genetic variants of recognized constituents.

# BIOLOGICAL ASPECTS

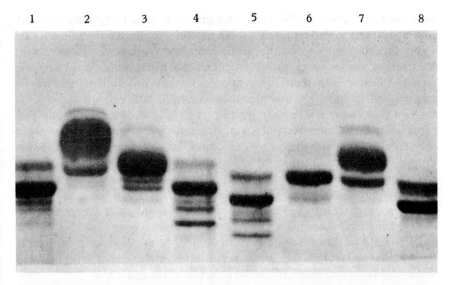

Fig. 2-6. Electrophoretogram of an acid gel to show the effects of dephosphorylation. Sample 2 was untreated ovalbumin B. Sample 3 was ovalbumin B treated with prostatic phosphatase. Sample 4 was ovalbumin B treated with both prostatic and intestinal phosphatase. Samples 7, 6, and 5 were corresponding ovalbumin A samples. Samples 1 and 8 were albumens (diluted five times), of Ov. type B and A respectively, treated with both enzymes. Reprinted from [141].

## BIOLOGICAL ASPECTS

BIOSYNTHESIS

The oviduct of the laying bird is a very active synthetic site for egg-white proteins. Because of this high activity and the relative ease with which tissue preparations from oviduct can be obtained, the hen's oviduct has been used by several investigators to study protein biosynthesis. Early work in Anfinsen's laboratory [147] on the biosynthesis of ovalbumin and more recently [148] on the biosynthesis of lysozyme are now part of the classical literature proving that biosynthesis of proteins proceeds from the amino terminal end to the carboxyl end.

The egg-white proteins are added to the egg near the terminal end of egg production. In contrast to the proteins of the egg yolk which are apparently synthesized elsewhere in the body, primarily in the liver, the present opinion is that the egg-white proteins are synthesized in the oviduct itself [35, 149]. In one of the few experiments directly related to the subject, Mandeles and Ducay [149] used two different approaches. In one they injected radioactive lysine-1-$^{14}$C and glutamic acid-1-$^{14}$C intravenously into a laying hen and examined the egg-white proteins for radioactivity. From this they concluded that ovalbumin and ovotransferrin were synthesized at similar rates but at a significantly higher rate than lysozyme was synthesized. In their other approach, they injected ovotransferrin labeled with lysine-1-$^{14}$C and glutamic acid-1-$^{14}$C intravenously into a laying hen and examined the egg-white proteins for radioactivity. The patterns of radioactivity in ovalbumin, ovotransferrin, and lysozyme were the same as obtained when the free amino acids were injected. This was considered to indicate that an extra-oviductal site for the formation of ovotransferrin was unlikely. These studies are of particular importance in view of the results of Williams [35] who has shown that the serum transferrin and the ovotransferrin of the chicken are identical except that the serum transferrin contains sialic acid. It seems probable that in the chicken both of these transferrins are under the same genetic control. It has not been proven, however, that injected serum transferrin (containing sialic acid residues) might not be transferred to the egg white.

Investigations of biosynthesis with tissue preparations of oviduct have been done in several different ways. Hendler [150] has studied the uptake of nucleic acid, protein, and carbohydrate precursors by the lipids of the hen oviduct. Carey [151] has studied amino acid incorporation and the synthesis of egg-white proteins by cell fractions from oviduct. He obtained particulate fractions high in RNA and other fractions active in incorporating amino acids into proteins.

That the egg is related to reproduction and that the biosynthesis of the egg-white proteins would have an hormonal relationship might well be expected, but a differential synthesis of one or more of the egg-white proteins might not be expected. Such a differential effect has been reported by O'Malley [152]. Progesterone added *in vitro* to minced chick oviduct in tissue culture medium induced the synthesis of a specific oviduct protein, avidin. This synthesis was inhibited by the addition of cycloheximide at any time during the incubation or by the addition of actinomycin D at zero time but not at 6 hours or later. The induction was specific, no increased synthesis of ovalbumin or lysozyme occurred during the incubation. These results may prove to be the first instance of an *in vitro* steroid-induced synthesis of a specific protein.

EMBRYOLOGICAL FUNCTIONS

In contrast to the extensive information on the physical and chemical properties of egg-white proteins and at least some information on the biosynthetic aspects, there is very little information on the function or importance of these proteins in embryological development. Unproven but possible functions for the egg-white proteins during the development of the embryo are the following:

1. A protective aqueous physical environment for the developing embryo on the surface of the egg yolk.
2. A source of water and protein to the developing embryo. The embryo apparently "drinks" the egg white during development.
3. A source of certain particular constituents, such as the ovotransferrin, for transfer to the blood.
4. A direct functional activity of a nutritional nature. One of these may be transport of calcium.
5. An antimicrobial barrier and protection for the embryo.

Direct evidence for any of these functions is apparently available. Some of the more sophisticated approaches to the problem have been in the laboratory of C. R. Grau [153]. These studies have involved a removal of the egg white during incubation of the embryo and its replacement by various substances. A preliminary interpretation of this data is that the egg white may prove to be not particularly important physiologically but may be utilized for its nutritional value.

The antimicrobial function has been the most popular. As is shown in the succeeding chapter, however, there seems to be no set patterns for the distribution of the antimicrobial constituents in the egg whites of different avian species. All of those proteins that show antimicrobial or antienzyme properties could be considered as potential microbial antagonists. These include lysozyme, ovoflavoprotein, avidin, ovotransferrin, and the inhibitors of proteolytic enzymes, ovomucoid, and ovoinhibitor. Lysozyme has received the most notoriety, and indeed, is bacteriacidal to a limited number of species. The factor almost entirely overlooked, however, is the antibacterial activity of the high pH of egg white during the first few days of its incubation of the egg. Nevertheless, ovotransferrin has been directly proven to be an important antimicrobial substance in a rather unique manner. It was shown that when eggs are washed for commercial sales, there was a large difference in the rate of spoilage which could not be explained on the basis of sanitation and source of the eggs [154]. It is necessary to wash eggs because most people will not buy them if they have the natural dirt and feces from the hen on them. The washing process, however, aids in transporting microorganisms through the thousands of pores of the egg shell. After a long period of detective work it was shown that the reason for the difference in subsequent spoilage of the eggs was directly related to the iron content of the water used for washing. The higher the iron content, the more the subsequent spoilage and growth of microorganisms in the egg white. An obvious conclusion was that the small amounts of iron gaining entrance into the egg white or in the pores of the shell was sufficient for microorganisms to initiate growth even in the presence of the high concentration of the iron chelating agent ovotransferrin.

## FOOD USES

Egg-white proteins contribute in an interesting way to the food uses of eggs, particularly insofar as the use of egg white for its foaming and coagulative properties in whipping techniques. Egg white is so unstable that even in the process of pouring it from one container to another denaturation of the ovalbumin may occur. This presents a real problem to the food processor. Considerable attention has actually been given to the foaming, whipping, and angel-cake-making capacities of egg white. MacDonnell et al. [155] reported on the contribution of the individual protein components to these properties. Ovalbumin by itself would whip well, but the globulins contributed significantly to the whipping rate and the ovomucin appeared to aid the stability of the foam. The addition of crystalline ovalbumin to unfractionated egg white gave a superior performing egg white. These studies also included studies with duck egg white. It was common knowledge that duck egg white produced a very unsatisfactory angel food cake. An often quoted explanation was that this was caused by contamination of the egg white with oily material from the shell. It was found that the addition of crude chicken globulins to duck egg white gave a satisfactorily performing product. A much simpler explanation and procedure for making duck egg white perform well was found later [156]. This resulted from observations of the titration curves of duck and chicken egg white and noting that duck egg white gave better coagulative properties when acidified. The solution was simply the addition of a small amount of acetic acid or vinegar to the meringue batter before whipping! There are, of course, numerous other references to the cookery properties of egg white. A recent interesting series of papers have been published by Nakamura [157] and Meehan [158].

## CHEMICAL AND PHYSICAL CHANGES IN EGGS OCCURRING ON INCUBATION

### Gross Changes

When the chicken egg is freshly laid the pH of the egg white is slightly below 7.6 and the pH of the yolk approximately 6.0. Immediately after the egg is laid, however, carbon dioxide escapes through the approximately ten thousand pores in the egg shell and the pH of the egg white increases. The rate of the increase is dependent on the temperature and the carbon dioxide tension. The pH will rise to above 8.0 within a matter of three or four hours at room temperatures. The reason for the high carbon dioxide content of the egg white is apparently due to a high carbon dioxide tension in the oviduct. The initial pH of 7.6 obviously closely approximates the pH of blood serum. After a few days at room temperature or a few weeks at refrigerated temperature, physical changes become evident. These changes are primarily a weakening and dissolution of the gel of the thick egg white and a weakening and eventual rupturing of the yolk membrane. The rate at which this reaction occurs depends on the pH of the egg white and the temperature of incubation. The physical changes themselves are well known to every housewife and, indeed, to anyone who has broken out a partially deteriorated egg and observed the watery white and perhaps rupturing of the yolk membrane.

The deterioration of the thick egg white and yolk membrane presents an interesting chemical and physical problem. The fundamental causes of these biophysical changes have been part of a long range study of our laboratory. Most investigators agree that the deteriorative process in the thick egg white involves changes of the ovomucin. Hoover [159] first observed that the addition of a mercaptan would "thin" egg white. It was later found by MacDonnell, Lineweaver, and Feeney [160] that thinning of the white and also weakening of the yolk membrane can be caused chemically by the addition of small amounts of a variety of reducing agents such as mercaptans or sulfite (Figs. 2-7, 2-8). The deterioration of the yolk membrane has been attributed to the same causes as those responsible for the thinning of the white, one involving changes in the ovomucin [160, 161]. Both changes, however, occur independently; the egg white thins in the absence of egg yolk, and the yolk membrane weakens when removed from the egg and placed in another environment [162]. These studies were extended to include incubation of pieces of "yolk membranes" (Fig. 2-9) [69, 163]. Several treatments that denature or inactivate protein prevented the changes. Another deteriorative change was found in our laboratory. This involved a number if not all of the egg-white proteins [59, 164]. The reaction in egg white was most evident by changes in the ovotransferrins (Fig. 2-10). Further investigations finally showed that the changes were due to interactions of the proteins with the glucose naturally present in the egg white. Such a glucose-protein reaction was not expected at room temperature or 37° and in relatively dilute solution. The situation in egg white is rather different because egg white is relatively alkaline and this glucose-protein reaction (Millard) is accelerated by alkaline conditions. Although it is possible that the glucose-protein reaction may contribute to the deteriorative thinning directly or indirectly by producing reductones, no direct causal relationship has been reported.

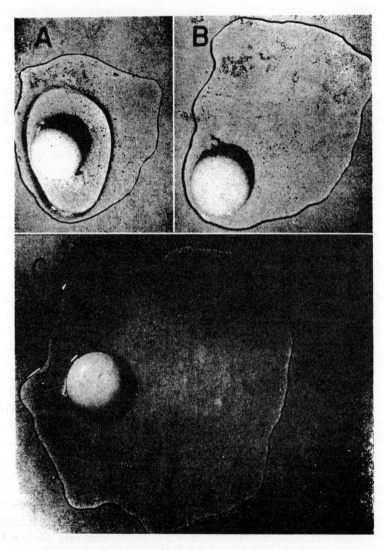

Fig. 2-7. Effect of adding dilute thioglycol to broken-out eggs (top view). (A) Control after 4 hours at 25°C; (B) 30 p.p.m. thioglycol after 4 hours at 25°C; (C) 300 p.p.m. thioglycol after 4 hours at 25°C. Essentially the same results were obtained by a 2-hour treatment with the same amounts of reducing agent. Reprinted from [160].

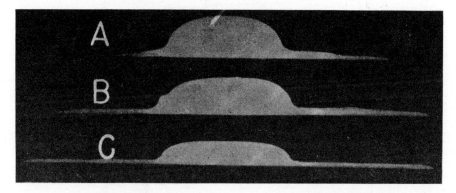

Fig. 2-8. Same as Fig. 2-7 (side view). Reprinted from [160].

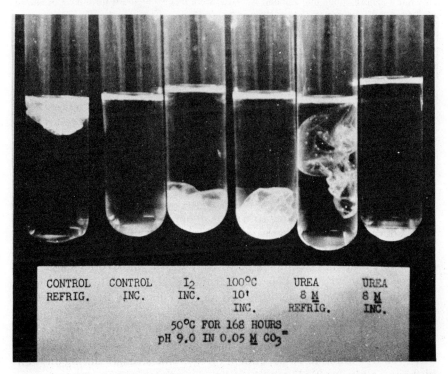

Fig. 2-9. Yolk membrane stability. Pieces of chicken egg-yolk membranes were incubated (INC.) in different media as indicated. Refrigerated (REFRIG.) samples were held for an equal time at approximately 4°C. Several of the treatments prevented the deterioration which occurred in the incubated control membrane (second tube from right) [163].

```
A  37°
B  Control
C  37°
D  Control
D  37°
E  Control
F  37°
```

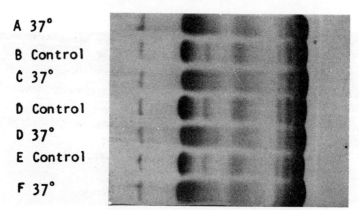

Fig. 2-10. Starch-gel electrophoretic patterns of incubated infertile eggs. Egg whites were all white Leghorn containing globulin $A_1$. Eggs were incubated at 37° for 6 days or stored at 2° for 6 days (controls). Letters refer to hen [59].

The following are several theories that have been proposed for the deteriorative mechanisms in egg white:

1. The ovomucin, a constituent supposedly responsible for the gel nature of the white, is reductively cleaved by reducing agents generated during the incubation of the white [160]. MacDonnell, Silva, and Feeney [71] showed that ovalbumin containing sulfhydryl groups could be titrated with *p*-chloromercuricbenzoate. It was calculated that a "denaturation" of less than 1% of the ovalbumin would be enough to supply the potential reducing agent required to thin the egg white. Changes have been reported in the chromatographic characteristics [158], in the solubility, and in resistance to denaturation [165] of ovalbumin.

2. There is a complex between the lysozyme and the ovomucin that dissociates during incubation [166].

3. There is no initial complex between lysozyme and ovomucin but one forms during incubation. This changes the gel characteristics and causes thinning [167]. The possibility given here is supported by observations that the enzymatic activity of lysozyme in egg white decreases during incubation [168]. The assayable lysozyme activities of egg white dropped extensively (up to 40%) during the thinning process. The activity actually observed was found to be dependent on the ionic strength during dilution of the egg white [168]. These observations indicate an ionic type of complex involving the lysozyme.

4. The solubility of the ovalbumin which represents over half of the protein changes during incubation [165].

5. The interaction of glucose with protein directly or indirectly causes thinning. The glucose interaction could cause thinning indirectly by generating reducing compounds [164].

None of these theories fits all the available data. Chicken, turkey *Meleagris gallopavo* [169], and probably penguin *Pygoscelis adeliae* eggs [170] deteriorate rapidly, while duck and goose eggs essentially show no deterioration. Duck eggs are relatively low in lysozyme and very low in sialic acid, and both of these conditions have been implicated as possibly related to deterioration. The possible role of lysozyme is obvious from the above described hypothetical mechanisms. Sialic acid has been considered because the ovomucin fractions contain sialic acid [69]. As the penguin egg contains even lower amounts of lysozyme (essentially none) and the highest amount of sialic acid observed in any bird, these latter mechanisms seem improbable. A reduction mechanism or a combination of several mechanisms may well be the cause of the changes finally expressed as a change in the colloidal distribution of the high molecular weight aggregates of ovomucin.

# 3
# Egg-White Proteins of Different Avian Species

Birds have been one of the primary orders employed for the study of evolution. One small group of birds, Charles Darwin's dull-looking finches in the Galapagos Islands, was a primary subject in Darwin's *The Origin of the Species* and influenced the consequent world-shaking developments in biological thinking. The thousands of diverse species and variants have continued to make birds a fertile group to study, not only by evolutionists but by scientists in many other fields. A condensed interpretation of the evolutionary and taxonomic relationships (Fig. 3-1) of birds has been presented as Welty's [171] "family tree of birds."

This chapter is concerned with the differences in the proteins of the egg whites of different birds. When these differences are used for considerations of evolutionary relationships, it must be remembered that we are dealing with the secretory proteins of only one organ. In addition we must coordinate our thinking much as David Lack [172] so admirably did when he was discussing and extending Darwin's studies of the finches and wrote, "This is a work of natural history, based on a study of living birds in the Galapagos and of dead specimens in museums. The evidence is circumstantial, not experimental, so that theories must be presented cautiously. They should not, however, be excluded." Many egg-white proteins are homologous with other proteins which have important functions elsewhere in the living animal.

Certain proteins may have special utility for determining evolutionary relationships. According to McCabe and Deutsch [173]:

"It seems possible that the physio-chemical character of an egg contains more of its incipient phylogeny than the more superficial aspects of the birds adult morphology . . . The selection for factors affecting egg-white proteins is probably slow, indirect, and less drastic. This, if true, would allow the various branches of a given phylogenetic stalk to retain a physiological niche with the parent stalk."

Sibley [174] has also supported these general ideas of McCabe and Deutsch and stated that more concentrated genetic information could be obtained from proteins than from any other known available source.

## GENERAL PROPERTIES OF EGGS OF DIFFERENT SPECIES

### GENERAL PHYSICAL CHARACTERISTICS

As is well known, eggs of various species vary tremendously in size, color, shape, etc. However, the physical structures of the contents appear quite similar to those shown in Fig. 2-1. Coloration of the egg whites is primarily yellow and is due to variable contents of riboflavin. All eggs have thick and thin egg whites in proportions similar to those found in chicken eggs; that is, fifty to sixty per cent thick egg white and forty per cent thin egg white. Chalazae exist in all egg whites, although their apparent sizes and opacities vary considerably; for example, the chalaza in the large cassowary egg is nearly transparent and even hard to locate. It could be described as being a slightly more dense portion of the thick egg white. Even the empirical physical characteristics employed by the commercial poultrymen for judging the quality of chicken eggs are in general similar. The yolk and white indices are the numerical ratios of the heights to the widths of the yolks and whites respectively, when the broken-out eggs are placed on a flat surface. The values are proportional to the relative strengths of the materials in the thick egg white and surrounding the yolk membrane. In one comparative experiment the yolk and white indices of several cassowary eggs were 0.35, and 0.09, respectively [77]. Average yolk and white indices of a series of chicken eggs were 0.42 and 0.08, respectively. There are several differences in the physical properties of various egg whites easily demonstratable. One difference is the relative rates of breakdown of the thick egg white and yolk membrane observed on incubation of eggs. Chicken, turkey, Japanese quail, guinea fowl, and Adelie penguin eggs all showed this deterioration [169]. Exactly comparable results were obtained when isolated segments of yolk membranes were incubated as described for chicken yolk membranes in Chapter 2 (Fig. 2-10) [69, 163]. With other treatments the yolk membranes and whites of different species were not affected in different ways. Both chicken and duck membranes were rapidly disintegrated in the presence of very small

Fig. 3-1. A hypothetical family tree of birds showing possible relationships. Some of the kinships indicated here will no doubt be more accurately determined as more precise methods of taxonomic research are developed. Corrections: "Doves and pigeons should be on the same small branch; cotingas should be on the same branch with the tyrant fly catchers and manakins." (Courtesy of Dr. Carl Welty, W. B. Saunders Co., 1968 [171].)

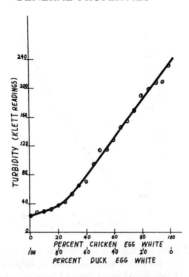

Fig. 3-2. Influence of percentage composition of mixtures of chicken and duck white on the turbidity obtained on dilution in water. Turbidity was produced by dilution with 5 volumes of water as described in text. Reproduced from [169].

concentrations of cupric ion. The addition of reducing chemicals also cause thinning of all egg whites examined.

Another easily seen difference in the properties of egg whites is the turbidity attained on dilution of the white with water. Some egg whites, such as chicken, become very turbid upon dilution with five volumes of water; others like duck, remain essentially clear. Figure 3-2 shows the turbidity obtained on dilution when chicken and duck egg white are mixed in various proportions and diluted. Figure 3-3 shows a similar effect when chicken lysozyme is added to duck egg white before dilution. It is evident that nearly a linear relationship was obtained between turbidity and the amount of lysozyme added, whether the lysozyme was added as chicken egg white or as purified chicken lysozyme. Although there may not be strict relationships between such turbidities and lysozyme contents in the whites of all avian species, there may be some fundamental importance between the physical interactions of lysozyme and the physical properties of avian egg whites.

## FLAVOR AND USE OF EGGS

The use of avian eggs for food has been a preoccupation of humans throughout recorded history. It was once thought that fishy flavors and other flavors unpleasant to humans were necessarily the properties of the

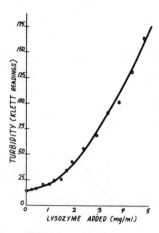

Fig. 3-3. Influence of addition of graded levels of lysozyme to duck white on the turbidity obtained on dilution in water. Turbidity was produced by dilution with 5 volumes of water as described in text. A blank consisting of equivalent amounts of lysozyme diluted in water was subtracted from each reading. (The blank reading subtracted was only 8-12 per cent of the total. Reproduced from [169].)

eggs themselves and perhaps even had some such teleological function as survival characteristics. Whereas these may exist it would not appear to be the general case. It has been found very difficult to distinguish between duck eggs and chicken eggs for many food purposes, particularly when the cooking of the duck eggs were modified to suit the human fancy [156]. The ducks and the chickens were fed the identical ration and no flavor differences whatsoever could be noted. Physical differences were minor but very real. The thick egg white became firmer on cooking so that boiled or fried eggs were frequently not acceptable to some individuals, because it was necessary to overcook the egg white to adequately cook the yolk. This is simply due to the fact that duck egg white coagulates at a lower temperature than chicken egg white. In addition there was an appreciable difference in the amount of sulfide production. A duck egg white produced significantly less sulfide on heating which was also easily evident by the difference in odors of the cooked eggs. Presumably this is due to the fact that duck ovalbumin contains different sulfhydryls than chicken egg white. There are various reports that [175] the egg whites of some species coagulate when cooked without becoming opalescent; for example, the egg white of the Adelie penguin do not become so opalescent as that of the chicken [134]. The penguin eggs are only slightly fishy in flavor and are palatable despite the marine organisms in the diets of the birds [170].

# AMOUNTS OF PROTEINS IN EGG WHITES

## Determination of Values

A major difficulty encountered in determining the contents of the homologous proteins in the different whites are their variations in properties [82, 93, 134]. The troublesome variations are not only in electrophoretic mobilities but also in the qualitative and quantitative activities of proteins.

The three proteins causing particular difficulties are ovalbumin, ovomucoid, and lysozyme. The ovalbumin is a problem because it is heterogeneous in most all species examined and possesses no biochemical activity to aid in its determination other than its content of sulfhydryl. The sulfhydryls are difficult to use since the sulfhydryl contents of the different ovalbumins vary [96]. In addition some whites contain a relatively high amount of another constituent containing a reactive sulfhydryl that will react with Ellmans reagent. Sulfhydryl groups reacting directly with $p$-chloro-mercuribenzoate but not with Ellman's reagent are, nevertheless, an indication of the relative amounts of ovalbumin.

The values for lysozyme content are all in doubt with the exception of chicken, turkey, duck, Adelie penguin, and the ratites (cassowary, emu, kiwi, ostrich, and rhea). These are separated in pure enough form to be able to estimate the amounts present in the original egg whites by comparisons with the purified lysozymes. Both the goose [176] and the kiwi [82] have specific activities considerably different from chicken lysozyme. All other values in Table 3-1 are subject to large errors until specific activities of each lysozyme are available.

The ovomucoids present still a different problem, one involving specificity and relative activity. Some ovomucoids are very poor inhibitors of trypsin but very good of chymotrypsin. It is absolutely essential to isolate the ovomucoid for each species and to determine its properties before estimating the contents in the egg whites. In Table 3-1 the values for the following species are consequently the most accurate: chicken, turkey, duck, Adelie penguin, ring-neck pheasant, golden pheasant, Lady Amherst pheasant, the ratites, and the tinamou.

## Comparisons of Contents

Table 3-1 shows large differences in the contents of the different proteins and no consistent differences in all the proteins in any of the groups. Nevertheless, as with the electrophoretic patterns whites of closely related species were usually more similar to one another than to whites of other species. A possible exception might be seen in the sialic acid contents of ratite and penguin egg whites. The sialic acid content of the egg

TABLE 3-1. PROTEIN COMPOSITION OF AVIAN EGG WHITES[a]

| Common Name | Scientific Name | Whole Egg Weight g | Dry Weight mg/ml | Ovotransferrin % | Ovomucoid % |
|---|---|---|---|---|---|
| | Order: Galliformes | | | | |
| Chicken | Gallus gallus | 60 | 121 | 12 | 11 |
| Red jungle fowl | Gallus gallus | 35 | 115 | 11 | – |
| Turkey | Meleagris gallopavo | 80 | 124 | 11 | 15 |
| Arizona scaled quail | Callipepla squamata pollido | 11 | 115 | 6 | – |
| California valley quail | Lophortyx californica | 9 | 108 | 5 | – |
| Texas bobwhite quail | Colinus virginianus texanus | 9 | 136 | 6 | – |
| Harlequin quail | Coturnix delegorguei | 8 | 111 | 15 | – |
| Philippine button quail | Coturnix c. lineata | 4 | 117 | 16 | – |
| Japanese quail | Coturnix c. japonica | 6 | – | 12 | – |
| Australian swamp quail | Coturnix ypsilophorus | – | – | 12 | – |
| Ring-necked pheasant | Phasianus colchicus | – | – | 12 | – |
| Reeves pheasant | Syrmaticus reevesi | 30 | 128 | 11 | – |
| Lady Amherst pheasant | Chrysolophus amherstiae | 30 | 108 | 16 | – |
| Blue eared pheasant | Crossoptilon auritum | – | – | 13 | – |
| Gray's francolin | Pternistes leucoscepus | 30 | 129 | 10 | – |
| Erckel's francolin | Francolinus erckeli | – | 128 | 12 | – |
| Helmeted guinea fowl | Numida meleagris | 40 | 134 | 9 | – |
| Green Java peafowl | Pavo muticus | 110 | 126 | 9 | – |
| Indian blue peafowl | Pavo cristatus | 100 | 105 | 10 | – |
| Indian chukor | Alectoris graeca | 35 | – | 8 | – |
| | Order: Anseriformes | | | | |
| Duck | Anas platyrhynckos | 60 | 132 | 2 | – |
| Goose | Anser anser | 150 | 133 | 4 | – |
| Black swan | Cygnus atratus | – | 135 | 4.5 | 16 |
| | Order: Rheiformes | | | | |
| Rhea | Rhea americana | 600 | 110 | 3 | 10 |
| | Order: Casuariiformes | | | | |
| Cassowary | Casuarius casuarius | 650 | 114 | 10 | 15 |
| Emu | Dromiceius novae-hollandiae | 650 | 101 | 10 | 20 |
| | Order: Struthioniformes | | | | |
| Ostrich | Struthio camelus | 1300 | 113 | 3 | 10 |
| | Order: Tinamiformes | | | | |
| Tinamou | Eudromia elegans | 30 | 118 | 12 | 20 |
| | Order: Columbiformes | | | | |
| Galapagos dove | Nesopelia galapagoensis | 7 | 125 | 9 | – |
| Pigeon | Columba livia | 18 | 101 | 9 | – |
| | Order: Psittachiformes | | | | |
| Masked lovebird | Agapornis personata | – | 126 | – | – |
| | Order: Sphenisciformes | | | | |
| Adelie penguin | Pygoscelis adeliae | 123 | 118 | 4.5 | 10 |
| Royal penguin | Eudyptes chrysolophus schlegeli | 130 | 110 | 5.0 | 10 |
| | Order: Procellariiformes | | | | |
| Albatross | Diomedea immutabilis | – | 115 | – | – |
| | Order: Apterygiformes | | | | |
| Kiwi | Apteryx australis mantelli | 350 | 123 | 10 | 8 |
| | Order: Charadriiformes | | | | |
| Banded plover | Zonifer tricolor | – | 107 | 3 | – |

TABLE 3-1 *(Continued)*

| Ovo-inhibitor % | Lysozyme % | Ovomacro-globulin % | Flavo-protein % | Avidin units/g | Ovomucin % | Sulf-hydryl μmoles/g | Sialic Acid % |
|---|---|---|---|---|---|---|---|
| 1.5 | 3.4 | 0.5 | 0.7 | 11.5 | 2.9 | 47 | 0.29 |
| – | 4.2 | – | – | – | – | – | 0.31 |
| 0.5 | 3.1 | 0.0 | 0.4 | 16.2 | – | 33 | 0.97 |
| – | 2.8 | – | – | – | – | 36 | 0.18 |
| – | 3.0 | – | – | – | – | 42 | 0.20 |
| – | 1.9 | – | 0.7 | – | – | 36 | 0.31 |
| – | 3.1 | – | – | – | – | 33 | 0.22 |
| – | 4.3 | – | – | – | – | – | – |
| – | – | – | – | – | – | – | – |
| – | – | – | – | – | – | – | – |
| – | 1.4 | – | 0.4 | – | – | 24 | 0.32 |
| – | 1.9 | – | 0.6 | – | – | 27 | 0.37 |
| – | – | – | – | – | – | – | – |
| – | 1.4 | – | – | – | – | 49 | 0.22 |
| – | 2.2 | – | – | – | – | – | 0.27 |
| – | 2.2 | – | 0.48 | – | – | 34 | 0.38 |
| – | 2.8 | – | 0.4 | – | – | 42 | – |
| – | 2.3 | – | 0.6 | – | – | – | 0.62 |
| – | – | – | – | – | – | – | – |
| – | 1.2 | – | 0.3 | 7.1 | – | 25 | 0.11 |
| – | 0.6 | – | 0.3 | 8.0 | – | 36 | 0.12 |
| – | 1.6 | 0.6 | – | – | – | 42 | 0.25 |
| 0.5 | 2.0 | 0.5 | 0.5 | – | 4.6 | 27 | 1.4 |
| 0.6 | 0.5 | 0.5 | 0.8 | – | 2.4 | 21 | 2.3 |
| 0.5 | 0.05 | 0.2 | 2.0 | – | – | 24 | 3.1 |
| 0.6 | 0.45 | 0.5 | 0.3 | – | 2.8 | 27 | 2.1 |
| 0.5 | 0.83 | 1.0 | 0.6 | – | 1.4 | 19 | 0.13 |
| – | 0.1 | – | – | – | – | 24 | 0.62 |
| – | 0.1 | – | – | – | – | – | 1.3 |
| – | <0.02 | – | – | – | – | – | 0.19 |
| – | 0.05 | 1 | 0.3 | 5.0 | – | 30 | 5 |
| – | <0.02 | 0.8 | – | – | – | 30 | 3.7 |
| 0.50 | – | – | – | – | – | – | – |
| 2.3 | 5 | 0.0 | 0.6 | – | – | 24 | 0.2 |
| – | 0.05 | – | 0.1 | – | – | – | 0.78 |

*a*Data from [77, 78, 62, 134, and 163].

white available from two species of penguins is the highest observed in any egg white. The sialic acid content of the egg white of all the ratites, with the exception of the kiwi, is also very high. As discussed at the end of this chapter, however, other evidence indicates only a distant relationship of the penguin to the ratites.

In all whites examined so far the sialic acid has been combined with protein [69]. Only low or insignificant amounts were found without preliminary hydrolysis by acid or by enzyme. Sialic acid was released from all whites examined at a similar rate by the enzyme neuraminidase. In addition the individual sialic acids prepared by enzymatic hydrolysis from the whites of chicken, emu, turkey, goose, cassowary, and guinea fowls had identical $R_f$ values when chromatographed on paper, indicating structural identity of the sialic acids.

Lysozyme contents varied from amounts which were undetectable in the lovebird and very low in the Adelie penguin to as high as 4.2% in the red jungle fowl egg white. These large differences in lysozyme content were not correlated with any other characteristic with the possible exception of the amount of turbidity occurring on dilution of the egg white with water. The higher content in red jungle fowl white as compared to the domestic chicken white is probably significant because the lysozyme content was found to vary approximately 20% in different breeds of chicken [138].

The concentration of ovotransferrin in the various whites did not vary as much percentagewise as the lysozyme activity varied, but the absolute differences were much greater for ovotransferrin than for lysozyme. In addition, all eggs examined were apparently devoid of any of the iron complex of the ovotransferrin as judged by the absence of a salmon-pink color.

The contents of ovoflavoprotein and the apoprotein vary not only as to the total amount of ovoflavoprotein but also as to the percentages existing as ovoflavoprotein and apoprotein. All species have relatively small amounts of ovoflavoprotein as compared to higher contents found for other proteins but the relative differences are approximately tenfold. In addition there are large differences in the percentages of ovoflavoprotein and apoprotein from a low of 14% flavoprotein for the emu to a high of approximately 50% for the guinea fowl. It has previously been noted, however, that the absolute amount of riboflavin in the egg white of the chicken depends directly on the diet. There are some species such as the duck, goose, and Adelie penguin which are apparently incapable of putting riboflavin in the egg white, regardless of the presence of high levels of riboflavin in the diet [57, 156, 170]. The flavoproteins all appear to be similar to chicken flavoproteins in that they are nonfluorescent and gave sharp end-points in the fluorescent titration. In addition the amount of yellow color in the egg white appeared to be approximately proportional to the amount of riboflavin found by chemical determination.

The ovomucoid content did not vary as much percentagewise as the lysozyme content but the absolute differences in content were as large as those of the ovotransferrins. It is critical, however, to emphasize the problems as regards the specific activity and specificity of the ovomucoids. Assays for ovoinhibitor for inhibition of fungal proteinase are included for only some of the species.

The data for avidin contents are limited, but relatively large differences are reported. For example, in studies from our laboratory, turkey, duck, and goose egg whites were found to contain approximately 300%, 40%, and 10% of the amount of avidin as in chicken egg white [77]. These values vary 30-fold. In addition to the limited data of Table 3-1 there are data from Jones and Briggs [177] and Hertz and Sebrell [178] (Table 3-2) which show an 80-fold variation.

TABLE 3-2.  AVIDIN CONTENT OF EGG WHITES

| Common Name | Scientific Name | Avidin Content (units/gram dried weight) |
|---|---|---|
| Chicken | Gallus gallus | 11.5 |
| Turkey | Meleagris gallopavo | 16.2 |
| Duck | Anas platyrhynchos | 7.1 |
| Goose | Anser anser | 8.0 |
| English Sparrow | Passer domesticus | 7.2 |
| Ostrich | Struthio camelus | 1.7 |
| Herring Gull | Larus argentatus | 0.1 |
| Great Cormorant | Phalacrocorax carbosinensis | 2.6 |
| Grackle | Quiscalus quiscula | 8.3 |
| Adelie Penguin | Pygoscelis adeliae | 5.0 |
| Robin | Tardus migratorius | 3.8 |
| European Coot | Fulica atra | 4.3 |
| Crow | Corvus brachyrhynchos | 7.5 |
| Agami Heron | Agamia agami | 2.8 |
| Red-throated Loon | Gavia stellata | 2.0 |
| Starling | Sturnus vulgaris | 6.1 |
| Zebra Finch | Taeniopygia castanea | 7.4 |
| Budgerigar | Melopsittacus undulatus | 6.8 |

Data from [177] and [178].

## IMMUNOLOGICAL COMPARISONS

Egg white and, in particular, chicken ovalbumin, have been used in many immunological studies. In 1938 Landsteiner [179] showed the immunological relationships of several avian species. Landsteiner's experiments are now classical and used both ovalbumins and hemoglobins of several species. More extensive data on egg whites using three different proteins, ovalbumin, ovotransferrins, and lysozyme came from Deutsch's laboratory at Wisconsin in the early nineteen fifties [180]. Turkey, guinea hen, pheasant, duck, and goose whites were analyzed with antibodies to the chicken proteins and were found to cross-react in approximately the order given (Table 3-3). These investigators also obtained evidence indicating that a major portion of the antibody response (immunogenicity) of rabbits to chicken egg white was directed against minor unidentified protein components. These minor components were present in the other egg whites examined and cross-reacted strongly.

Ovalbumins of different species have been compared immunologically by many workers [181–183]. Allan Wilson and co-workers [184] have studied the complement fixation by antiserum against chicken ovalbumin with a large number of different egg whites in twenty-two taxonomic groups. A summary of his findings are given in Table 3-4. These immunological titrations appear to agree well with most taxonomic classifications, although there are a few irregularities. This type of study has now been extended to lysozyme. Table 3-5 illustrates the "indeces of similarity" reported by Arnheim and Wilson [185].

Recent studies in our laboratory under the leadership of Dr. Herman Miller [60] have employed immunoelectrophoresis (Figs. 3-4 to 3-6) and immunodiffusion (Figs. 3-7, 3-8) for comparative immunological studies. Antibodies were prepared in the rabbit against chicken, Japanese quail, duck, and cassowary egg whites, as well as to purified chicken ovalbumin and ovotransferrin and cassowary ovotransferrin. Again, the degrees of cross-reactions agreed with most taxonomic classifications with the exception that comparisons of the cross-reactions to ovotransferrins and ovalbumin showed no parallelism between the cross-reactions of the two proteins. In confirmation of the observations of Deutsch's laboratory [186] a minor component was found to be strongly antigenic and to show extensive cross-reactions between all species examined. This component was identified as component 18 (now named ovomacroglobulin) noted on starch-gel electrophoresis [58, 93]. Figure 3-4 identifies this component. The cross-reactivity of the ovomacroglobulins was so extensive that antibodies to either cassowary or chicken egg white reacted strongly with the ovomacroglobulins of thirteen other species studied.

TABLE 3-3. CROSS-REACTIONS OF VARIOUS EGG WHITES WITH RABBIT ANTIBODY TO CHICKEN CONALBUMIN, OVOMUCOID, AND LYSOZYME

| Egg White Source of Protein | Percentage Cross-reaction | | |
|---|---|---|---|
| | Conalbumin | Ovomucoid | Lysozyme |
| Chicken | 100 | 100 | 100 |
| Turkey | 36 | 18 | 70 |
| Guinea hen | 27 | 37 | 49 |
| Pheasant | | 23 | 33 |
| Duck | 8 | 20 | Very weak |
| Goose | 8 | 21 | Negative |

Reproduced from [180].

TABLE 3-4. QUANTITATIVE IMMUNOLOGICAL COMPARISON OF BIRD OVALBUMINS[a]

| Taxonomic Group | Index of Dissimilarity[b] |
|---|---|
| Order: Galliformes | |
| Domestic chicken, *Gallus gallus* | 1.00 |
| Burmese red jungle fowl, *Gallus gallus* | 1.05 |
| Domestic turkey, *Meleagris gallopavo* | 2.2 |
| Ruffed grouse, *Bonasa umbellus* | 2.2 |
| Domestic guinea fowl, *Numida meleagris* | 3.2 |
| Japanese quail, *Coturnix coturnix* | 3.8 |
| Ring-necked pheasant, *Phasianus colchicus* | 7.5 |
| Order: Anseriformes | |
| Domestic duck, *Anas platyrhynchos* | 7.0 |
| Mallard duck, *Anas platyrhynchos* | 7.0 |
| Canada goose, *Branta canadensis* | 6.9 |
| Order: Charadriiformes | |
| Murre, *Uria lomvia* | 11 |
| Gull, *Larus argentatus* | 13 |
| Order: Columbiformes | |
| Pigeon, *Columba livia* | 80 |

[a] Data provided by A. C. Wilson and D. Wachter, Biochemistry Department, University of California, Berkeley [184].
[b] An antiserum (No. 94B3) was prepared by injecting a rabbit with five-times crystallized chicken ovalbumin. The quantitative microcomplement fixation method was then used to detect reactions between this antiserum and egg whites of other species. Most of the egg whites employed in this work were supplied by Dr. C. G. Sibley. The term index of dissimilarity (ID) refers to the relative concentration of antiserum required to produce a given amount of complement fixation. The larger the ID the weaker the cross-reaction (cf. Arnheim and Wilson, 1967) [185].

Wachter (unpublished) has found that the phosphate groups on ovalbumin have a slight effect on its immunological reactivity. Hence the ID values given here may be influenced by variation between species in the number of phosphate residues on ovalbumin. It is conceivable that if variation also occurs in the carbohydrate residue on ovalbumin, this could affect the immunological results as well.

TABLE 3-5. CROSS-REACTIONS OF LYSOZYMES FROM VARIOUS SPECIES WITH ANTI-CHICKEN LYSOZYME

| Species | Index of Dissimilarity[a] | |
|---|---|---|
| | Antiserum A24 | Antiserum A54 |
| Fowl | | |
| Domestic chicken | 1.00 | 1.00 |
| Burmese red jungle fowl | 1.00 | 1.00 |
| Partridges | | |
| Sharp's francolin | 1.02[b] | 1.07 |
| Chukar | 1.13[b] | 1.10 |
| American quails | | |
| Bobwhite quail | 1.04[b] | 1.01 |
| California quail | 1.08[b] | 1.09 |
| Other gallinaceous birds | | |
| Domestic guinea fowl | 1.33 | 1.45 |
| Ruffed grouse | 1.35 | 1.32 |
| Blue-eared pheasant | 1.37 | 1.60 |
| Lady Amherst's pheasant | 1.44 | 1.60 |
| Blue peafowl | 1.52 | 1.31 |
| Golden pheasant | 1.63 | 1.48 |
| Domestic turkey[c] | 1.66 | 1.16 |
| Swinhoe pheasant | 1.81 | 1.87 |
| Japanese quail | 1.86 | 1.81 |
| Reeves's pheasant | 1.88 | 1.70 |
| Ring-necked pheasant | 2.20 | 1.70 |

[a]Average for at least two experiments. Reproduced from [185].
[b]Experiments with a third antiserum confirm that these lysozymes are immunologically distinct from chicken lysozyme, the index of dissimilarity for all four species being between 1.1 and 1.2.
[c]Experiments with three other antisera gave the following values for turkey lysozyme: 1.33, 1.37, and 1.63.

# COMPARATIVE ELECTROPHORETIC STUDIES

## Earlier Studies

One of the simpler techniques for comparative studies is the determination of the electrophoretic mobility of the individual proteins or a mixture of the proteins. Deutsch showed that the electrophoretic technique was discriminatory and descriptive of the general proteins in the egg whites of thirty-seven different species [173], [187]. These earlier data of Deutsch's laboratory are also valuable, because the free boundary electrophoretic technique allowed Deutsch to calculate mobilities of individual components which are more difficult by procedures employing fixed supports. No patterns were found completely identical and the electrophoretic patterns showed similarities between most species that were taxonomically related (Fig. 3-9). There were, nevertheless, some discrepancies that pointed up the necessity for more discriminatory physical and chemical procedures such as electrophoresis to aid in taxonomy. In these studies free boundary electrophoresis was employed, which is not so discriminatory and much more difficult than some of the newer and simpler gel-electrophoretic techniques.

Nearly a decade later Sibley [174] published his paper electrophoretic results. Sibley determined the electrophoretic profiles of the egg-white proteins of three-hundred-fifty-nine species of nonpasserines and at least three hundred species of passerines. Figure 3-10 is a reproduction of one of Sibley's figures selected to show the differences and similarities. The Lady Amherst and the golden pheasants are very closely related and interbreed readily. In fact there are some that consider them breeds. Their paper electrophoretic profiles are nearly identical whereas those of the pheasants have markedly different patterns. The patterns of the domestic fowl and Sonneret's jungle fowl show a similar relationship. Sibley attempted to summarize his data by comparing the peak distances of electrophoretic mobilities (Fig. 3-11). Such paper electrophoretic methods are being displaced by more definitive methods. Sibley himself stated, "Before this paper appears in print it is expected that more sensitive methods

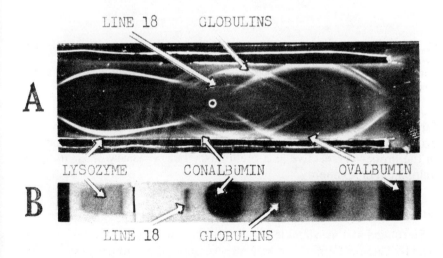

Fig. 3-4. Comparisons of immunoelectrophoretic and starch-gel patterns of who' chicken egg white. Antichicken egg-white serum used for development of immunoele' phoresis. Reproduced from [60], *Academic Press*, 1964.

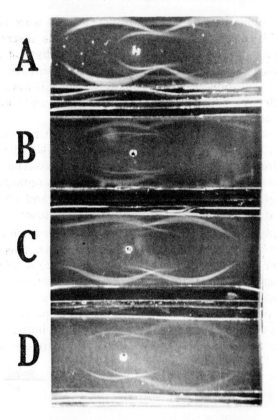

Fig. 3-5. Immunoelectrophoresis of ratite egg whites. Anticassowary white used for development in both top and bottom troughs. (A) Cassowary, (B) rhea, (C) emu, and (D) ostrich followed by immunodiffusion against cassowary egg-white antiserum in both troughs. Reproduced from [60], *Academic Press*, 1964.

for the detection of the evolutionary and genetic information contained in protein structure will be in use. The present procedure will soon be seen in the proper perspective as a preliminary but necessary phase in the development of taxonomy."

GEL ELECTROPHORETIC STUDIES

Starch-gel and acrylamide-gel electrophoresis have proven to be powerful techniques for performing general comparative studies of egg whites and have been used in most of the studies in our laboratory. Egg whites of all the different avian species examined to date have shown different and distinguishable patterns in starch-gel electrophoresis. Starch-gel electrophoresis gave much better definition than paper electrophoresis and consequently a separation of constituents in a manner that made it possible to count and even estimate the amounts of some of the different constituents.

Figures 3-12 and 3-13 are reproductions of examples of starch gels with several different avian species. Figure 3-14 shows planar sections of the same gel used to locate and identify the ovotransferrins by staining and radiography. In these experiments $^{59}$Fe was added to the egg white prior to electrophoresis. A horizontal section of the gel was used to make the radioautograph (Fig. 3-14B) whereas a second section of the gel was stained for the gel pattern (Fig. 3-14A). Most of the egg whites of the species listed in Table 3-1 have been studied by starch-gel electrophoresis

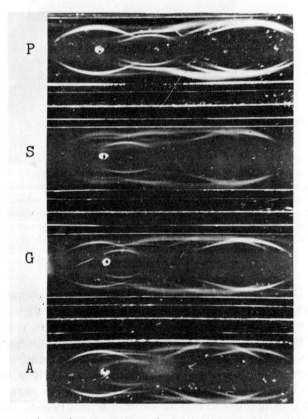

Fig. 3-6. Immunoelectrophoretic patterns of whole avian egg whites. Antisera in both top and bottom troughs were anti-penguin egg-white sera. Egg whites electrophoresed in center. (P) penguin egg white; (S) pink-footed shearwater egg white; (G) grebe egg white; (A) albatross egg white. Reproduced from [134], *Pergamon Press,* 1964.

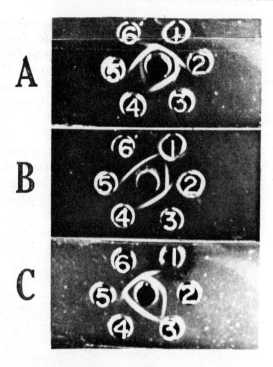

Fig. 3-7. Immunodiffusion against anti-cassowary conalbumin serum. The antigens were: (A): 1, cassowary conalbumin; 2, acetylated cassowary conalbumin; 3, rhea egg white; 4, duck egg white; 5, ostrich; and 6, emu egg white; (B): 1, quail conalbumin; 2, acetylated cassowary conalbumin; 3, emu egg white; 4, golden pheasant egg white; 5, tinamou egg white; and 6, cassowary conalbumin; (C): 1, acetylated cassowary conalbumin (ppt.); 2, acetylated cassowary conalbumin; 3, duck egg white; 4, emu egg white; 5, quail conalbumin; and 6, cassowary conalbumin. Reproduced from [60], *Academic Press*, 1964.

[78, 82]. Differences were noted between all species examined. The division of the ovalbumin area into several components, as exist in chicken, are quite evident, but the proportions of the components vary. New proteins, or homologous proteins very different from those found in the chicken are evident in many species. The most extensive differences were observed in the ovotransferrin area, particularly with the ratite group. As is evident from the radioautograph of Fig. 3-14 the ratites can be arranged in an order so that the ovotransferrins become progressively more alkaline as judged by their mobilities; in addition the ovotransferrins separated into several multiple forms.

Many other laboratories have made important contributions to the characterization of the egg-white proteins of different avian species. Stratil and Valenta [188] have shown different genetic loci in goose egg whites (Fig. 3-15). Baker and co-workers [140], [189–191] have examined many different species and variants within species. The occurrences of different ovalbumin types in populations of two closely related pheasants as illustrated by Baker are given in Table 3-6. Lush [142] has also studied the ovalbumin locus in various species. Sibley has recently repeated his previous studies but using starch-gel electrophoresis (Fig. 3-16).

Such electrophoretic techniques show differences only in electrophoretic migrations and are probably very useful for demonstrating taxonomic relationships. Although these procedures are powerful adjunctive tools, more definitive studies at a molecular level are required to provide information about the architecture and function of the molecules.

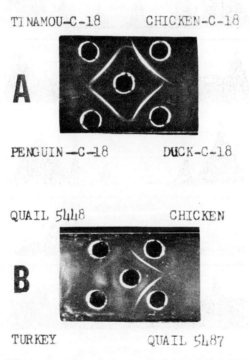

Fig. 3-8. Immunodiffusion patterns of (A) purified component 18 from four species against anti-chicken component 18; and (B) various whole egg whites against anti-chicken component 18. Rabbit antiserum to chicken C-18 was in the center cells. Reproduced from [61].

## PROPERTIES OF PROTEINS OF DIFFERENT SPECIES

The physical and chemical properties of the egg-white proteins of birds other than the chicken are just beginning to receive general attention. However, most of the studies from laboratories other than our own have so far been with lysozymes.

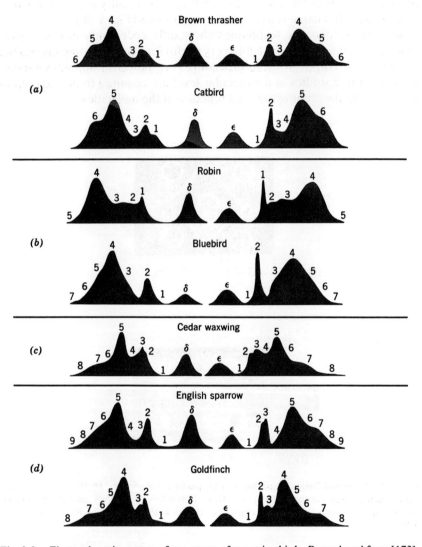

Fig. 3-9. Electrophoretic patterns for a group of passerine birds. Reproduced from [173].

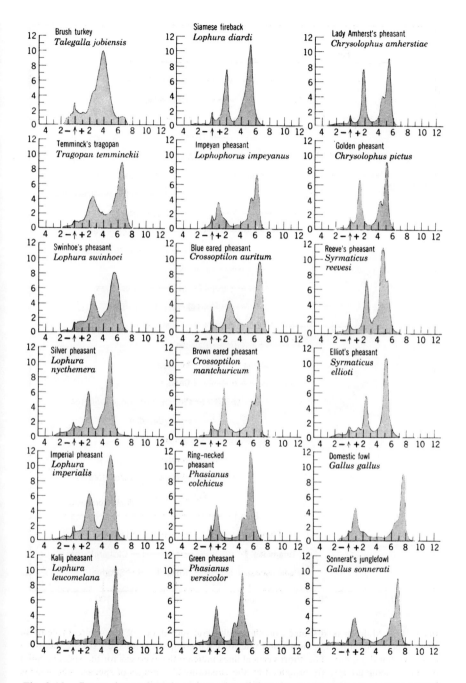

Fig. 3-10. Paper electrophoretic patterns for Galliformes (part). Reproduced from [174].

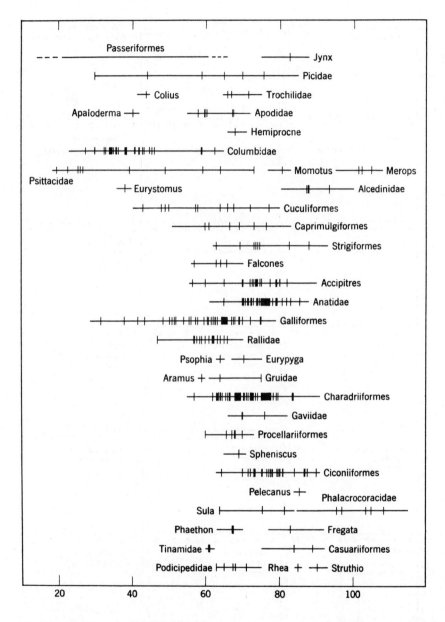

Fig. 3-11. Main peak distances of electrophoretic movement of egg-white proteins of some groups of birds. The horizontal line for each group includes the extreme measurements within the group. The short vertical lines indicate the averages for the species. Solid blocks, in some groups, are produced by the clustering of averages of species. The scale is in millimetres from the application point. Reproduced from [174].

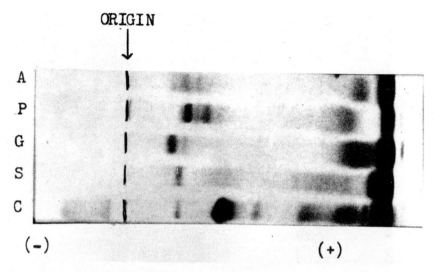

Fig. 3-12. Starch-gel electrophoretic patterns of various species of egg white. Performed at pH 8.6. Samples are: (A) albatross; (P) penguin; (G) grebe; (S) Pink-footed shearwater; (C) chicken. Reproduced from [134], *Pergamon Press*, 1966.

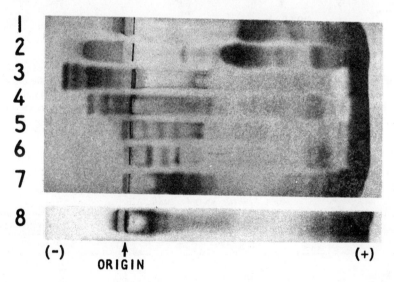

Fig. 3-13. Starch-gel electrophoretic patterns of egg whites. The discontinuous buffer system at pH 8.6 was used: Tris-citric acid (0.076 M Tris) with 2 M urea in the gel buffer; boric acid-NaOH(0.3 M boric acid) in the bridge buffer. The run was conducted at room temperature for 18 hours with 15 mA and 7 V/cm, and stained with aniline blue-black. The egg-white samples were 1, turkey; 2, chicken; 3, cassowary; 4, emu; 5, ostrich; 6, rhea; 7, tinamou; 8, kiwi. Reproduced from [82], *Academic Press*, 1968.

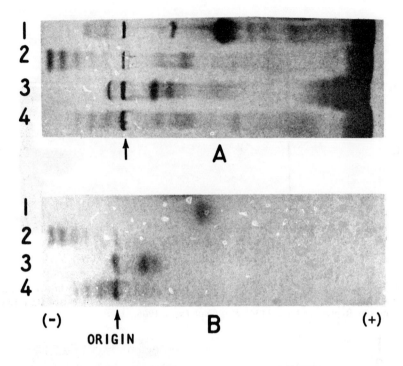

Fig. 3-14. Starch-gel electrophoretic experiment of $^{59}$Fe-treated egg whites. Electrophoretic conditions were the same as described in Figure 3-13. A and B are from the same gel. A, stained by aniline blue-black, and B, X-ray film exposed to gel. The samples were 1, chicken; 2, cassowary; 3, kiwi; and 4, emu. Reproduced from [82], *Academic Press*, 1968.

TABLE 3-6. THE OCCURRENCE OF THE OVALBUMIN TYPES
OF *C. pictus* AND *C. amherstiae*

| Population | *C. pictus* Ovalbumin Types | | | *C. amherstiae* Ovalbumin Types | | |
|---|---|---|---|---|---|---|
| | Fast | Slow | Both | Fast | Slow | Both |
| Savoy | — | 3 | 3 | 1 | — | 1 |
| Yorkville | — | 10 | 6 | 1 | 2 | 5 |
| Total | — | 13 | 9 | 2 | 2 | 6 |

Reproduced from [190], Pergamon Press, 1965.

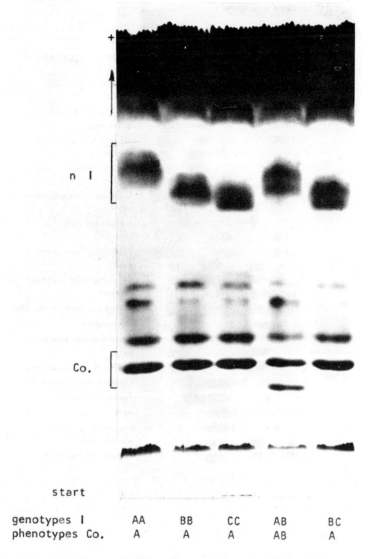

Fig. 3-15. Starch-gel electrophoresis patterns of goose egg whites. Two polymorphic regions, (1) and conalbumin (Co.). Reproduced from [188].

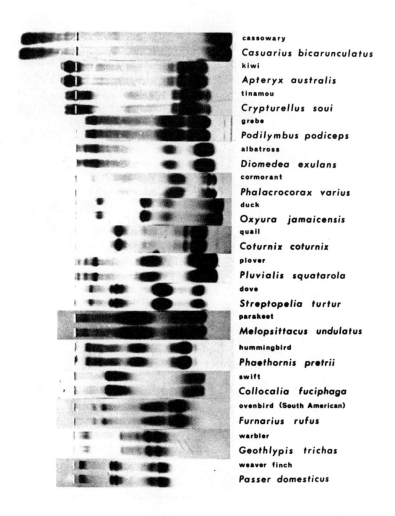

Fig. 3-16. Examples of the starch-gel electrophoretic patterns of avian egg-white proteins. Reproduced from [18].

## Preparation of Proteins

Many experiments have been done on the fractionation of proteins of the different species in our laboratories. The most detailed studies have been made with the egg whites of the turkey, duck, Adelie penguin, cassowary, emu, kiwi, ostrich, and rhea [77, 78, 81, 99, 134, 192]. Other species as shown in Table 3-1 were also studied. These egg whites were fractionated by modification of the methods used for chicken egg white. The modifications depended on the particular egg white and on the properties of the individual constituents of each egg white.

## Ovotransferrins

The ovotransferrins were probably the easiest to study in a general comparative way because of the simplicity of identifying them in a gel by radioautography (Fig. 3-14). The ovotransferrins also gave strong precipitin lines with heterologous sera in immunoelectropherograms (Figs. 3-4 to 3-6). All ovotransferrins apparently exist in multiple forms. In some species five or six forms have been observed. They vary the greatest in isoelectric points of all of the egg-white proteins. The range of isoelectric points was from pH 6.0 for the chicken to slightly above pH 9.0 for the cassowary [78].

The metal binding properties of all ovotransferrins are nearly the same [78], [192]. Two atoms of iron or copper can be chelated to produce red or yellow complexes, respectively. Electron spin resonance studies of the complexes of chicken, turkey, and Japanese quail ovotransferrins gave identical results [192]. Small differences in absorption maxima of iron complexes were observed.

Compositions of a series of ovotransferrins are compared in Table 3-7. Many similarities are obvious, particularly in certain residues such as histidine.

## Ovomucoids

The ovomucoids have shown the greatest physical, biochemical, and chemical differences. They all existed in multiple forms that differed in some species by their contents of sialic acid [99].

The most striking differences between ovomucoids from different species were their different specificities for inhibiting proteolytic enzymes [81, 82, 93, 95, 193]. These different specificities are listed in Table 3-8: some inhibit trypsin (chicken); some inhibit $\alpha$-chymotrypsin (golden pheasant); and some inhibit both trypsin and $\alpha$-chymotrypsin (turkey). Penguin ovomucoid has still a further type of activity. It is a strong inhibitor of the bacterial proteinase, subtilisin, and a good inhibitor of trypsin. But it is only a poor inhibitor of chymotrypsin. The use of the homologous ovomucoids as tools for studying the mechanism of interaction between the ovomucoids and the proteolytic enzymes has been actively pursued in our laboratory [193-196]. One recent pertinent finding is that only one of twelve ovomucoids tested had inhibitory activity against a partially

TABLE 3-7. CHEMICAL COMPOSITION OF OVOTRANSFERRINS

| Constituents | Ovotransferrins | | | | |
|---|---|---|---|---|---|
| | Chicken[a] | Cassowary[b] | Tinamou[a] | Kiwi[a] | Penguin[a] |
| | Residues per Mole (76,000 g)[c] | | | | |
| Alanine | 54.5 | 51.8 | 45.8 | 45.8 | 51.1 |
| Arginine | 29.6 | 28.4 | 21.8 | 23.6 | 20.4 |
| Aspartic acid | 76.0 | 71.6 | 68.0 | 63.8 | 70.1 |
| CySH | 24.9 | 22.9 | 25.9 | 24.1 | 25.9 |
| Glutamic acid | 69.1 | 68.6 | 59.1 | 69.1 | 65.1 |
| Glycine | 52.1 | 59.7 | 59.7 | 56.2 | 52.5 |
| Histidine | 11.9 | 10.9 | 11.2 | 12.5 | 13.3 |
| Isoleucine | 25.9 | 28.8 | 23.3 | 28.6 | 27.4 |
| Leucine | 49.9 | 46.7 | 51.1 | 45.8 | 50.5 |
| Lysine | 57.8 | 62.1 | 60.6 | 54.3 | 57.8 |
| Methionine | 10.7 | 9.6 | 9.7 | 8.8 | 7.8 |
| Phenylalanine | 25.8 | 22.0 | 23.8 | 22.4 | 23.6 |
| Proline | 28.1 | 21.8 | 24.9 | 26.4 | 22.6 |
| Serine | 46.0 | 38.5 | 45.1 | 44.4 | 51.7 |
| Threonine | 36.6 | 36.1 | 39.7 | 36.5 | 35.0 |
| Tryptophan[d] | 11.7 | 12.0 | 12.8 | – | 11.6 |
| Tyrosine | 19.8 | 25.5 | 31.2 | 23.4 | 23.3 |
| Valine | 51.4 | 35.6 | 38.5 | 37.0 | 41.9 |
| Sialic acid | 0.0 | 1.1 | 0.2 | 0.0 | 0.4 |
| Hexose | 4.3 | 13.3 | 12.2 | – | 11.3 |
| Glucosamine[e] | 4.6 | 10.3 | 8.5 | 7.6 | 7.1 |
| N-Terminal | Alanine | Alanine | Alanine | Alanine | Alanine |
| Total recovery | 76,800 g | 76,500 g | 75,600 g | 73,200 g[f] | 75,900 g |

[a]Values for amino acids obtained as described in Table 3-9. Published data from [82].
[b]Values for amino acids obtained as described in Table 3-9 with exceptions that only 22-hour acid hydrolysates used.
[c]The value of 76,000 g was used for the molecular weight of chicken ovotransferrin [40] and assumed for the other proteins.
[d]The values of chicken ovotransferrin for tryptophan are lower than values of 18 residues reported by other workers [40].
[e]Values for glucosamine obtained as described in Table 3-9.
[f]Values for tryptophan of 12 and for hexose of 12 were assumed in the calculations of total recovery for kiwi ovotransferrin.

## TABLE 3-8. PROPERTIES OF OVOMUCOIDS [82], [193]

| Ovomucoids | $S_{20,w}{}^a$ | Isoionic pH[b] | Inhibitory Activity[c] | | |
|---|---|---|---|---|---|
| | | | Bovine Trypsin | Bovine Chymotrypsin | Subtilisin |
| Chicken | 2.3 | 4.48 | +++ | 0 | 0 |
| Turkey | 2.2 | 4.28 | +++ | +++ | + |
| Duck | 2.1 | 4.28 | +++[d] | +++ | ++ |
| Cassowary | 2.1 | 3.90 | ++ | 0 | 0 |
| Emu | 2.2 | 3.78 | +++[d] | + | ± |
| Kiwi | 2.2 | nd[e] | ++ | ± | nd |
| Ostrich | 2.2 | 3.97 | ++ | ± | 0 |
| Rhea | 2.1 | 3.82 | +++ | ± | ± |
| Tinamou | 2.2 | 4.73 | + | +++ | ++ |
| Penguin | 2.1 | 3.62 | ++ | + | +++ |

[a] 1.0% solutions in 0.05 $M$ phosphate buffer at pH 7.0. Data summarized from [82] and [193].
[b] Value for kiwi ovomucoid not determined.
[c] Ovomucoids that have a weak affinity for an enzyme will vary in the apparent inhibitory activity depending on assay procedure. Inhibitory activities were estimated from results of several different procedures. Ovomucoids with the greatest inhibitory activity are marked +++ (strong inhibitor); those with less activity are marked ++ (intermediate), + (weak), ± (very weak), and 0 (no inhibitory activity).
[d] Emu ovomucoid appears to be like duck ovomucoid in inhibiting more than one trypsin depending on the assay conditions.
[e] Not determined.

purified preparation of human trypsin (196a). The ovomucoid exhibiting this inhibitory activity was from Japanese quail egg white. Chicken ovomucoid did not inhibit human trypsin! These studies are discussed in more detail in Chapter 8.

Compositions of several ovomucoids are given in Table 3-9. Large variations in two amino acids, arginine and methionine, are evident. These variations have also been useful as tools in general protein chemistry. Liu et al., [196] have studied the chemical modifications of arginine residues in proteins. The use of ovomucoids with different contents of arginine and without arginine was valuable in determining the specificity and accuracy of the methods.

## Ovomacroglobulin

A primary property of the ovomacroglobulin was its extensive immunological cross-reactivity (Fig. 3-4 to 3-8). Miller and Feeney [61] have found many similar physical and chemical properties in the ovomacroglobulins of four distantly related avian species: chicken, duck, Adelie penguin, and tinamou. Slight differences were shown by comparisons of electrophoretic mobility and by lines of partial identity in immunodiffusion tests (Fig. 3-8). As can be seen from the compositions of four ovomacroglobulins (Table 3-10) only small differences in the percentages of amino acids were noted.

Ovomacroglobulin was found widely distributed in the various flocks of chickens at the University of California at Davis. It was absent from many genetic lines of Japanese quail and not found in twenty turkey egg whites selected to obtain a population sample [197] (see Figure 3-8b).

TABLE 3-9. CHEMICAL COMPOSITION OF OVOMUCOIDS

| Constituents | Ovomucoids[a] | | | | | | | | |
|---|---|---|---|---|---|---|---|---|---|
| | Chicken | Turkey | Cassowary | Emu | Ostrich | Rhea | Tinamou | Duck | Penguin |
| | Residues per Mole (28,000 g)[b] | | | | | | | | |
| Alanine | 11.7 | 8.6 | 7.3 | 7.0 | 5.6 | 5.2 | 8.5 | 8.4 | 5.4 |
| Arginine | 6.3 | 5.8 | 1.0 | 0.0 | 3.4 | 0.0 | 3.5 | 1.2 | 3.1 |
| Aspartic acid | 31.9 | 27.2 | 25.8 | 27.1 | 27.3 | 25.7 | 30.8 | 33.0 | 28.0 |
| CySH | 17.5 | 16.7 | 19.9 | 16.7 | 19.9 | 17.4 | 19.6 | 20.2 | 18.8 |
| Glutamic acid | 14.9 | 19.0 | 17.7 | 16.6 | 18.0 | 20.1 | 18.1 | 20.1 | 15.7 |
| Glycine | 16.1 | 17.3 | 14.3 | 14.9 | 17.0 | 14.6 | 17.6 | 19.2 | 14.7 |
| Histidine | 4.3 | 5.2 | 3.1 | 3.0 | 2.2 | 4.1 | 3.4 | 3.5 | 2.2 |
| Isoleucine | 3.2 | 4.4 | 6.0 | 6.0 | 4.5 | 3.7 | 4.7 | 2.6 | 4.1 |
| Leucine | 12.2 | 13.5 | 12.5 | 13.4 | 15.3 | 12.6 | 9.4 | 13.5 | 12.6 |
| Lysine | 13.6 | 11.2 | 16.5 | 16.2 | 14.8 | 15.8 | 17.1 | 17.2 | 13.5 |
| Methionine | 1.9 | 1.8 | 1.0 | 0.9 | 1.0 | 0.9 | 0.0 | 7.9 | 2.0 |
| Phenylalanine | 5.3 | 3.2 | 3.1 | 3.1 | 3.3 | 5.0 | 4.7 | 4.7 | 2.5 |
| Proline | 7.7 | 8.8 | 10.4 | 8.9 | 11.4 | 9.7 | 14.0 | 10.4 | 9.5 |
| Serine | 12.5 | 10.0 | 14.2 | 13.4 | 18.3 | 16.7 | 12.5 | 13.3 | 13.4 |
| Threonine | 14.6 | 14.2 | 11.8 | 12.2 | 17.0 | 16.2 | 14.0 | 19.9 | 16.0 |
| Tryptophan | 0.0 | 0.0 | 0.0 | 0.0 | 0.0 | 0.0 | 0.0 | 0.0 | 0.0 |
| Tyrosine | 6.7 | 6.8 | 10.6 | 6.4 | 11.1 | 8.2 | 10.4 | 11.5 | 10.1 |
| Valine | 16.0 | 15.7 | 15.4 | 15.1 | 16.9 | 17.9 | 17.6 | 17.4 | 19.4 |
| Sialic acid | 0.3 | 2.2 | 4.7 | 8.9 | 2.9 | 5.3 | 0.1 | 0.0 | 6.8 |
| Hexose | 16.7 | 18.0 | 16.4 | 15.6 | 15.2 | 16.9 | 15.9 | 12.8 | 14.8 |
| Glucosamine[c] | 21.0 | 18.5 | 16.2 | 15.1 | 16.0 | 14.5 | 16.4 | 10.2 | 12.8 |
| N-Terminal | Alanine | Valine | Valine | Valine | Valine | Valine | Valine | Valine | Valine |
| Total recovery | 27,400 g | 27,100 g | 27,300 g | 26,900 g | 28,200 g | 27,200 g | 27,500 g | 28,000 g | 27,100 g |

[a]Values for amino acids are averages of duplicate analyses of 22-hour and 72-hour acid hydrolysates with following exceptions: values for valine, isoleucine, and leucine were averages of duplicate analyses of 72-hour hydrolysates only; values for CySH, threonine, and serine were obtained by extrapolations to zero time; values for tyrosine and tryptophan were determined spectrophotometrically on unhydrolyzed proteins. Published data from [82] and [193].
[b]The value of 28,000 g was used for the molecular weight of chicken ovomucoid (81,53) and assumed for the molecular weights of the other proteins.
[c]The value for glucosamine was obtained by ion-exchange chromatography of 13-hour hydrolysate at 95°.

TABLE 3-10. AMINO ACID ANALYSIS OF OVOMACROGLOBULIN RESIDUES PER 10,000 $g^a$

| Amino Acid | Chicken | Duck | Penguin | Tinamou |
|---|---|---|---|---|
| Aspartic acid | 6.50 | 5.85 | 7.65 | 6.41 |
| Threonine | 4.55 | 4.21 | 4.42 | 3.62 |
| Serine | 5.05 | 5.21 | 5.53 | 4.83 |
| Glutamic acid | 7.82 | 8.11 | 9.05 | 7.21 |
| Proline | 3.65 | 3.75 | 3.65 | 2.91 |
| Glycine | 3.49 | 3.81 | 3.80 | 3.22 |
| Alanine | 3.95 | 4.01 | 4.26 | 3.84 |
| Valine | 5.41 | 5.11 | 6.12 | 5.57 |
| CySH | 1.15 | 1.47 | 0.60 | 1.24 |
| Methionine | 1.43 | 1.61 | 1.05 | 1.44 |
| Isoleucine | 4.42 | 3.52 | 4.02 | 3.73 |
| Leucine | 6.25 | 5.61 | 5.76 | 5.52 |
| Tyrosine | 2.68 | 2.68 | 2.20 | 2.53 |
| Phenylalanine | 3.45 | 3.34 | 3.64 | 3.42 |
| Lysine | 4.11 | 4.05 | 4.63 | 4.04 |
| Histidine | 1.25 | 1.92 | 1.00 | 0.81 |
| Arginine | 2.60 | 2.61 | 2.22 | 2.47 |
| Tryptophan[b] | 0.5 | 0.5 | 0.5 | 0.5 |

Published data from [61].
[a] All values, except those for tryptophan, are averages of duplicate determinations by ion-exchange chromatography on all four proteins after 22-hr hydrolysis and also of duplicate determinations of the chicken and tinamou protein after 72-hr hydrolysis.
[b] Tryptophan values were estimated from ultraviolet absorption spectra and are only approximate.

It was also absent from the white of the single kiwi egg available for study [82].

OVOMUCIN

This constituent is difficult to study because of its physical properties and absence of criteria for determining purity. Nevertheless, comparisons of composition show many similarities (Table 3-11).

LYSOZYMES

Definitive studies are available in comparatively great depth for chicken [86], [198], duck [199], and goose [176] lysozymes (Table 3-12), but few definitive studies have been reported for other lysozymes. All egg-white lysozymes have strongly alkaline (pH 9) isoelectric points as shown by electrophoretic migrations. All appear to have molecular weights near that of chicken lysozyme. But important differences have been noted in specific activities and in the type of catalytic activities.

TABLE 3-11. CHEMICAL COMPOSITION OF OVOMUCINS

| Constituents | Ovomucins (Residues per 10,000 gm)[a] | | | | |
|---|---|---|---|---|---|
| | Chicken | Cassowary | Ostrich | Rhea | Tinamou |
| Alanine | 4.18 | 4.29 | 4.35 | 4.29 | 4.15 |
| Arginine | 2.80 | 2.55 | 2.56 | 2.63 | 2.67 |
| Aspartic acid | 7.49 | 7.52 | 7.60 | 7.72 | 6.91 |
| CySH | 4.72 | 4.82 | 4.75 | 4.30 | 3.80 |
| Glutamic acid | 8.60 | 8.34 | 8.67 | 8.65 | 7.73 |
| Glycine | 4.70 | 5.36 | 5.28 | 5.18 | 5.00 |
| Histidine | 1.63 | 1.53 | 1.66 | 1.82 | 1.60 |
| Isoleucine | 3.58 | 3.58 | 3.84 | 4.14 | 3.50 |
| Leucine | 5.38 | 5.18 | 5.51 | 5.97 | 5.43 |
| Lysine | 4.68 | 4.55 | 4.54 | 4.69 | 4.44 |
| Methionine | 1.50 | 1.26 | 1.35 | 1.42 | 1.21 |
| Phenylalanine | 3.15 | 2.99 | 3.20 | 3.40 | 2.68 |
| Proline | 4.19 | 4.59 | 4.55 | 4.60 | 3.93 |
| Serine | 6.14 | 5.90 | 5.92 | 5.50 | 5.35 |
| Threonine | 5.41 | 5.30 | 5.36 | 5.59 | 5.42 |
| Tryptophan | – | – | – | – | – |
| Tyrosine | 3.33 | 2.44 | 2.58 | 2.59 | 2.62 |
| Valine | 4.99 | 5.42 | 6.02 | 5.79 | 5.15 |
| Sialic acid | 0.61 | 1.05 | 0.72 | 0.54 | 0.37 |
| Hexose | 2.84 | 3.19 | 2.92 | 2.80 | 2.88 |
| Glucosamine[b] | 2.79 | 2.51 | 2.49 | 2.06 | 1.98 |
| Galactosamine[b] | 0.75 | 0.74 | 0.52 | 0.46 | 0.66 |
| Recovery (gm) | 9665 | 9603 | 9680 | 9633 | 8858 |

[a]Values for amino acids were obtained as described in Table 3-9. Data from [82].
[b]Values for glucosamine and galactosamine were averages of duplicate analyses on 22-hour acid hydrolyzates. Analyses were done by ion-exchange chromatography.

Canfield [176] has reported that goose lysozyme has several-fold the activity of chicken lysozyme against a synthetic substrate. A somewhat similar relationship has been found with kiwi lysozyme by Osuga and Feeney [82]. In this case, the kiwi lysozyme was much more active (4 to 5 times the rate of chicken lysozyme) but it caused incomplete reduction of the turbidity of suspensions of *Micrococcus lysodeikticus*.

OVALBUMIN

Ovalbumin has been the most neglected protein in studies of chicken egg white and it has been even more neglected in comparative biochemical studies of other egg whites. Feeney and co-workers [77] have studied partially purified samples from twelve different species and found large differences (nearly two-fold) in electrophoretic migrations and differences in contents of sulfhydryl groups. In further investigations of the sulfhydryl contents [82, 96, 134] it was confirmed that chicken ovalbumin is distinctive in that it contains an additional sulfhydryl group which can be titrated only after denaturation (Table 2-4). Values of 2 or 3 groups per mole have been found in each of the other approximately fifty species studied. The fact that the red jungle fowl was found to have the additional unreactive fourth sulfhydryl group is further confirmation of the identity of the species in *Gallus gallus*.

The ovalbumins appear to have similar molecular weights. One per cent solution of chicken, duck, and turkey ovalbumin were reported to have $S_{20,w}$ values of 2.8, 3.0, and 3.0, respectively [77].

## Other Constituents

Very little information on comparative properties of other constituents is available. Limited information is available on the ovoflavoprotein and ovoinhibitor of the Adelie penguin [134] and the ratites [82]. No comparative information appears to have been published on avidin. A constituent of highly reactive sulfhydryl groups existing in small amounts in chicken

TABLE 3-12. AMINO ACID COMPOSITIONS OF DIFFERENT EGG-WHITE LYSOZYMES

| Amino Acid | Residues/Mole | | |
|---|---|---|---|
| | Chicken [86] | Duck [199] | Goose [176] |
| Ala | 12 | 11 − 12 | 10 |
| Arg | 11 | 13 ± 1 | 6 − 7 |
| Asp | 21 | 19 ± 1 | 13 − 14 |
| CySH | 8 | 8 | 3 − 4 |
| Glu | 5 | 5 | 10 |
| Gly | 12 | 12 | 14 |
| His | 1 | 0 | 3 − 4 |
| Ileu | 6 | 5 − 6 | 9 |
| Leu | 8 | 8 | 4 − 5 |
| Lys | 6 | 6 | 11 |
| Met | 2 | 2 | 2 |
| Phe | 3 | 1 | 2 |
| Pro | 2 | 2 | 2 − 3 |
| Ser | 10 | 10 − 11 | 6 − 7 |
| Thr | 7 | 7 | 8 − 9 |
| Trp | 6 | 5 − 6 | 2 − 3 |
| Tyr | 3 | 4 − 5 | 6 |
| Val | 6 | 6 | 7 |

Data from [86], [199], and [176].

egg white [77] was found in much greater amounts in Adelie penguin egg white [134]. The very high content of sialic acid present in penguin and emu whites (Table 3-1) was not reflected by a high content in the ovomucin fraction, but it was high in other fractions such as the ovomucoid [99, 134, 182]. A constituent found in several species has not been found in chicken white. This is a "doublet protein" with a very alkaline isolectric point (pH 9) easily seen on electrophoresis of Adelie penguin egg white.

Other uncharacterized constituents have also been reported in the whites of various species. Although some of these are doubtlessly present as trace constituents in chicken egg white, perhaps the best way to study them would be to use the whites of the species from which they can be easily obtained.

## TAXONOMIC RELATIONSHIPS

The immunological studies on the cross-reactions of avian egg-white proteins have supplied important information for taxonomic relationships. Newer immunological techniques should continue to be used in conjunction with physical and chemical procedures. The recent studies from Allan Wilson's laboratory [185] on the quantitative immunological comparison of bird lysozymes are good examples of the continuing value of immunological methods. These workers found that quantitative microcomplement fixation could be used to distinguish between the lysozymes of all species tested (chicken, partridges, quails, and pheasants) with the exception of the chicken and the jungle fowl. The similarity of these two lysozymes is in agreement with the similarity of the ovalbumins of these birds as described above. We do not attempt here to correlate all of these studies. Such an undertaking would merit an independent monograph.

The various studies from our laboratory are still not sufficiently complete to warrant any further rearrangement of classifications. Of our studies, those on the ratites and the penguins have been the most thorough. Results of the immunochemical and chemical studies indicate relationships among the five species of birds presently classified as members of the ratite group. The tinamou gave relatively strong immunochemical reactions with all the ratites, although it appeared more distantly related to any of the ratites than the ratites were to one another. The biochemical results of this study also showed that the tinamou was generally different from the ratites. The kiwi has been depicted by some investigators as an off-shoot in the evolutionary scheme between the rhea and the ostrich on one branch and the cassowary and the emu on the next branch [171]. In immunological studies, the emu and cassowary proteins gave the greatest degree of cross-reaction of the ratites. The kiwi proteins reacted strongly with both cassowary and emu proteins. The cassowary and emu ovotransferrins migrated as alkaline proteins in the starch-gel electrophoretic experiments, and egg whites of these species have the highest concentration of sialic acid among ratites. The apparent alkalinity of the ovotransferrins would place the kiwi near the ostrich-rhea pair, although the content of sialic acid would indicate a remoter relationship. The ovotransferrins of penguins and gulls [82] have also similar alkaline characteristics. Such characteristics are therefore possibly useful to show differences rather than close relationships. The immunochemical studies, however, demonstrate that the kiwi is without question a ratite. The tinamou may not be closely related to the general ratite group but it is not distant. Our studies on the ratites appear to confirm the classification of these paleognathous birds as monophyletic [200, 201].

The possibility that penguins may have originated from nonmarine flightless birds might suggest relationships between the penguins and the ratites. Immunochemical comparisons of the egg-white proteins of the ratites and the Adelie penguin showed no close relationships. In fact, the penguin egg-white proteins showed a much stronger cross-reaction with the egg-white proteins of the duck *Anas platyrhynchos* than with the proteins of the ratites. Our data show no relationships between the ratites and the penguins and agree with a suggestion of Simpson [202] that penguins evolved from flying ancestors.

# 4
# Milk Proteins

Each of the three fluids discussed in this book (milk, blood serum, and egg white) has its own "built-in" interest factor based on its origin and function. But milk is probably the most revered of the three. Milk is considered by many as "the most nearly perfect food." We drink it as a fluid and eat it in many forms. We even call cow's milk, *milk,* and human milk, *mother's milk!* Indeed, until only a few years ago most all information available on milk was for cow's milk, but this is now rapidly changing.

Most of the early studies on bovine milk proteins concerned their nutritional value — an obvious consequence of the early observations of its high nutritional quality and its extensive utility for infant feeding. Casein was noted long ago, and an albumin and globulin were separated from the whey proteins by salt fractionations approximately 80 years ago. But it wasn't until the advent of the Tiselius electrophoretic apparatus and the Svedberg ultracentrifuge that recognition of the multiplicity of components was realized. Knowledge of milk proteins has thus paralleled the knowledge of blood serum proteins in this respect.

The complexity of bovine milk, particularly its colloidal characteristics, dissuaded many protein chemists from studying milk proteins for many years. The last thirty years have seen the identification of numerous minor components as well as the reclassification and subdivision of previously known protein fractions. In 1939 Mellander [203] classified casein into three groups, $\alpha$, $\beta$, and $\gamma$ in order of decreasing electrophoretic mobility in moving boundary electrophoresis at pH 8.6. Since that time, and especially with the advent of various gel-electrophoretic techniques, the heterogeneity of these fractions and the presence of various minor constituents have been demonstrated. Genetic polymorphism in most of the major constituents also has been found and studied. There was once little information, for example, on the milks of different species, but now with the use of semimicro electrophoretic procedures and other techniques, the comparative biochemistry of milks is beginning to be a significantly investigated subject.

No attempt is made in this chapter to present a comprehensive review of milk or milk proteins. The reader is referred to the general text of Webb and Johnson [204] on bovine milk and the more recent extensive review of milk proteins under the authorship of H. A. McKenzie [205]. In this chapter we only attempt to cover the general properties of the major milk proteins, but we emphasize those proteins that exist as homologs in milk, blood, and eggs.

## COMPOSITION

The composition of bovine milk has been the subject of literally thousands of publications. Many of these had to do with the effect of diet, environment, season, time after parturition, and breed of cow on milk composition. The economics involved have been the driving force in most all of these investigations, and many of them were consequently related to the development of practical tests for finding differences that would increase the value of the commodity. Nevertheless, these studies have been valuable in the eventual understanding of the composition of milk as we now know it.

Milk is considered one of the classical examples of a biological colloid due to the presence of micelles of the caseins, the main protein group of milk. Other proteins are complexed or aggregated in the casein micelles and not all of the casein is in the micelles. The formation and structure of the micelles is not understood even today.

### General Classification

The proteins of milk have been divided into two general groups, the caseins and the remainder of the proteins which are termed whey proteins. Most fractionation procedures begin with the precipitation of casein from skimmilk by the addition of acid. Other methods, however, are used for separation of casein from whey proteins. The action of the digestive enzyme rennin will cause milk to clot and precipitate casein. There are approximately 30 grams of protein per liter of milk and about 25 grams of this is casein. Almost all of the noncasein proteins are in the whey, but there are some proteins associated with the lipids. In fact a subject of current active interest is the so-called proteins of the milk fat globule membrane. Table 4-1 is an abbreviated list of the protein composition of bovine milk. In this list the caseins are divided into four classifications: $\alpha_s$-

casein, κ-casein, β-casein, and γ-casein. This nomenclature arose originally from the observations of Mellander described above. The current classification does not correspond exactly to that of Mellander but the nomenclature has held with some modifications. Waugh and von Hippel [207] showed that there were several proteins in the casein micelle, one of which they called κ-casein. The κ-casein had been isolated along with the α-casein fraction by previous workers. Removal of κ-casein from this fraction left most of the remaining casein sensitive to calcium, that is, most of the α-casein was precipitated by calcium. This fraction is now referred to as $\alpha_s$-casein.

The main proteins of milk whey are the β-lactoglobulin and the α-lactalbumin. β-Lactoglobulin has been the subject of numerous physical-chemical investigations and exists as polymers and genetic variants.

Included in Table 4-1 are a number of proteins rightfully called blood serum proteins. They probably originate in the blood serum and therefore not only have the same genetic patterns but the identical biosynthetic sites in the animal. Studies of these proteins in milk have been and should continue to be necessary because they might be changed during passage through the mammary gland into the milk. These include the blood serum albumin and immunoglobulins.

There is a large number of proteins in milk, many present only in trace amounts. The proteins of direct interest to this study are the lactotransferrins, the lysozymes, the α-lactalbumins, and the inhibitors of proteolytic enzymes ("colostrum" inhibitors). The enzyme activities that have been identified in milk are included in Table 4-2. Many of these enzymes may either come from the blood or be present in the milk as by-products of the biosynthetic processes occurring in the mammary glands.

## PROPERTIES OF PROTEINS

### CASEINS

As discussed before, caseins are involved in complex interactions in the micelle and many of them tend to associate with themselves or other proteins. Harsh conditions have often been used to separate the individual components but recently more gentle procedures have been developed. Fractionation of the caseins by gel filtration and the corresponding gel electrophoretic patterns are given in Fig. 4-1. The reader is referred to the review by McKenzie [205] for the considerable information on the association characteristics of the caseins. Of more pertinence here are the interesting biological relationships and the genetic polymorphisms of the caseins.

TABLE 4-1. PROTEIN FRACTIONS OF COW'S SKIMMILK AND SOME OF THEIR PROPERTIES[a]

| Classical nomenclature | Contemporary nomenclature | Approximate per cent of skimmilk protein | Other characteristics |
|---|---|---|---|
| Casein (precipitated from skimmilk by acid at pH 4.6) | | 81 | |
| | $\alpha$-casein | 54 | Contains 1% phosphorus. Consists of a mixture of proteins including $\alpha_s$- and $\kappa$-casein. Formed in the udder. |
| | $\beta$-casein | 24 | 0.6% phosphorus. Formed in udder. |
| | $\gamma$-casein | 4 | 0.1% phosphorus. Preformed from blood. |
| Noncasein proteins | | 19 | |
| Lactalbumin (Soluble in ½ saturated $(NH_4)_2SO_4$ soln.) | $\beta$-lactoglobulin A | — | Associates in pH range 3.7 to 5.3. Formed in udder. |
| | $\beta$-lactoglobulin B (Mixed A and B) | 9 | Exists principally in monomeric form. Formed in udder. |
| | $\beta$-lactoglobulin C | — | Exists principally in monomeric form. Formed in udder. |
| | $\alpha$-lactalbumin (B-form) | 3 | 7% tryptophan. Formed in udder. |
| | Blood serum albumin | 1 | Apparently identical to bovine serum albumin. Preformed from blood. |
| | Immunoglobulins | 2 | High in colostrum. |
| | Others (Lactotransferrin, Lactolein) | 5 | |

[a]Data from [205] and [206].

## $\alpha_s$-CASEIN

Three genetic variants in $\alpha_s$-casein were identified [210] following the classical work of Aschaffenburg in 1961 [211] on the genetic variants of $\beta$-casein. Figure 4-2 illustrates gel electrophoretic patterns of the genetic variants of $\alpha_s$-casein, $\alpha_{s,1}$-A, $\alpha_{s,1}$-B, and $\alpha_{s,1}$-C. The various forms are controlled by three allelic autosomal genes with no dominance [212]. The

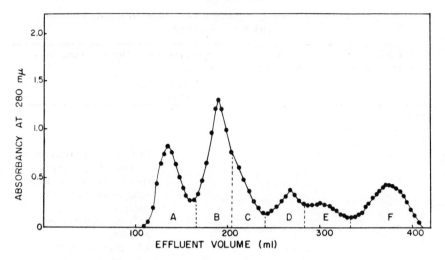

Fig. 4-1a. Gel filtration of skimmilk (3 ml) on a 2.5- by 80-cm column of Sephadex G-200 at 4°C, Eluant: 0.02 M Na-phosphate, pH 7.0. Reproduced from [209].

genes occur with different frequencies in various breeds; the B allele is the most common and the A allele is rare. The amino acid analyses of the $\alpha_s$-casein genetic variants are presented in Table 4-3. There is some discrepancy between the results reported from various laboratories. Examination of the results of Gordon, Basch, and Thompson [213] indicates a glutamic

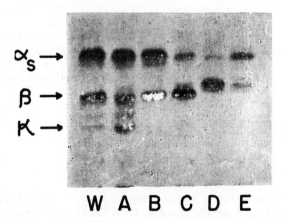

Fig. 4-1b. Starch-gel electrophoresis (pH 8.6, 7.0 M urea, 2-mercaptoethanol) and whole casein (W). Reproduced from [209]. Patterns A–E were obtained from the corresponding fraction of Fig. 4-1a.

TABLE 4-2.  MILK ENZYMES AND THEIR SERIAL CLASSIFICATION NUMBERS ACCORDING TO SUGGESTIONS OF THE COMMITTEE ON ENZYMES OF THE INTERNATIONAL UNION OF BIOCHEMISTS

| Enzyme | Classification No. | Enzyme | Classification No. |
|---|---|---|---|
| Aldolase | 4.1.2.7 | Lysozyme | 3.2.1.17 |
| $\alpha$-amylase | 3.2.1.1 | Peroxidase (incl. lactoperoxidase) | 1.11.1.7 |
| $\beta$-amylase | 3.2.1.2 | Phosphatase (acid) | 3.1.3.2 |
| Carbonic anhydrase | 4.2.1.1 | Phosphatase (alkaline) | 3.1.3.1 |
| Catalase | 1.11.1.6 | Protease (quite specific) | 3.4.-.- |
| Cytochrome C reductase[a] | 1.9.3.1 | Rhodanese | 2.8.1.1 |
| Diaphorase | 1.6.4.3 | Ribonuclease | 2.7.7.16 |
| Esterase | 3.1.1.1 | Salolase (arylesterase) | 3.1.1.2 |
| Lactase[b] ($\beta$-galactosidase) | 3.2.1.23 | Xanthine oxidase | 1.2.3.2 |
| Lipase | 3.1.1.3 | | |

[a]May be the same as Cytochrome C oxidase. Reproduced from [208].
[b]Galactose synthetase could have the same classification number as $\beta$-galactosidase (3.2.1.23) since it also catalyzes the galactose transferase reaction.

TABLE 4-3.  AMINO ACID COMPOSITION OF BOVINE $\alpha_{s,1}$-CASEINS (RESIDUES PER MONOMER M.W. 27,000)[a]

| | Variant | | |
|---|---|---|---|
| | $\alpha_{s,1}$–A | $\alpha_{s,1}$–B | $\alpha_{s,1}$–C |
| Gly | 10.3 | 10.1 | 11.1 |
| Ala | 9.5 | 10.2 | 10.2 |
| Ser | 17.2 | 16.3 | 16.6 |
| Thr | 6.5 | 5.7 | 5.8 |
| Pro | 19.9 | 19.2 | 19.2 |
| Val | 11.5 | 12.6 | 12.8 |
| Ileu | 13.1 | 12.4 | 12.5 |
| Leu | 16.7 | 19.2 | 19.3 |
| Phe | 7.2 | 9.1 | 9.2 |
| Tyr | 11.7 | 10.9 | 11.0 |
| Try | 2.7 | 2.5 | 2.6 |
| CySH | 0.0 | 0.0 | 0.0 |
| Met | 5.7 | 5.4 | 5.4 |
| Asp | 16.2 | 17.1 | 17.2 |
| Glu | 44.9 | 43.8 | 42.9 |
| $NH_3$ | 26.1 | 29.4 | 28.0 |
| Arg | 6.0 | 6.8 | 6.8 |
| His | 6.0 | 5.8 | 5.8 |
| Lys | 17.4 | 16.0 | 16.0 |

[a]Data from [213].

acid residue of B is substituted by a glycine in C. The analysis of the A variant shows many differences from the B and C variant. Peptide maps of the three genetic variants suggest that the A variant is devoid of a portion of the molecule [214]. This would explain the larger variations in amino acid composition of A when calculated on the basis of the same molecular weight as B and C.

Many properties of caseins in solution resemble those of denatured proteins, that is, they exist in a disordered configuration or a random coil. The susceptibility of caseins to attack by proteolytic digestive enzymes is thought to be a consequence of this lack of secondary and tertiary structure and an advantage in the use of milk as a food for infants. A feature of $\alpha_s$-casein seen in Table 4-3 is the high proline content. It has been suggested that this is sufficient to prevent helix formation [205].

## $\beta$-CASEINS

$\beta$-casein apparently exists as coiled polymers [205]. The molecular weight of the monomer is around 20,000 g and this is the form in which $\beta$-casein exists at low temperature. As the temperature is raised it slowly polymerizes. The $\beta$-caseins have also been found to have genetically interesting polymorphisms. Three variants with differing electrophoretic mobilities were found in five major British breeds. It was through a study of the distribution of these variants that Aschaffenburg [215] showed differences in the derivation of Guernsey and Jersey breeds.

## $\kappa$-CASEIN

The $\kappa$-casein is now considered to be the primary fraction acted upon by the enzyme rennin, releasing a glycopeptide and forming para-$\kappa$-casein. Two genetic variants have been reported for $\kappa$-casein. Figures 4-2 and 4-3 illustrate gel-electrophoretic patterns demonstrating these variants. Mackinlay and Wake [216] concluded from their investigations of the variants that any amino acid replacements responsible for the variants must reside in the glycopeptide portion of the molecule that is split off by the action of rennin. This observation has been substantiated by amino acid analysis of the isolated glycomacropeptides from $\kappa$-casein A and B [217].

Studies on $\kappa$-casein have been complicated by the presence of sulfhydryl or disulfide groups and a polymerization of $\kappa$-casein through disulfide linkages. A molecular weight of 125,000 g has been reported for $\kappa$-casein in 67% acetic acid and 5.0 M guanidine hydrochloride by Swaisgood and co-workers [218]. They concluded the smallest particle had a molecular weight of 56,000 under these conditions, but they obtained a value of 28,000 g for the molecular weight of mercaptoethanol-reduced $\kappa$-casein.

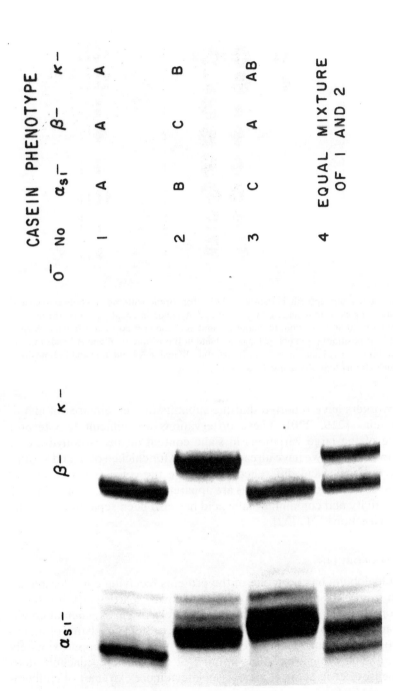

Fig. 4-2. Acrylamide-gel electrophoresis patterns of caseins isolated from the milk of cows representing different phenotypes. Performed on 7% acrylamide at pH 9.1, Tris-$Na_2$ EDTA, $H_3BO_3$ buffer. (Courtesy of Dr. M. P. Thompson.)

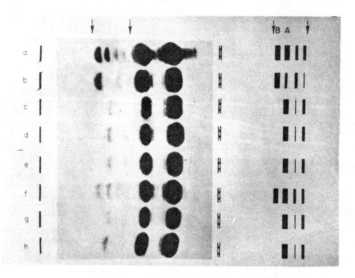

Fig. 4-3. Urea mercaptoethanol-starch gel electrophoretic patterns of whole acid caseins from milk samples of individual cows: (a) B and A bands in equal amounts; (b) major B band, minor band in A position; (c) major A band, no band in B position; (d) major A band, no band in B position; (e) major A band, no band in B position; (f) B and A bands in equal amounts; (g) major A band, no band in B position; (h) major A band, no band in B position. Reproduced from [205], *Academic Press*, 1967.

Some workers have reported that the subunits of $\kappa$-casein are not identical proteins [219, 220]. These observations are difficult to interpret because of the large variations in sialic content of the carbohydrate of $\kappa$-casein [216]. As we have already discussed for chicken ovo- and serumtransferrin, sialic acid differences can cause the gel patterns to appear different even though the proteins are apparently identical. A protein with lipase activity and containing sialic acid has also been separated from the $\kappa$-casein fraction [221, 222].

### $\beta$-LACTOGLOBULIN

$\beta$-lactoglobulin has been one of the proteins receiving considerable attention from both protein and physical chemists, particularly insofar as the properties of polymerization and the effect of environmental conditions on this phenomena. $\beta$-Lactoglobulin is a protein consisting of identical subunits, the molecular weight of the monomer is approximately 17,000 g. The protein exists as a dimer at intermediate concentrations near the isoelectric point. It may dissociate at more extremes of temperature, pH, etc. Bovine β-lactoglobulin A forms an octamer in acid solution and low temperatures. Of the various animal species studied only bovine β-lactoglobulin A has been found to undergo this formation of a higher polymer.

β-Lactoglobulin contains two poorly reactive sulfhydryl groups per dimer (one per monomer). Table 4-4 shows the reactivity of the sulfhydryl groups of β-lactoglobulin with four different types of reagents with and without denaturation. This should be compared to Table 2-5 in which chicken ovalbumin was tested using identical reagents and conditions. With both N-ethylmaleimide and DTNB the sulfhydryls of β-lactoglobulin react very poorly, but will react with prolonged incubation of the reaction mixture. In fact the value given for DTNB is intermediate and this will progressively increase to the value obtained with denaturation with prolonged incubation.

Table 4-5 lists the amino acid composition of the genetic variants of β-lactoglobulin A, B, and C. It can be seen that there are only very small differences in the amino acid content. These small differences have been found to be very important to the association-dissociation behavior of the bovine β-lactoglobulin subunits.

## α-LACTALBUMIN

The recent discovery that α-lactalbumin is the B protein of lactose synthetase [223] and the somewhat astounding discovery by Brew and coworkers [224] that α-lactalbumin appears to be a homolog of lysozyme is currently setting off a tremendous surge of investigations on α-lactalbumin. It must also be remembered that milk contains a lysozyme, as shown

TABLE 4-4. COMPARISONS OF SULFHYDRYL DETERMINATIONS ON BOVINE β-LACTOGLOBULIN

| Procedure | | Sulfhydryls Found (Moles /Mole of Dimer) | |
|---|---|---|---|
| Reagent | Condition | Native | Denatured[a] |
| p-Chloromercuri-benzoate | pH 4.6 | 1.9 | — |
| Iodine | pH 6.7 | 2.1 | — |
| N-Ethylmaleimide | pH 7.0 | 0.0 | 1.9 |
| 5,5'-Dithiobis-(2-nitrobenzoic acid) | pH 8.0 | 1.0[b] | 1.8 |

[a] Denaturant was 0.48% dodecyl sulfate. Data from [96].
[b] Reaction was very slow with both N-Ethylmaleimide and 5,5'-Dithiobis (2-nitrobenzoic acid). It was faster with 5,5'-Dithiobis (2-nitrobenzoic acid) but incomplete in two hours.

TABLE 4-5. PROBABLE NUMBER OF AMINO ACID RESIDUES PER MONOMER
OF $\beta$-LACTOGLOBULIN

| Species: | Bovine | | | Ovine | | Caprid |
|---|---|---|---|---|---|---|
| Variant: | A | B | C | A | B | |
| Gly | 3 | 4 | 4 | 5 | 5 | 5 |
| Ala | 14 | 15 | 15 | 15 | 15 | 15 |
| Ser | 7 | 7 | 7 | 6 | 6 | 6 |
| Thr | 8 | 8 | 8 | 8 | 8 | 8 |
| Pro | 8 | 8 | 8 | 8 | 8 | 8 |
| Val | 10 | 9 | 9 | 10 | 10 | 10 |
| Ileu | 10 | 10 | 10 | 9 | 9 | 9 |
| Leu | 22 | 22 | 22 | 20 | 20 | 20 |
| Phe | 4 | 4 | 4 | 4 | 4 | 4 |
| Tyr | 4 | 4 | 4 | 4 | 3 | 4 |
| Try | 2 | 2 | 2 | 2 | 2 | 2 |
| Cystine | 2 | 2 | 2 | 2 | 2 | 2 |
| Cysteine | 1 | 1 | – | – | – | 1 |
| Met | 4 | 4 | 4 | 4 | 4 | 4 |
| Asp | 16 | 15 | 15 | 15 | 15 | 15 |
| Glu | 25 | 25 | 24 | 24 | 24 | 24 |
| $NH_3$[a] | 15 | 15 | 14 | 16 | 16 | 15 |
| Arg | 3 | 3 | 3 | 3 | 3 | 3 |
| His | 2 | 2 | 3 | 2 | 3 | 2 |
| Lys | 15 | 15 | 15 | 14 | 14 | 15 |

[a]Estimate from values obtained from amino acid analyzer. Reproduced from [205], Academic Press, 1967.

in Table 4-2. To a certain extent it is this discovery of the role of $\alpha$-lactalbumin as a subunit in an enzyme and, more particularly, its probable homologous relationship with lysozyme that adds fascination and general interest to the discussion of milk proteins in this book.

Figure 4-4 illustrates the fascinating homologous relationship between $\alpha$-lactalbumin and chicken egg-white lysozyme. The size, shape, and other physical properties of the molecule are very similar, and the homologous sequences in various areas are obvious. The biochemist interested in protein evolution has here before him an exciting history to trace. It can only be hoped that $\alpha$-lactalbumins or lysozymes that show other relationships and possibly even pathways of divergence will be found in other animal species. It is always possible that the similar sequences and properties of these two proteins are due to convergent rather than divergent evolution. But we prefer to think that here is another example of the use of previously available genetic information to develop a new protein having a completely different function. The lysozymes, as is discussed in

Chapter 7, must be "more ancient" proteins and if there is eventual proof for divergent evolution of the α-lactalbumins, they must have originated from a gene pool of lysozymes.

It is interesting to look at the mechanism of action of lactose synthetase (UDP-galactose: D-glucose 1-galactosyltransferase; EC 2.4.1.22) in light of the relationship of the B protein with lysozyme, which is another enzyme with a carbohydrate substrate. The A protein has been shown to be a galactosyltransferase that catalyzes the reaction:

(1) $UDP$-galactose + $N$-acetylglucosamine $(NAG) \rightarrow N$-acetyllactosamine $+ UDP$.

α-Lactalbumin changes the specificity of the A-protein so that it can catalyze the reaction:

(2)     $UDP$-galactose + glucose → lactose + $UDP$ [225].

A molecular symbiosis appears to exist here. The B-protein, α-lactalbumin, controls the A-protein's function in lactose synthetase. Much work is still necessary to describe the means by which this control is effected; α-lactalbumin inhibits reaction 1 and allows the synthesis of lactose in the presence of glucose, but the relationship is not simple in that the inhibition of reaction 1 does not parallel the stimulation of reaction 2. The situation becomes even more complicated when the affect of NAG concentration is considered. The importance of this compound in lysozyme activity is discussed in Chapter 7. At low concentrations of NAG, reaction 1 is actually stimulated by α-lactalbumin, but at higher concentrations of NAG the reaction is inhibited [225]. These relationships can be seen in Fig. 4-5 and 4-6.

The milk of two lactating sea lions *Zalophus californianus* was examined and found to contain no lactose [226]. This is apparently the only mammal whose milk has been reported to contain no lactose. It should be interesting to investigate the presence or absence of the A and B proteins of lactose synthetase in this species and to make a comparative study of the proteins if homologous forms are found to be inactive.

LACTOTRANSFERRINS

As is discussed in Chapter 6 nomenclature of the iron-binding proteins in milk is presently confusing. There may be more than one iron-binding protein in milk. Table 4-6 lists the amino acid compositions of human, rabbit, and bovine lactotransferrins; there is considerable difference between the compositions of the iron-binding proteins of these species. The components called bovine lactotransferrin and bovine red protein demonstrate that considerable differences in composition may exist between

A. H$_2$N-Lys-Val-*Phe*-Gly-*Arg*-Cys-Glu-*Leu*-Ala-Ala-Ala-*Met*-Lys-Arg-His-Gly-Leu-Asp-Asn-Tyr-Arg-Gly-Tyr-Ser-Leu-Gly-*Asn*-

B. H$_2$N-Glu-Gln-*Leu*-Thr-*Lys*-Cys-Glu-*Val*-Phe-Arg-Glu-*Leu*-Lys-▨-Asp-Leu-Lys-Gly-Tyr-Gly-Val-Ser-Leu-Pro-*Glu*-

A. Trp-Val-Cys-*Ala*-*Ala*-Lys-Phe-Glu-▨Ser-Asn-*Phe*-*Asn*-Thr-*Gln*-Ala-Thr-Asn-Arg-*Asn*-Thr-*Asp*-Gly-Ser-Thr-Asp-Tyr-Gly-

B. Trp-Val-Cys-*Thr*-*Thr*-▨Phe-His-Thr-Ser-Gly-*Tyr*-*Asx*-Thr-*Glx*(Ala, Ile, Val, Glx)*Asx*▨-*Asx*(Glx, Ser, Thr, Asx)Tyr-Gly-

A. *Ile*-*Leu*-Gln-Ile-Asn-Ser-*Arg*-Trp-Trp-Cys-*Asn*-Asp-Gly-Arg-Thr-Pro-Gly-Ser-Arg-*Asn*-*Leu*-Cys-Asn-Ile-Pro-Cys-Ser-Ala-

B. *Leu*-*Phe*(Glx,Ile,Asx,Asx)*Lys*-Ile-*Trp*-Cys-Lys-*Asx*-Asx-Asx-Glx-Asx-Pro-His-Ser-*Asx*-*Ile*-Cys-Asn-Ile-Ser-Cys-Asp-Lys-

A. *Leu*-*Leu*-Ser- Ser- *Asp*-*Ile*-Thr- Ala-Ser-Val-Asn-Cys-Ala-Lys-Lys-Ile-*Val*-Ser-Asp-Gly-Asp-Gly-*Met*-Asn-Ala-Trp-*Val*-Ala-

B. *Phe*-Leu-Asx-Asx-*Asx*-*Leu*-Thr(Asx,Asx,Ile) Met-Cys-Lys-Val-Lys-Ile-*Leu*-▨-Asp-Lys-Val-Gly- *Ile*-Asn-Tyr-Trp-*Leu*-Ala-

A. Trp-*Arg*-Asn-Arg-Cys-Lys-Gly-Thr-Asp-Val-Gln-Ala-Trp-*Ile*-Arg-Gly-Cys-▨-*Arg*-Leu-COOH

B. His-*Lys*-Ala-Leu-Cys-Ser-Glu-Lys-Leu-Asp-Gln-▨Trp-Leu-▨-Cys-Glu-*Lys*-Leu-COOH

iron-binding proteins. In fact, in the original isolations of the proteins from milk they were isolated in different ways. As discussed in Chapter 6 there is confusion on the iron-binding proteins of some milks. In the case of the rabbit, however, only one milk iron-binding protein has been reported and it was identical with the serum transferrin [230].

IMMUNOGLOBULINS

Immunoglobulins in milk were early considered as milk constituents rather than blood constituents. We agree with the suggestion of McKenzie [205] that the terms, euglobulin and pseudoglobulin should no longer be used for this group and that the nomenclature proposed by the World Health Organization [231] be used. In the case of the bovine, immunoglobulins exist in by far the highest concentration in colostrum. In bovine colostrum the protein content is as high as 200 g/liter and over half of this protein may be immunoglobulins.

A study on the proteins of porcine colostrum, milk, and blood serum has demonstrated the presence of IgG, IgM, and IgA common to the three sources. IgG predominated in colostrum but IgA and IgM predominated in mature milk [232].

COLOSTRUM INHIBITOR

Also present in colostrum is an inhibitor of proteolytic enzymes, colostrum inhibitor [194], [233]. This protein inhibits the enzyme trypsin and is discussed further in Chapter 8. It is one of the milk proteins that has been crystallized. On the basis of its combining capacity with trypsin, it is a relatively small protein with a molecular weight between 5,000 and 10,000 g. Of particular interest is the fact that it exists in much higher concentrations in colostrum than in normal milk.

---

Fig. 4-4. The partial amino acid sequence of bovine α-lactalbumin (B) compared with the sequence of hens egg white lysozyme (A) (8, 9). Those portions of the sequence of α-lactalbumin designated by the *arrows* refer to the tryptic ($T_p$), chymotryptic ($C_p$), or peptic ($P_p$) peptides which were used to establish the sequence. Two of the four disulfide bonds in α-lactalbumin are formed between residues 6 and 120 and 28 and 111. Tryptic peptides terminating in half-cystine were obtained from S-aminoethyl α-lactalbumin. Those residues which are identical in α-lactalbumin and lysozyme are indicated in **boldface**; those residues which are similar structurally are given in *italics*. The gaps which are required to give maximum homology between the two proteins are indicated by the *hatched bars*. Asx and Glx indicate aspartic acid or its amide and glutamic acid or its amide. The presence or absence of an amide has not been established in these cases. Reproduced from [224].

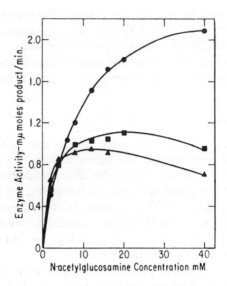

Fig. 4-5. Effect of α-lactalbumin (α-LA) on N-acetyllactosamine (NAL) synthetase at different N-acetylglucosamine (NAG) concentrations. The symbols correspond to: —●—, no α-LA: —■—, 5 μg α-LA per assay mixture: —▲—, 10 μg α-LA per assay mixture. Reproduced from [225].

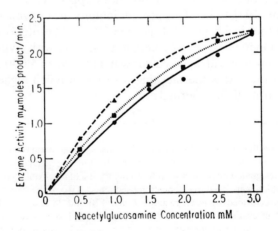

Fig. 4-6. Effect of α-LA on NAL synthetase at low NAG concentration. The A protein concentration is three times that in Fig. 4-5. The symbols are the same as in Fig. 4-5. Reproduced from [225].

TABLE 4-6. AMINO ACID COMPOSITION OF TRANSFERRINS

| | Lactotransferrins | | | | | | | | Serum Transferrins | | | | |
|---|---|---|---|---|---|---|---|---|---|---|---|---|---|
| | Human | | | | Bovine | | | | Bovine (T₂a) | | Rabbit | | |
| | [227] | | [228] | | [228] | | "Red Protein" [229] | | [229] | | [230] | | |
| | Res/ 95,000g | Res/ 10,000g | Res/ 88,000g | Res/ 10,000g | Res/ 80,000g | Res/ 10,000g | Res/ 86,100g | Res/ 10,000g | Res/ 86,100g | Res/ 10,000g | Res/ 70,000g | Res/ 10,000g | Res/ 70,000g | Res/ 10,000g |
| Lys | 54 | 5.7 | 56 | 6.4 | 60 | 7.5 | 49 | 5.7 | 57 | 6.6 | 51.0 | 7.3 | 51.7 | 7.4 |
| His | 14 | 1.5 | 12 | 1.4 | 11 | 1.4 | 10 | 1.2 | 15 | 1.7 | 16.0 | 2.3 | 16.4 | 2.3 |
| Arg | 53 | 5.6 | 53 | 6.0 | 39 | 4.9 | — | — | 23 | 2.7 | 59.6 | 3.4 | 24.0 | 3.4 |
| Asp | 82 | 8.7 | 81 | 9.2 | 68 | 8.5 | 64 | 7.4 | 82 | 9.5 | 24.0 | 10.0 | 70.2 | 10.0 |
| Thr | 41 | 4.3 | 35 | 4.0 | 35 | 4.4 | 34 | 3.9 | 35 | 4.1 | 70.1 | 2.9 | 20.3 | 2.9 |
| Ser | 63 | 6.6 | 52 | 5.9 | 44 | 5.5 | 40 | 4.6 | 45 | 5.2 | 20.4 | 4.8 | 33.6 | 4.8 |
| Glu | 84 | 8.9 | 88 | 10.0 | 74 | 9.3 | 66 | 7.7 | 58 | 6.7 | 33.5 | 8.7 | 60.9 | 8.7 |
| Pro | 42 | 4.4 | 42 | 4.8 | 35 | 4.4 | 32 | 3.7 | 33 | 3.8 | 60.7 | 4.7 | 32.5 | 4.6 |
| Gly | 61 | 6.4 | 66 | 7.5 | 50 | 6.3 | 48 | 5.6 | 47 | 5.5 | 32.8 | 6.3 | 45.1 | 6.4 |
| Ala | 75 | 7.9 | 72 | 8.2 | 66 | 8.3 | 64 | 7.4 | 50 | 5.8 | 44.6 | 7.1 | 51.2 | 7.4 |
| CySH | 31 | 3.3 | 32 | 3.6 | 34 | 4.3 | 36 | 4.2 | 17 | 2.0 | 49.5 | 4.5 | 32.2 | 4.6 |
| Val | 50 | 5.3 | 50 | 5.7 | 41 | 5.1 | 44 | 5.1 | 41 | 4.8 | 42.0 | 6.0 | 42.3 | 6.0 |
| Met | 4 | 0.42 | 6 | 0.68 | 3 | 0.37 | 5 | 0.58 | 11 | 1.3 | 5.4 | 0.8 | 5.6 | 0.8 |
| Ileu | 18 | 1.9 | 20 | 2.3 | 14 | 1.8 | 16 | 1.9 | 20 | 2.3 | 15.2 | 2.2 | 15.1 | 2.2 |
| Leu | 65 | 6.9 | 66 | 7.5 | 64 | 8.0 | 62 | 7.2 | 50 | 5.8 | 57.4 | 8.2 | 58.4 | 8.3 |
| Tyr | 24 | 2.5 | 28 | 3.2 | 28 | 3.5 | 20 | 2.3 | 23 | 2.7 | 22.8 | 3.3 | 23.2 | 3.3 |
| Phe | 37 | 3.9 | 37 | 4.2 | 30 | 3.8 | 26 | 3.0 | 29 | 3.4 | 24.2 | 3.5 | 24.7 | 3.5 |
| Try | — | — | 1 or 0 | 0.11 or 0 | 15 | 1.9 | 15 | 1.7 | — | — | 10.6 | 1.5 | 9.4 | 1.3 |
| NH₃ | | | | | | | | | | | 59.6 | 8.5 | 61.2 | 8.7 |

## BIOSYNTHESIS

The biosynthesis of milk has fascinated biologists for a long time. It is nearly a sacred subject when one considers the human aspects and the psychology connected with nursing of human infants. Nevertheless, even without this emotional relationship the synthesis of milk protein in mammary glands, as the synthesis of egg-white proteins in the oviduct, is an important biochemical subject. Although many animals produce materials of nutritional importance for their newborn, the mammals have developed this function to a high degree of perfection. A related subject is that of the formation of "crop milk" by certain birds. "Crop milk" is probably the product of autolyzed cells lining the crop and is thus regurgitated rather than secreted, but the initial release of the cells is apparently also under hormonal control.

The reader is referred to reviews for the general synthesis of milk protein [234, 235]. It is important in this discussion, however, to differentiate between proteins synthesized in the mammary gland and those which come from the blood. Information on these human proteins is given in Table 4-7. Figure 4-7 shows immunoelectrophoretic patterns of bovine blood serum using antisera made against milk and blood serum. As can be seen by comparison of those patterns approximately half of the blood serum proteins appear related to milk proteins.

TABLE 4-7. SERUM PROTEIN PRESENT IN HUMAN COLOSTRUM AND MILK[a]

| Protein | Colostrum | Human Milk |
|---|---|---|
| Prealbumin | (+) | (+) |
| Albumin | ++ | + |
| $\alpha_1$-acid glycoprotein (orosomucoid) | (+) | (+) |
| $\alpha_1$-3.5S glycoprotein | (+) | (+) |
| $\alpha_2$-macroglobulin | − | − |
| $\alpha_2$-lipoprotein | (+) | ((+)) |
| Ceruloplasmin ($\alpha_2$-globulin) | (+) | ((+)) |
| Haptoglobulin ($\alpha_2$-globulin) | (+) | (+) |
| Transferrin | (+) | (+) |
| Fibrinogen | (+) | − |
| $\beta_1$-lipoprotein | ((+)) | ((+)) |
| $\beta_{1A}$-globulin | + | + |
| $\gamma A(\gamma_1 A; \beta_2 A)$-globulin | + | ++ |
| $\gamma M(\gamma_1 M; \beta_2 M)$-globulin | ++ | + |
| $\gamma G(\gamma ss; 7S\gamma)$-globulin (rapidly migrating) | ++ | + |
| $\gamma G(\gamma ss; 7S\gamma)$-globulin (slowly migrating) | (+) | ((+)) |

[a]Modified from von Muralt et al. [236]; (+) = trace, ((+)) = smallest trace only detected by enrichment procedure. Reproduced from [205], Academic Press, 1967.

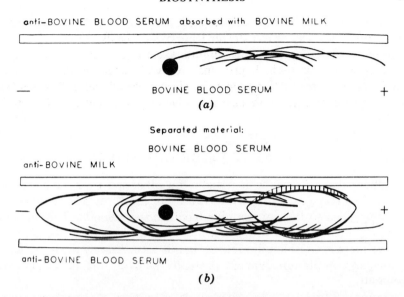

Fig. 4-7. (a) Diagram of the immune electrophoretic analysis of bovine blood serum by means of anti-bovine blood serum immune serum absorbed with milk; (b) diagram of the immune electrophoretic analysis of bovine blood serum by means of anti-bovine milk and anti-bovine blood serum immune sera. Reproduced from [237].

Investigators of this subject generally agree that evidence obtained using radioisotopes and immunological procedures supports the belief that the mammary gland synthesizes caseins, $\beta$-lactoglobulin and $\alpha$-lactalbumin [235]. As we have mentioned earlier the origin of many of the other proteins are inconclusively established. The interpretation of results from immunological and isotopic tracer studies have been limited for several reasons:

1. A blood protein synthesized by the liver may be modified in the mammary gland before it is secreted into the milk.
2. The mammary gland may synthesize a protein identical to a blood protein that was synthesized elsewhere in the body.
3. The mammary gland may synthesize a protein differing from a similar blood protein only in carbohydrate composition.
4. A combination of these situations may exist.

The actual site of milk immunoglobulin synthesis may in some cases be a problem of semantics. Plasma cells associated with the lymphatic system present in the mammary gland may synthesize immunoglobulins that find their way into the milk. However, it has been established that circulating antibodies may be transferred from blood to milk [238, 239].

Cell-free extracts and tissue culture experiments may provide clear evidence concerning which proteins are synthesized by the mammary gland. Experiments of this type have shown the effects of various hormones on the syntheses of caseins, $\alpha$-lactalbumin, and $\beta$-lactoglobulin [240–242].

## COMPARATIVE BIOCHEMISTRY

Although the milk of the goat and a few other species have been used by man for many years, only within the last decade or two has much interest been given to the comparative biochemistry of milk. The situation here is analogous to that with egg white. Now the interest factor is changing and there is considerable activity in this area. Among the most active in this area is Robert Jenness presently studying the milks of many species.

Figures 4-8 to 4-10 belong to a series of tracings of paper electrophoretic patterns of various species of milk proteins. These patterns of the casein and whey proteins show large differences as well as some obvious relationships within certain groups. Studies of this type have also been done using gel-electrophoresis as shown in Fig. 4-11 for the comparison of cow and sow milk proteins.

The study of $\beta$-lactoglobulins of different species has unearthed some interesting relationships; for example, $\beta$-lactoglobulin appears to be totally absent from human milk [245]. A matter of considerable interest is the large variation in the content of $\alpha$-lactalbumin. It may prove to be the most prevalent of the noncasein proteins. When one reflects that human milk does not contain $\beta$-lactoglobulin, it is obvious that there are large differences in the proportions of proteins as has also been found in the different avian egg whites. An homologous relationship involving bovine $\beta$-lactoglobulin and an "unidentified" protein in swine milk has recently been suggested [246]. The swine protein has several different physical properties from $\beta$-lactoglobulin and no immunological cross-reactions were noted. Homology was suggested on the basis of: (a) many similarities in the amino acid composition; (b) approximately similar molecular weights. Although no sequences were available and these are critical in establishing homology, this suggestion of homology is worthy of careful examination. How many similar examples may be found in the future which are now overlooked because of differences in gross properties?

The colostrum trypsin inhibitors from bovine, porcine, and human colostrum have been studied [247]. All three inhibitors are relatively small proteins. Maximum inhibitory activities in the milks of all three species are present in the first day and almost disappear by the fifth day. The quantities in the milks of the three species, however, are very different. In relative quantities the following amounts were found: human, 1; bovine, 10; and porcine, 67.

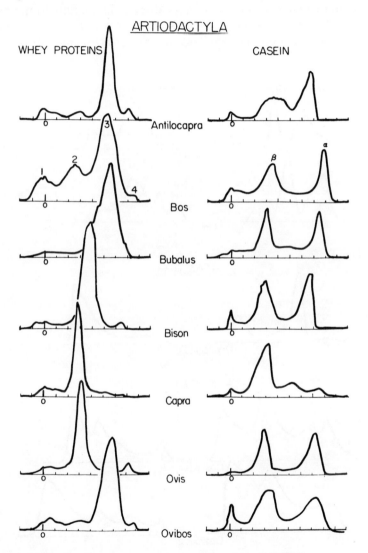

Fig. 4-8. Electrophoretic patterns of milk proteins of 7 artiodactyls (in whey protein pattern of *Bos*, 1 = immune globulins, 2 = $\alpha$-lactalbumin, 3 = $\beta$-lactoglobulin, and 4 = bovine blood serum albumin). Reproduced from [243], Pergamon Press, 1961.

Comparative studies of bovine and human milks have shown large variations in the amounts of lysozyme, lipase, and alkaline phosphatase. However, large variations of enzyme contents were also found to exist within the milks of each species. These intra-species variations are probably partially due to the genetic differences between individual milk donors rather than to a large variation of enzyme levels in the milk from one donor as the lactation period progresses. Table 4-8 indicates that human milk and colostrum contain similar amounts of lysozyme but also pro-

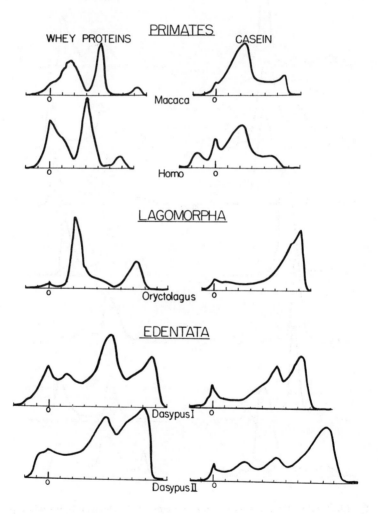

Fig. 4-9. Electrophoretic patterns of milk proteins of 2 primates, 1 lagomorph, and 1 edentate. Reproduced from [243], Pergamon Press, 1961.

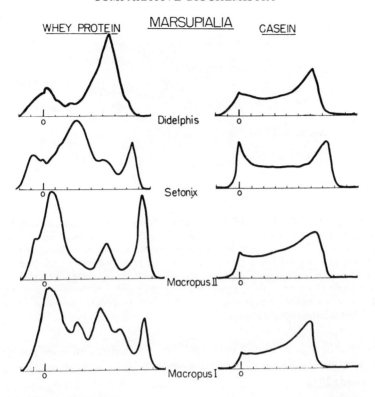

Fig. 4-10. Electrophoretic patterns of milk proteins of 3 marsupials. Reproduced from [243], Pergamon Press, 1961.

vides evidence for wide ranges in the lysozyme content of the individual samples possibly due to genetic differences. The average lysozyme content of human milk is about 3000 times that of bovine milk [248]. The lipase content of human milk is also several times higher than that of bovine milk. Freshly drawn human milk intended for storage has been shown to require pasteurization prior to cooling to prevent extensive hydrolysis of milk fat and the consequent rancidity [249]. Wide variation in alkaline phosphatase occurs from mother to mother and during the period of lactation of a given individual. Analyses of 199 samples of human milk gave an average alkaline phosphatase level of 147 units per ml. The range was from 30 to 540 units per ml [250]. The activity of alkaline phosphatase is important due to its use as a test for pasteurization because this enzyme is destroyed more slowly by heat than many bacteria. However, some 40-fold variation in the content of bovine milk samples from the same cow at

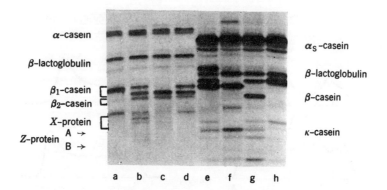

Fig. 4-11. Photograph of the electrophoretic resolution of sow's milk, diluted 1:2 with 6·6 M solution of urea (a−d), and undiluted cow's milk (e−h), in starch gel with urea and mercaptoethanol, *Tris*-citrate and borate discontinuous system of buffers, pH 8·6. This system detects a polymorphic component, X-protein, present in sow's milk (seen only in b). No analogous protein was detected in cow's milk. A second polymorphic protein, Z-protein, formed one or two very weak bands near the origin. This designation of the sow's milk proteins is a modification of the original identification (personal communication from Dr. V. Glasnak). Reproduced from [244].

different milkings has been reported. The concentration does not appear to depend on the breed of cow but seems to be in inverse relationship to milk yield [251].

## BIO-UTILIZATION

The nutritional aspects of milk have been investigated for many years. It is well recognized that milk proteins are highly nutritious and there will be no attempt here to argue the pros and cons of this point. The following four areas show interesting biochemical relationships.

TABLE 4-8. LYSOZYME CONTENTS OF HUMAN MILK AND COLOSTRUM[a]

| Sample | No. of Samples | Lysozyme (mg/100 ml milk) | | |
|---|---|---|---|---|
| | | Minimum | Maximum | Average |
| Human milk | 105 | 3 | 300 | 39 |
| Human colostrum | 12 | 9 | 102 | 46 |

[a]Reproduced from [248].

## The Action of Rennin on Milk

The action of rennin on milk has also been the subject of many investigations. From an evolutionary standpoint it is of particular interest that there is an enzyme in mammalian digestive systems whose function is to act on a particular constituent in the colloidal system of milk disrupting the colloidal system and changing the physical state, thereby enhancing digestion. This latter statement obviously contains considerable teleology and may indeed be untrue. It is known, nevertheless, that the action of rennin does greatly affect the digestibility of milk proteins.

## Immunology of Milk

The transfer of an immune defense system to the newborn is of great importance. In the case of the bovine species and a few other species, the antibodies of the newborn are transported via the colostrum into the digestive tract of the newborn animal. In contrast, in some other species including man the antibodies are transported via the blood and placenta before birth. This is discussed more fully in Chapter 5.

The review of Campbell and Peterson [252] has indicated that the presence of antibodies in bovine milk may have pharmaceutical use in the treatment of animal and human disease. A cow can be caused to synthesize immunoglobulin against many antigens and allergins related to human disease. Some of these immunoglobulins have been reported in the milk. Several reports have supported the idea that the mammary gland itself can be immunized without affecting the normal appearance of immunoglobulins in the blood serum [253, 254]. Immunization via the teat canal was proposed as a method for eliciting maximum antibody production in the milk [255]. According to this idea the antibodies to antigens related to human disease could be introduced into a human who would drink the milk as an aid in the control of disease. Most attention has apparently been paid to rheumatoid arthritis and to hay fever. Limited success has been claimed in treatment of the former by feeding milk from cows immunized against a certain pathogenic microorganism [256]. The idea of controlling disease in human and other animals by this means is obviously intriguing but it has not gone unchallenged. Lascelles [257] concluded:

1. The increase of antibody against antigen introduced through the teat canal may be explained by an increased permeability of the glandular epithelium to serum protein.
2. There is little evidence to support the idea that mature animals including man can absorb antibody from the gut in significant amounts.

3. Large-scale, well-controlled, clinical trials have not been made to ascertain the feasibility of treating diseases in this manner.

## Presence of Enzymes or Enzyme Inhibitors in Milk

The presence of small amounts of enzymes in milk may be advantageous and a consequence of the biosynthetic process in the mammary gland. They might be nothing more than by-products of the biosynthetic process lost or sloughed-off into the milk. On the other hand, the future may show that some of these may be of either physical importance to the milk or have a functional importance in nutrition. There is the possibility that the colostrum inhibitor may be present in order to retard digestion of the $\gamma$-globulins during the early stages of absorption of the $\gamma$-globulin [247]. This has not been substantiated.

Another protein that may have a function is the iron-binding protein. As is discussed in Chapter 6 there is evidence of an increase in iron-binding capacity in milk and the iron content of milk shortly after parturition.

## Milk Allergy

Allergic reactions of infants to milk other than human have been known for many years. The actual process by which an infant develops an allergy to milk is not known and indeed controversy exists concerning the effect of such an allergy on the health of an infant. For example, there is an increase in the number of agglutinating antibodies to bovine milk proteins in the sera of newborn infants fed cow's milk [258]. The normal immune response of the infant may be overwhelmed by the antigens of bovine milk in the first few days following birth.

Efforts directed at determining which of the milk proteins are the most powerful allergins have been inconclusive. One of the proteins most antigenic to children is bovine serum albumin present in very low concentration in bovine milk [259]. Circulating antibodies in the sera of infants have also been demonstrated against casein, $\alpha$-lactalbumin, $\beta$-lactoglobulin, and other milk proteins [258, 260, 261]; but a correlation of these circulating antibodies to allergic symptoms has not been observed by all investigators [205].

# 5
# Blood Plasma Proteins

The blood serves as the principal transport medium of the body and plays many important roles in co-ordinating the individual body cells into a complex organism. Its importance is further evident by the relatively large amount in an animal, 5 to 7% of the body weight in case of the human. Since blood has numerous functions its composition is consequently complex. The plasma proteins are a complicated mixture containing constituents with very different properties. One big advantage in the use of blood for genetic studies is simply that both the male and the female have a phenotype, which is obviously not the case with milk or eggs.

The proteins of blood sera were studied over one hundred years ago and one of the proteins, horse serum albumin, was crystallized before 1900. Definitive characterization of the blood plasma proteins did not come until the development of more discriminatory procedures such as moving boundary electrophoresis and analytical ultracentrifugation in the 1930's. Indeed, it was not until 1937 that Tiselius [262] separated and named the $\alpha$-, $\beta$-, and $\gamma$-globulin fractions electrophoretically.

World War II supplied the impetus for the development of fractionation methods better than the salting techniques used up to that time. This led to the development of general techniques for the preparation of many fractions from blood plasma by the low dielectric solvent procedure developed in the Harvard laboratory of E. J. Cohn and his colleagues [263]. This development made possible the separation and study of a variety of minor components of human blood serum which were previously unobtainable in sufficient quantity or purity. Notable among more recent advances in the fractionation of blood plasma were the development of the cellulose ion-exchangers by Sober and Peterson [264, 265] and the introduction of gel filtration [266].

Smithies' [267] development of the technique of starch-gel electrophoresis provided a rapid micromethod for identifying constituents during fractionation as well as for the identification of genetic differences of minor components in human and other animal blood plasma. In fact it opened the way to a whole new field in genetics and evolutionary biology. Immunoelectrophoresis has added antibody specificity to the discriminatory powers of electrophoretic identification of plasma proteins. The resolution obtained with each of these electrophoretic techniques is illustrated in Fig. 5-1. Data from ultracentrifugation studies demonstrating the proteins of different molecular weights which comprise the various fractions separated by free boundary electrophoresis are summarized in Fig. 5-2. Another electrophoretic technique receiving wide application is disc electrophoresis shown in Fig. 5-3. Many of these techniques were developed for plasma proteins and subsequently adapted to the proteins of other systems. The identification of analogous and homologous protein systems has progressed with refinements in methods for characterizing the individual proteins.

Fortunately there exist extensive reviews on the physical and chemical aspects of the plasma proteins [268], [271] as well as on their comparative biochemistry and embryology [272] so it is not necessary to present the same material in depth in the following sections. We have tried to select the material necessary to emphasize possible evolutionary and genetic relationships.

## COMPOSITION AND PROPERTIES OF CONSTITUENTS

### General Comparison of Plasma

Table 5-1 lists the principal proteins of human blood plasma, their general properties, and approximate per cent of the total serum protein. The serum albumin is the principal component and of particular interest in this discussion because it is apparently identical to the albumin found in milk. It also might be considered to be the counterpart of ovalbumin, the major protein in egg white. But it is neither an analog nor homolog of ovalbumin. Their structure and any possible functions (unknown) appear to be quite different. The $\gamma$-globulins are of interest because they contain the antibodies and also appear in colostral milk. They are of additional biochemical significance because their content varies over wide ranges as a result of genetic differences and disease. Particularly pertinent for discussion in this book are the transferrins and the $\alpha_1$- and $\alpha_2$-globulins. Many genetic variants of transferrins have been identified and homologs have been identified in egg white and milk. The $\alpha_1$- and $\alpha_2$-globulins discussed in

# COMPOSITION AND PROPERTIES

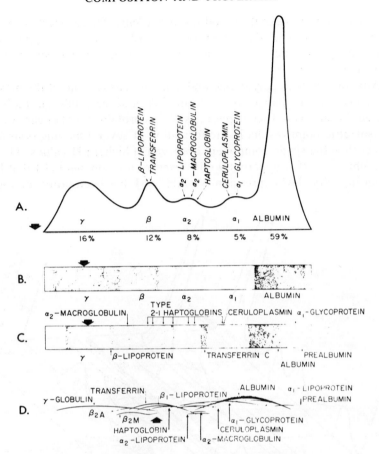

Fig. 5-1. Schematic representation of the electrophoretic pattern of normal human serum in pH 8.6 buffer as obtained by four methods: (*a*) Tiselius or free boundary electrophoresis; (*b*) paper electrophoresis; (*c*) starch-gel electrophoresis; (*d*) immunoelectrophoresis. The broad vertical arrow indicates the starting point in each case. $\beta_{2M}$-Globulin remains in the starting slot in starch-gel electrophoresis but moves in the $\gamma$- to $\beta$-range in other methods. Reproduced from [268], Academic Press, 1965.

Chapter 8 are of interest because trypsin inhibitors reside in these fractions.

## Albumin

Serum albumin has a molecular weight reported between 65,000 and 70,000 g and consists of a single polypeptide chain; it is defined operationally by its high electrophoretic mobility in Veranol buffer at pH 8.6 in most all descriptions of the blood serum proteins. This property is observable in Fig. 5-1. The major physiological roles of albumin appear to be maintaining osmotic pressure and transporting low molecular weight substances.

Albumin tends to polymerize and denature easily. One of the most extensively studied changes is the formation of several different molecular species below pH 4. Electrophoresis of this protein at a pH near or below its isoelectric point results in the formation of several moving boundaries rather than the single component observed at higher pH values. The normal albumin (N) is apparently converted to a component (F) that has a faster mobility. The expression $N + qH^+ \rightleftharpoons F$ has been proposed for this

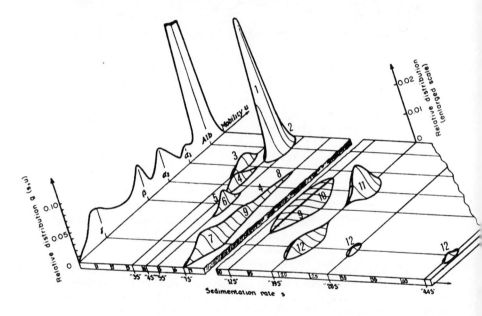

Fig. 5-2. Three-dimensional diagram showing the electrophoretic and ultracentrifugal components of normal lipid-free human serum (adapted from [269]). The sedimentation rate (s) and the relative electrophoretic mobility of each fraction is indicated. Observe that the electrophoretic γ-globulin fraction contains four components of different molecular weight. Also note the enlarged scale in the right-hand part of the diagram. Key: (1) albumin; (2) prealbumin; (3) 3.5-S $\alpha_1$-glycoproteins; (4) haptoglobin; (5) hemopexin; (6) transferrin; (7) 7-S γ-globulin; (8) ceruloplasmin; (9) $\gamma_{1A}$-($\beta_{2A}$)-globulin; (10) $\alpha_2$-mucoprotein; (11) $\alpha_2$-macroglobulin; (12) $\gamma_{1M}$-($\beta_{2M}$)-macroglobulin. Reproduced from [268], Academic Press, 1965.

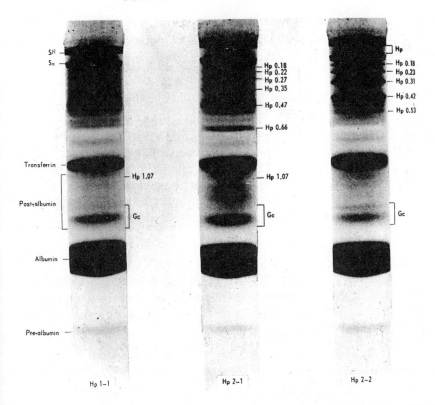

Fig. 5-3. Disc electrophoretic patterns in polyacrylamide gel of normal adult human serums representative of haptoglobin types 1-1, 2-1, and 2-2. The identification of a few bands are noted: (S$\beta$) slow beta globulin; (S$\alpha$) slow alpha globulin; (Gc) group-specific components; (Hp) haptoglobin bands; transferrin; post-albumin region; albumin; and pre-albumin. Reproduced from [270].

equilibrium [274]. The normal (N) form is converted to the fast (F) form by addition of a constant (q) number of protons. This change probably involves a shift in the conformation of the molecule.

The shape of the serum albumin molecule has been proposed to be like two spheres connected by a string. The molecule appears to have two globular regions connected to each other by a polypeptide chain [275, 276].

## HAPTOGLOBIN

Haptoglobin, a hemoglobin binding protein, is found in the $\alpha_2$-electrophoretic region (Fig. 5-1). Peroxidase activity is known to be associated with this protein when it binds hemoglobin [277]. This protein has been studied in many mammalian species because of genetic polymorphisms and biochemical interest in its primary structure.

## METAL-BINDING PROTEINS

In 1948 Holmberg and Laurell [278] proposed the name ceruloplasmin for a copper-binding protein that they had discovered because of its blue color. On starch-gel and immunoelectrophoresis this protein is located in the $\alpha_2$-region as in Fig. 5-1. The physiological importance of an oxidase activity associated with this protein is not completely evaluated. It has been estimated that nearly all of the copper of plasma is bound by this protein. The bound copper may be present in the cuprous or cupric state [279].

Heterogeneity has been noted by some laboratories in preparations of ceruloplasmin from human sera [280]. One group reports the isolation of two components, C-C and C-D, having different physical properties. The molecular weight of C-C was 148,000 g and the molecular weight of C-D was 125,000 g; these components differed also in carbohydrate content and electrophoretic properties [281].

TABLE 5-1. MAJOR HUMAN BLOOD SERUM PROTEINS[a]

| Protein | Approximate % of Total Protein | Properties |
|---|---|---|
| Prealbumin | 0.1–0.5 | High electrophoretic mobility |
| $\alpha_1$-glycoprotein | 0.5 | |
| Albumin | 55–64 | Single polypeptide chain; contains a reactive sulfhydryl; denatures easily |
| $\alpha_1$-lipoprotein | 3 | |
| Ceruloplasmin | 0.2–0.5 | Binds copper |
| $\alpha_2$-macroglobulin | 1.5–4.5 | Inhibits proteolytic enzyme trypsin |
| $\alpha_2$-lipoprotein | — | |
| Haptoglobin | — | Binds hemoglobin; genetic variants |
| $\beta$-lipoprotein | 4–14 | |
| Transferrin | 1–3 | Binds iron; genetic variants |
| IgM | 1.1 | Contains antibodies, molecular wt. over 900,000 g/mole |
| IgA | 1.6 | Contains antibodies, principal antibody associated with mucous membranes and gastrointestinal tract |
| IgG | 10 | Major antibody containing fraction of serum |

[a] Data from [268], [271], and [273].

Work is currently underway on variants of the protein from several species. Physiochemical studies have indicated that it is a relatively labile protein. The electrophoretic properties are easily altered in the presence of many of the commonly employed buffers. This complicates the interpretation of results and the identification of naturally occurring variants by electrophoretic methods. A colorless component containing only 4 atoms of copper per molecule has also been isolated and studied. Copper is proposed to play an important role in the secondary and tertiary structure of ceruloplasmin [282].

Serum transferrin is the iron-transporting protein of blood serum. The electrophoretic mobility of transferrin is indicated in Fig. 5-1 as being in the $\beta$-region. Chapter 6 discusses this protein in more depth.

## Fibrinogen

At the time of this writing the chemistry and biochemistry of blood clotting is one of the most complex and most actively studied areas of medical science. Fibrinogen, one of the largest proteins in plasma, has a molecular weight of 340,000 g [283, 284]. The protein has been isolated from many mammalian species and found to be composed of three pairs of polypeptide chains [285]. The role of fibrinogen in clot formation was postulated as early as 1905 [286] to be the source of fibrin by the conversion of fibrinogen to fibrin by thrombin. The enzyme thrombin converts fibrinogen through limited proteolysis, to a fibrin monomer that rapidly polymerizes, an important step in clot formation. The limited proteolysis releases two peptides A and B from fibrinogen. These have provided a source of material for fruitful studies on biochemical evolution based on amino acid sequence [5].

## Gamma Globulins

Gamma globulins have been the subject of intensive investigations by many disciplines in the past few years because of the importance of the immunoglobulins — members of this group of proteins [287–289]. Table 5-2 is a summary of the nomenclature of immunoglobulins composed of 2 types of polypeptide chains in a basic unit consisting of a pair of each type. The electrophoretic position of several immunoglobulins is shown in Fig. 5-1. A schematic representation of IgG is seen in Fig. 5-4. This molecule is the most prevalent immunoglobulin in mammalian plasma. The heavier chains of each immunoglobulin class are distinctive to the individual classes. The amino terminal end of the heavy and light chains have variable sequences but the carboxy terminal ends have a relatively constant sequence for a given class or subclass of light or heavy chains. Two classes of light chains, $\kappa$ and $\lambda$, and five classes of heavy chains, $\gamma$, $\alpha$, $\mu$, $\delta$, and ϵ, have been identified. In addition several of the heavy chains have been further distinguished into subclasses (Table 5-2).

TABLE 5-2. NOMENCLATURE OF IMMUNOGLOBULINS[a]

| Subunits | | |
|---|---|---|
| Light Chains | Heavy Chains | Immunoglobulins (Ig) |
| κ | γa, γb, γc, γd | $(κγa)_2 / (λγa)_2$ = IgGa, etc. |
| λ | αa, αb | $(καa)_2 / (λα)_2$ = IgAa, etc. |
| | μ | $(κμ)_{10} / (λμ)_{10}$ = IgM |
| | δ | $(κδ)_2 / (λδ)_2$ = IgD |
| | ϵ | $(κϵ)_2 / (λϵ)_2$ = IgE |

[a]Reproduced from [289].

The immunoglobulins are distinctive among the blood proteins in their biological function, heterogeneity, and biosynthesis. Immunoglobulins are formed with high specificities for combining with the particular structure of antigens. The high specificity is due to differences in structures of the antibodies caused by the biosynthesis of different primary sequences of amino acids. The heterogeneity is therefore related to the differences in specificities, as well as to different inherited phenotypes.

## Plasma Enzymes

Blood plasma contains trace amounts of many enzymes which have been of interest not only to the biochemist studying protein structure and distribution in the animal body but also to the clinician interested in diag-

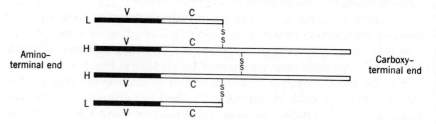

Fig. 5-4. A schematic representation of an immunoglobulin molecule. The variable and constant portions are labeled $V$ and $C$ respectively. It is not known whether $V$ is the same length in $H$ and $L$ chains. Reproduced from [289].

nosis of disease states in man and other animals. An altered level of activity of a plasma enzyme is often indicative of a specific disorder; an example is the level of acid phosphatase in human plasma [290]. A high acid phosphatase level in man suggests cancer of the prostate gland. Altered levels of other enzymes often indicate a disorder in a tissue which can serve as a source of an enzyme similar to one normally present in serum. Most enzymes present in man have at one time or another been found in plasma. This includes enzymes associated with cytoplasm and the subcellular organelles. One of the plasma enzymes leucine aminopeptidase is discussed in the following section.

## COMPARATIVE BIOCHEMISTRY AND GENETICS

Discussion of comparative aspects of the serum proteins in this chapter is directed at vertebrates and more particularly toward mammalian species, for it is here that the greatest amount of information is available on the structure and function of the plasma proteins. With the exception of insects the plasma of invertebrates have fewer types of proteins than the plasma of vertebrates. It is difficult to establish, however, that a homologous protein, e.g. $\gamma$-globulin, is not functioning in invertebrate plasma in some form. That is, even though a $\gamma$-globulin-like protein is not isolated or demonstrated electrophoretically an ancestral peptide of the mammalian, $\gamma$-globulin may be performing the similar physiological function of immunity by being associated with a cellular immunity phenomenon. The problem of identifying serum albumin in fish, amphibia, and reptiles by means of the existing definition which is based on such properties as electrophoretic mobility and molecular size has been recognized by many researchers [272]. This problem occurs whenever the plasma proteins of lower vertebrates and invertebrates are compared with those of higher vertebrates. The protein concentration of plasma has apparently increased progressively with evolution; the number of distinguishable proteins has also increased at a comparable rate. These observations are only a reflection of the extensive mutation and probably gene duplication which have produced proteins with a myriad of physiological functions.

The large differences encountered in electrophoretic patterns of distantly related species are shown in Fig. 5-5. We have been using the method of disc electrophoresis (Fig. 5-6) to compare the blood serum proteins of penguins to each other and to the more familiar patterns obtained with chicken and human sera. The patterns of distantly related species are quite different but differences in electrophoretic patterns can be obtained from closely related species. Investigations in this laboratory on the serum proteins of several species of Antarctic fish have not only revealed

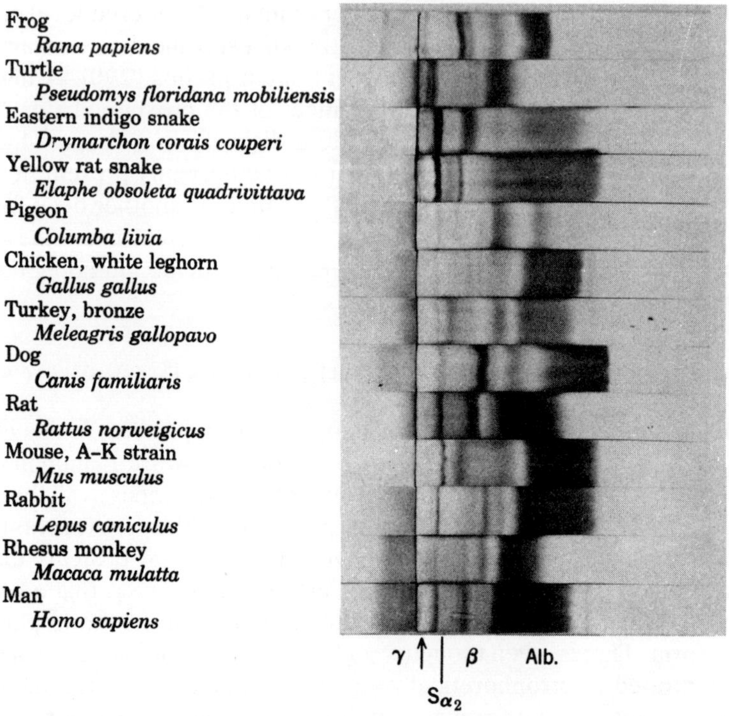

Frog
  *Rana papiens*
Turtle
  *Pseudomys floridana mobiliensis*
Eastern indigo snake
  *Drymarchon corais couperi*
Yellow rat snake
  *Elaphe obsoleta quadrivittava*
Pigeon
  *Columba livia*
Chicken, white leghorn
  *Gallus gallus*
Turkey, bronze
  *Meleagris gallopavo*
Dog
  *Canis familiaris*
Rat
  *Rattus norweigicus*
Mouse, A-K strain
  *Mus musculus*
Rabbit
  *Lepus caniculus*
Rhesus monkey
  *Macaca mulatta*
Man
  *Homo sapiens*

$\gamma \uparrow \mid \beta$ Alb.
$S_{\alpha_2}$

Fig. 5-5. Comparative electrophoretic patterns of serum proteins of some amphibia, reptiles, birds, and mammals, Position at which samples were applied is indicated by arrow below figure. Cathodic area is left of origin and anodic area is toward the right. Reproduced from [272], Academic Press, 1960.

variations between closely related species but have also revealed many differences within some species (Fig. 5-7). In a comparative study of homologous and analogous proteins it is difficult to consider the serum proteins of distantly related species without considering evidence for structural and functional similarities. No less difficult are comparisons of very closely related species for which the sensitivity of available techniques for detecting small changes in many instances often is the limiting factor.

PLASMA ALBUMIN

An amino acid analysis of the plasma albumin of several species is presented in Table 5-3; it can be seen that albumin contains a large proportion of acidic and basic amino acids and also a very low amount of tryptophan. This appears to be true of all species so far studied. Attempts at establishing the complete amino acid sequence of human and bovine albumin have not been reported but the sequences of isolated peptides are being studied. Each molecule contains 17 or 18 disulfide linkages and one reactive sulfhydryl group. This sulfhydryl may be blocked with cysteine or glutathione [293]. Thus the impression may result from sulfhydryl content that there is less than one free sulfhydryl per molecule or that two molecular species occur in a purified albumin sample.

Comparative amino acid analysis of the sequence around the free sulfhydryl in human and bovine albumins has revealed the sequences to be identical [294]. It may well be that this group plays a role in the physiolog-

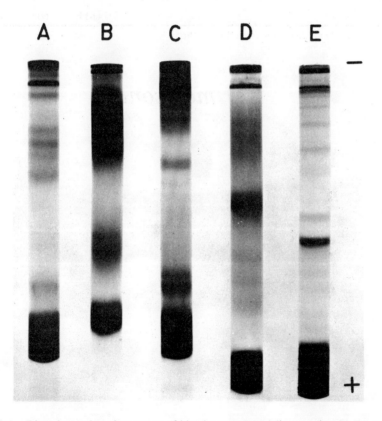

Fig. 5-6. Disc electrophoretic patterns of blood sera: (A) Adelie penguin; (B) Emperor penguin; (C) Humboldt penguin; (D) chicken; (E) human sera. Reproduced from [291]. Academic Press, 1968.

TABLE 5-3. COMPARISON OF THE NUMBER OF AMINO ACID RESIDUES PER MOLECULE IN THE ALBUMINS OF DIFFERENT SPECIES[a]

| Amino acid | Human | Bovine | Canine | Amino acid | Human | Bovine | Canine |
|---|---|---|---|---|---|---|---|
| Lysine | 59 | 61 | 62 | Alanine | 60 | 49 | 64 |
| Histidine | 16 | 18 | 13 | Half-cystine | 36 | 38 | 35 |
| Arginine | 25 | 24 | 24 | Valine | 46 | 35 | 43 |
| Aspartic acid | 55 | 57 | 55 | Methionine | 6 | 4 | 4 |
| Threonine | 29 | 34 | 24 | Isoleucine | 9 | 14 | 6 |
| Serine | 25 | 28 | 27 | Leucine | 64 | 65 | 68 |
| Glutamic acid | 83 | 79 | 88 | Tyrosine | 18 | 20 | 23 |
| Proline | 31 | 29 | 33 | Phenylalanine | 33 | 28 | 32 |
| Glycine | 15 | 17 | 24 | Tryptophan | 1 | 2 | 1 |
| | | | | | 611 | 602 | 626 |

[a]Reproduced from [268], *Academic Press*, 1965.

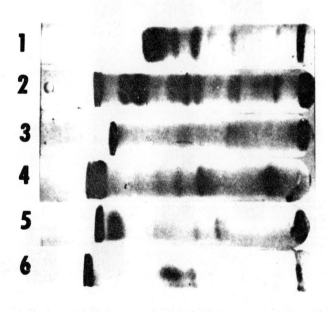

Fig. 5-7. Disc electrophoretic patterns of *Dissostichus mawsoni* seri from six different specimens [292].

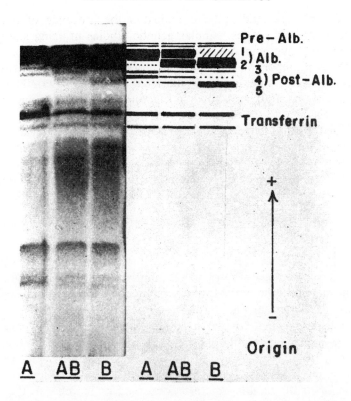

Fig. 5-8. A starch gel on 3 samples of plasma, all from horses of transferrin phenotype FF, which illustrates the resolution of the 3 albumin phenotypes A, AB, and B. Zones 1 and 2 in the albumin (Alb.) region and zones 3, 4, and 5 in the post-albumin (Post-Alb.) region indicate the positions of the diagnostic bands. Stippled lines indicate extremely faint bands which may or may not be seen on individual gels. Cross-hatched lines in zone 1 in phenotype B represent diffuse but rather densely staining material which often obscures the outlines of the thick band in zone 2. Reproduced from [296].

ical functions of serum albumin and that the amino acid sequence around it is essential to this role.

Although some authors feel that serum albumin can be considered as a single pure protein from a given species, polymorphisms exist which are under genetic control and not a result of polymerizations or electrophoretic anomalies. Such polymorphisms have been reported in cattle [295], horses [296], chickens [297], and turkeys [298], and is probably present in other species. Figure 5-8 illustrates the presence of three albumin phenotypes in horse serum. These phenotypes correlate with three postalbumin phenotypes enabling the phenotype to be recognized by examining either pattern. Inheritance studies have indicated the presence of two co-dominant autosomal alleles. Similar genetic studies utilizing starch-gel electrophoresis have been done on the albumins of other species.

An interesting report on the inheritance of albumin phenotypes resulting from crosses of chickens and quails has appeared [299]. The quail phenotypes of this study are shown in Fig. 5-9a. Three phenotypes of the quail appear similar to the three observed in the case of horse serum. The phenotypes resulting from crossing a homozygous chicken (C-C) with a heterozygous female quail ($Q_1$-$Q_2$) are seen in Fig. 5-9b. The hybrid receives one allele from each parent. Another example of this type has been reported. The existence of three albumin phenotypes, A, B, and AB in a breeding flock of turkeys was interpreted as having arisen from the cross of two species of turkeys each introducing one of the alleles [298].

## HAPTOGLOBINS

Smithies [267] described the electrophoretic mobilities of several phenotypes of haptoglobin found in human plasma. The starch-gel electrophoretic patterns of the three common phenotypes, Hp1-1, Hp2-1, and Hp2-2 are shown in Fig. 5-10.

The molecular weight of the hemoglobin-free haptoglobin has been reported as approximately 85,000 for Hp1-1 [301], and the average molecular weights of Hp2-2 and 2-1 have been reported as 400,000 and 200,000, respectively [302, 303]. Sedimentation studies demonstrated (very similar to the banding patterns observed on starch gel) that only Hp1-1 sedimented as a pure protein whereas Hp2-1 and Hp2-2 appeared to contain fractions of multiple molecular weight.

Electrophoresis following reductive cleavage in urea has permitted the three major phenotypes to be further divided into six types, Hp 1F-1F, 1F-1S, 1S-1S, 2-1S, 2-2S, and 2-2. It has been noted that reductive cleavage of haptoglobin yields two polypeptide chains, $\alpha$ and $\beta$. Only the $\alpha$-chains are seen in the electrophoretic patterns of the haptoglobin subtypes in Fig. 5-10. The $\beta$-chain does not migrate from the origin. It has been postulated that all haptoglobins contain a common $\beta$-chain. But the $\alpha$-chain varies and can be one of three types, 1F$\alpha$, 1S$\alpha$, and 2$\alpha$, which are the expressions of three alleles [268], [304]. The difference between the $\alpha$-polypeptide chains 1F$\alpha$ and 1S$\alpha$ may be as small as the replacement of a single lysine residue in 1F$\alpha$ with an acidic amino acid in 1S$\alpha$. However, the $\alpha$-chain, 2$\alpha$, appears to be twice the molecular weight (17,300 g) of either $\alpha$-chains, 1S$\alpha$ or 1F$\alpha$ (8860 g). From studying the results of partial amino acid sequence and peptide maps, Smithies et al. [305] determined that the $\alpha$-chain, 2$\alpha$, represents a combination of the $\alpha$-chains,

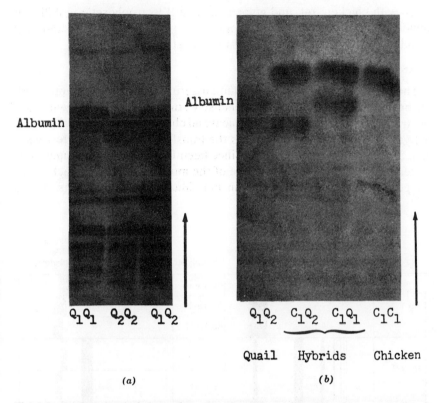

Fig. 5-9. (a) Starch gel of plasma from three Japanese quail, and the three serum albumin phenotypes; (b) starch gel of plasma from a homozygous chicken male ($C_1C_1$), a heterozyquos quail female ($Q_1Q_2$), and their hybrid offspring $C_1Q_1$ and $C_1Q_2$. Reproduced from [299].

$1F\alpha$ and $1S\alpha$. They also predicted the existence of two $2\alpha$ polypeptide chains each composed of two $1F\alpha$ or two $1S\alpha$ chains. A number of other haptoglobin variants have been found in human sera and reviewed in detail [306]. Two variants are shown in Fig. 5-10.

CERULOPLASMIN

Immunochemical studies have been interpreted as showing the existence of four antigenic determinants in human ceruloplasmin. The presence of a protein in the sera of other species containing one or more of these determinants has been demonstrated. Rabbit antisera to human ceruloplasmin cross-reacted strongly with a component in the sera of monkey, four species of deer, and one goat. Only a slight cross-reaction was noted with chicken, rat, and guinea pig sera. Electrophoretic mobility as demonstrated by starch-gel electrophoresis (Fig. 5-11) had no apparent correlation with the cross-reactions obtained [307].

## Transferrins

The identification of 17 or 18 human serum transferrins is apparently the beginning in a long and comprehensive study on the human genetics of the transferrins as well as the physical and chemical properties of individual variants. Figure 5-12 diagrams the transferrins as known a few years ago. Several more variants have since been found. The determination of patterns of transferrins may be one of the more important characteristics for genetic comparisons of human individuals and populations. So far

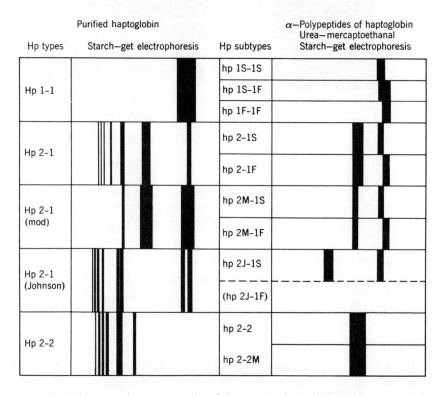

Fig. 5-10. Diagrammatic representation of the starch-gel electrophoretic patterns of the haptoglobin phenotypes and their respective subtypes. Those examples of Hp 2-1 (Johnson) which have so far been subtyped have had the α-polypeptides hp 2J and 1S. The starch-gel pattern associated with hp 2J and hp 1F may vary from Hp 2-1 (Johnson). Reproduced from [300].

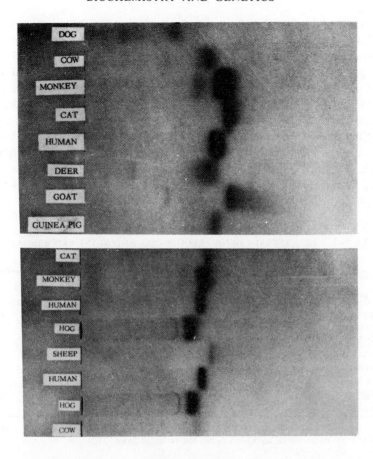

Fig. 5-11. Starch-gel electrophoretograms of various animal sera stained with benzidine to show the oxidase activity of ceruloplasmin. Reproduced from [307].

there have been no pathologies traced to variants of serum transferrins or any correlations with other genetically transmitted differences. The genetic polymorphism is not due to differences in sialic acid content, although the sialic acid contents also vary [309]. Transferrin variants appear in high frequency in all populations, whereas particular variants are restricted to certain races and others to certain close kinship. As an example, $B_2$ has been found only in Caucasians, $D_{Chi}$ only in Chinese, $B_{0-1}$ only in Navajo Indians and $D_1$ only in Negroes. All D variants have electrophoretic mobilities less than that of the common variant C, whereas all B variants have mobilities greater than that of C. Several variants have been shown to be modified by the action of enzymes hydrolyzing the carbohydrate prosthetic groups [310].

Considerable diversity of serum transferrins occurs within the primates [311]. With the exception of the chimpanzee the transferrin phenotypes of the nonhuman primates have variants analogous to those of man. Principal studies with other mammals have been with cattle, sheep, and deer. Many variants have been found in sheep. Table 5-4 tabulates the transferrin gene frequencies in Merino sheep from different areas in Australia and Tasmania. In contrast to this variation, Braend and Stormont [313] found no difference in the transferrin patterns of 113 samples of bison blood. Also no differences were found in the hemoglobin patterns. Although the sample was considered possibly too small for generalization, it was suggested that genetic variation in these traits was lost during the drastic reduction of the bison population in the nineteenth century.

TABLE 5-4. TRANSFERRIN GENE FREQUENCIES AND STANDARD ERRORS FOR VARIOUS SHEEP POPULATIONS STUDIED

| Population | No. of Animals | $Tf^F$ | $Tf^G$ | $Tf^A$ | $Tf^H$ | $Tf^J$ | $Tf^K$ |
|---|---|---|---|---|---|---|---|
| Tasmanian fine-woolled Merinos | 210 | 0·012 ±0·005 | — | 0·314 ±0·023 | 0·279 0·019 | 0·371 ±0·028 | 0·024 ±0·007 |
| Peppin Merinos from Central Western Queensland | 298 | 0·181 ±0·016 | 0·059 ±0·010 | 0·149 ±0·015 | 0·200 0·016 | 0·391 0·020 | 0·020 ±0·006 |
| Peppin Merinos from Badgery's Creek, N.S.W. | 71[a] | 0·112 ±0·037 | 0·084 ±0·033 | 0·075 ±0·031 | 0·187 ±0·046 | 0·458 ±0·059 | 0·084 ±0·033 |

[a]Excluding inbred animals, and counting maternally bestowed genes only in the group of offspring from ewes mated to inbred rams. Reproduced from [312].

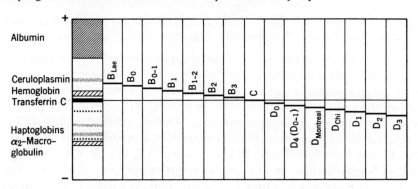

Fig. 5-12. Diagrammatic representation of known genetic variations (electrophoretic) of human transferrin. Reproduced from [308].

In a study of the serum transferrins of 150 species of reptiles and amphibians [314] only one iron-binding protein was found in the plasma of most individuals, but two were often present and three sometimes were found. Transferrins of closely related taxonomic groups tended to cluster within limited mobility ranges and some taxonomic categories could be differentiated on this basis. In addition there appeared to be much less variability in mobilities of transferrins of some species of amphibians, turtles, crocodelians, and lizards.

These iron-binding proteins are commonly identified by using radioactive iron ($^{59}$Fe). Figure 5-13 shows the results of such an experiment; it demonstrates the existence of multiple bands in sera of several species of penguins. In a limited survey no evidence has been found for genetic variation within the individual species. Differences in the electrophoretic mobility of the chicken ovotransferrin and serum transferrin can be seen in Fig. 5-13. This difference has been proposed as due to altered carbohydrate content. A similar difference does not appear in a comparison of Adelie penguin ovotransferrin and serum transferrin. Figure 5-13 also demonstrates the higher electrophoretic mobility of human serum transferrin in comparison to the avian species.

### Fibrinogen

Sequence study of fibrinopeptides A and B has indicated a large degree of variation during vertebrate evolution [315, 316]. A recent study of fibrinopeptides A and B from 18 members of the Order Artiodactyla has indicated that changes in the amino acid sequence of these peptides are generally consistent with previously established taxonomic relationships [317]. Figure 5-14 is a summary of the findings of this study showing the relationships established by amino acid sequence studies. The fibrinopeptides are changing so rapidly that they are particularly useful for establishing detailed relationships between closely related species where differences in many proteins (e.g., cytochrome and hemoglobin) are small.

### Immunoglobulins

The evolutionary, taxonomic, and biosynthetic aspects of immunoglobulins are fundamental to many problems of modern biology and, most certainly, to an understanding of the biochemical evolution of proteins. Here is a protein(s) apparently evolved in response to the need for a mechanism of protection against foreign substances invading the more complicated systems of higher animals. The molecular structure itself shows this evolution. Homologous sequences have been found in the light and heavy

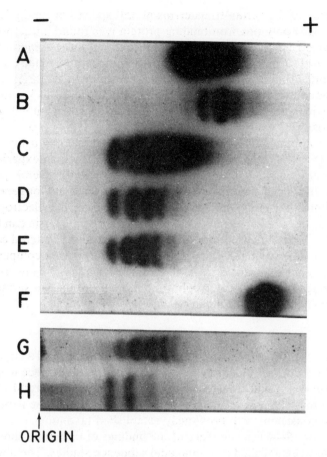

Fig. 5-13. Autoradiograms of starch-gel electrophoretic patterns demonstrating iron-binding proteins in egg whites and sera. (A) Chicken egg white; (B) chicken serum; (C) Adelie penguin egg white; (D) Adelie penguin serum; (E) Emperor penguin serum; and (F) human serum. In a separate experiment: (G) Emperor penguin serum and (H) Humboldt penguin serum. Reproduced from [291], Academic Press, 1968.

chains of immunoglobulins. The ancestral structure appears to have been a protein with a molecular weight of 12,000 to 13,000 g. Variations of this unit structure constitute the building blocks of the light (L) chains with molecular weights of approximately 25,000 g and the heavy (H) chains with molecular weights of approximately 50,000 g (Fig. 5-4).

Two types of variation in immunoglobulin primary structures exist within a species: one is the common phenotypic variation as found in most all proteins; the other is directly related to the biosynthesis resulting in specificity of the immunoglobulin for interaction with particular antigens. The mechanism controlling this latter variation is under intensive study in many laboratories. The entire subject is currently very complicated. Lennox and Cohn [289] have recently summarized in a concise manner what they believe as currently established [318-320]. We do not believe we can improve on the summary of Lennox and Cohn as follows:

"(a) immunoglobulins are made up of nonidentical subunits, light chains, L, and heavy chains, H, and there are several classes of light and heavy chains each of which has many members; (b) each subunit has a contiguous amino-terminal sequence of amino acids which varies from one member of the class to another and a contiguous carboxy-terminal sequence that is almost invariant in the class but which is different from

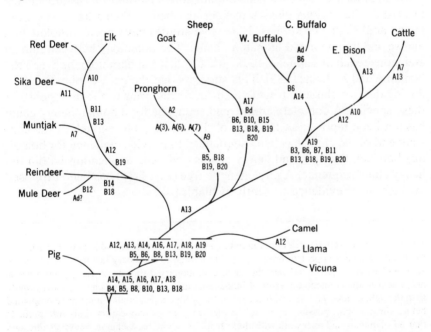

Fig. 5-14. Classical relationship of 18 artiodactyls and a mutational scheme consistent with the present day fibrinopeptide amino acid sequences. Letter-number designation indicates position in fibrinopeptide A or B in which an amino acid replacement has occurred on that branch of the phylogenetic tree. Deletions are designated by "d." On the branch leading to the pronghorn antelope three of the changes are noted in parentheses. At least three such amino acid replacements have occurred in this region of the fibrinopeptide A, but their exact positions are still uncertain. Reproduced from [317], Academic Press, 1967.

class to class; (c) the number of variable sequences associated with any one class is known to be greater than $10^2$ and is probably greater than $10^4$; (d) the variable sequence, V, determines the antibody specificity and the almost invariant sequence, C, characteristic of the species, determines the class of immunoglobulin as well as polymorphism within the class; (e) higher vertebrates show two, possibly three, classes of light chains and about ten classes of heavy chains, but lower vertebrates seem to have one light and one heavy chain class [321-323]; (f) any light chain can be linked by disulfide bridges to any heavy chain to give the basic monomer, LH; certain classes of immunoglobulin molecules are in the form of the dimer $(LH)_2$ with the heavy chains linked by a disulfide bridge (this structure is shown in Fig. 5-4); however, light and heavy chains can interact in solution to form $(LH)_2$ without formation of a covalent bond."

Much of the evidence for gene duplication of a primordial sequence that has led to the various classes of light and heavy chains has come from amino acid sequence studies. Extensive studies have been completed on mouse and human $\kappa$ and human $\lambda$ chains. The homology between human and mouse $\kappa$ chains appears to be greater than the homology between the human $\kappa$ and $\lambda$ chains [289]. This suggests that the gene duplication producing the two classes of light chains occurred prior to the divergence of these species. A complete amino acid sequence for a human heavy chain has not been reported at the time of this writing, but enough of the constant end of the chain has been sequenced to provide evidence for homology between the light and heavy chains and gene duplication within the heavy chain sequence. A phylogenetic tree (Fig. 5-15) has been proposed on the basis of evidence currently available [8].

---

Fig. 5-15. Phylogenetic tree of the immunoglobulins. The mouse and human $\kappa$-chains and the human $\lambda$-chains appear to have a perfectly normal evolutionary history. Both the constant and variable halves indicate the same evolutionary distances. A rough time scale is given next to the common ancestor of human and mouse. The heavy chain is more remote than the light chains. The special mechanism for producing variability must have originated before the $\kappa$-$\lambda$ split, possibly even before the heavy chain duplication. This tree predicts that all mammals had ancestors with these features: $\kappa$-chains, $\lambda$-chains, heavy chains, and the special mechanism for producing variability. The two halves of the 220 carboxyl-terminal links of the rabbit heavy chain are clearly related. Dayhoff and Eck have placed their common ancestor near the time of the light-heavy chain gene duplication. Because of evolutionary distances involved, they propose that the immunoglobulin heavy and light chains found in sharks have really arisen by an independent gene duplication in the shark line. (Reproduced from the *Atlas of Protein Sequence and Structure 1967-68*, Margaret O. Dayhoff and Richard V. Eck, National Biomedical Research Foundation, Silver Spring, Maryland (1968) [8, p. 25].)

At this time very little is known about the proteins acting in defense systems of lower vertebrates and invertebrates. The evidence indicates that most of the defense is nonadaptive (i.e., there are no large increases in a particular protein with specific combining properties like antibodies, in response to an antigen) and frequently a form of phagocytosis. However, even for phagocytosis a recognition of an unwanted foreign substance is necessary.

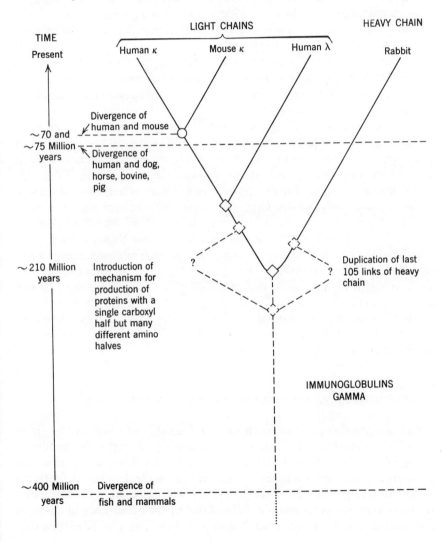

## Leucine Aminopeptidase

A review of Fishman [290] has described the presence of over three dozen enzymes in human plasma encompassing acid and alkaline phosphatases, glycolytic enzymes, choline esterase, lipase, amylase, transaminases, and peptidases. One of these, leucine aminopeptidase (LAP), we have chosen to illustrate several typical aspects of plasma enzymes. We acknowledge that this enzyme is neither the best chemically or physically characterized nor the most widely studied in different species.

The level or even the presence of a given plasma enzyme varies from species to species. Within a species variation occurs between individuals due to genetic differences. The levels of many enzymes are subject to hormonal control but levels also vary as a result of injury or disease.

A useful assay for the presence of leucine aminopeptidase detects the position of the enzyme on a starch gel following electrophoresis by means of its enzymatic activity. The substrate employed for this enzyme is L-leucyl-$\beta$-napthylamide-HC1 which, when hydrolyzed in the presence of an azo dye, fixes the position of the enzyme in the gel. An assay of normal human serum by this means is shown in Fig. 5-16$a$. Alteration of this pattern is known to occur during certain types of cancer and other diseases. An interesting aspect of this enzyme in humans is the appearance of multiple bands of enzyme activity in serum of a woman during pregnancy. Figure 5-16$b$ demonstrates the appearance of these bands after several months of the pregnancy period. The origin of such multiple bands or the enzymes of altered mobility is being investigated in many enzyme systems. It might be suspected that these enzymes arise from pre-existing forms present in the tissues of various organs, that new enzymes are being synthesized or that the normal enzyme is being altered. It has been reported that leucine aminopeptidase in serum from patients with infectious mononucleosis becomes inactivated.

## Constituents Peculiar to Physiological Condition

Certain physiological or environmental conditions may greatly influence the composition of blood. The process of egg formation in the female is one of the physiological states with which extensive changes in serum proteins are effected. Changes are readily seen in electrophoretic patterns of sera of laying birds and gravid fish [292]. The association of blood composition with an environment is exemplified by the occurrence of a unique constituent in the sera of some Antarctic fishes. The sera of certain (but

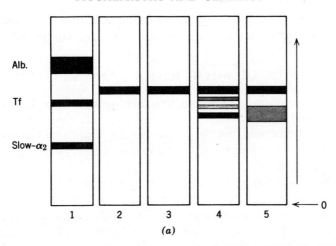

Fig. 5-16a. Schematic picture of starch-gel electrophorogram showing the electrophoretic patterns of serum LAP in normal adult males (2); normal adult females (3); pregnant women at term (4); and a rare type found in a few newborn children (5). For comparison, the electrophoretic mobilities of albumin (Alb.), transferrin (Tf) and the slow $\alpha_2$-macroglobin are given at the left (1). The arrow indicates the direction of migration towards the anode, and (O) indicates the point of origin (insertion of samples). Reproduced from [324].

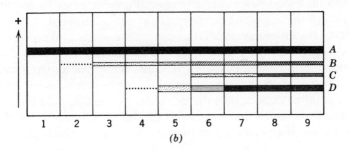

Fig. 5-16b. Schematic zymogram showing the differential time of onset of the pregnancy LAP isozymes (B, C, D). (A) is the normal LAP zone. The numbers indicate the period of pregnancy in months when the samples were taken and examined. Arrow indicates direction of migration. Degree of shading indicates relative intensities. Reproduced from [324].

not all) fishes living in the cold Antarctic waters (−1.86°C) contain a glycoprotein not found in temperate water fishes. This constituent has the important property of lowering the freezing point of these fishes, and it has been aptly termed a "freezing point depressant glycoprotein [324a]."

Many other physiological conditions are likely to influence the composition of sera. Ureotilic and uricotilic animals differ in the amounts of urea and uric acid in their sera. Some associated effects on other blood constituents might also be expected, but extensive studies of this type have not been reported.

## RELATIONSHIP OF MILK, BLOOD, AND EGG PROTEINS

The existence of proteins having similar properties in milk, blood, and egg white has been discussed in this and previous chapters. Some are found in all three fluids, but the reason for the distribution of some proteins in these three systems and the exclusion of others from this distribution is not understood.

The albumins of bovine blood serum and milk appear to be identical but the albumins of chicken blood serum and egg white appear very different. On the other hand, the transferrin of rabbit blood sera and milk are apparently identical (see Chapter 6) and the transferrins of chicken serum and chicken egg white are identical except for carbohydrate residues [35]. The blood proteins should more properly be compared with the egg-yolk proteins than the egg-white proteins. However, the absence of a riboflavin-binding protein from the blood of genetically selected strains of chickens was shown to correlate with the absence of flavoprotein from the egg white and yolk [132]. The blood flavoprotein is synthesized in the liver. A portion of the blood flavoprotein is normally incorporated into yolks but the source of the egg-white flavoprotein is not established.

Probably the best example of the appearance of a serum protein in milk and eggs is given by the $\gamma$-globulins. The immunologically active fraction of the blood serum proteins, the $\gamma$-globulins, are transferred to the circulation of young of all birds and mammals by various mechanisms. The three routes are: (a) passage across the placenta; (b) passage into the yolk sac membrane from which they are absorbed into the circulation; and (c) secretion via the mammary gland into the milk from which they are ab-

TABLE 5-5. RELATIONSHIP OF PLACENTAL STRUCTURE AND MODE OF PASSIVE TRANSFER OF IMMUNOGLOBULINS TO THE FETUS[a]

| Species | Tissue Layers between Maternal and Fetal Circulation at Term | Placental or Amniotic Transmission | Importance of Colostrum |
|---|---|---|---|
| Pig | 6 | − | +++ |
| Ruminants | 5 | − | +++ |
| Carnivores | 3 | ± | + |
| Rodents | 2 | + (yolk sac) | +−++ |
| Ape, man | 2 | +++ (placental) | − |

[a]Reproduced from [325], *Academic Press*, 1964.

sorbed across the intestinal wall into the circulation. The yolk sac is the sole means observed in avian species by which young receive parental immunoglobulins. This route also is important to rodents. Table 5-5 shows the correlation of placenta structure, and the transmission of immune globulins. The primates are the only species that appear not to absorb immune globulins from colostrum. As the number of tissue layers between maternal and fetal circulation increases the relative importance of colostrum increases [325].

The general aspects of the relationships between the proteins of blood plasma, milk, and eggs should become an increasingly fertile field for studies. As minor components are isolated and identified and as sequences become available, perhaps many homologies will be found. The $\alpha$-lactalbumin-lysozyme relationships may prove to be only one of many examples.

# 6
## The Transferrins

The transferrins are a family of homologous proteins that bind iron strongly. They are found in quantity in egg whites and blood sera, in smaller amounts in milks, and in trace amounts in other vertebrate fluids. They are thus an excellent example of a widely dispersed series of homologous proteins that exist in many animal species in three different, very important, fluids. Two of these fluids, milk and egg white, are produced by two of the most rapid animal biosynthetic systems, the mammary gland and oviduct. The biosynthetic origins of these proteins and their evolutionary, taxonomic, and genetic relationships are currently stimulating much interest and research. The mechanism of transport of iron in animal blood serum, the chemistry of the metal complexes of transferrins, and the ligands involved in the chelation are currently being studied by several groups. It seems probable that the transferrins will become one of the more extensively investigated proteins and will have continued importance to several fields of science.

All known transferrins bind two atoms of iron per protein molecule forming a salmon-pink colored complex with an absorption maximum near 470 m$\mu$. The two most studied transferrins have been chicken ovotransferrin and human serum transferrin. These and several other transferrins have been shown also to bind copper to form a yellow complex with an absorption maximum near 440 m$\mu$. Colored complexes result from a specific conformation and arrangement of the amino acid side chains in the protein molecule that interact with the metal ion. The metal-binding property of the transferrins is intimately related to the maintenance of their tertiary structures. There are no specialized prosthetic groups for complexing metals such as the heme found in hemoglobin or the cytochromes. The absence of specialized binding groups may have been responsible for the relatively late discovery and identification of transferrins. The specific requirements for metal binding by the protein itself, rather than by a prosthetic group, may have favored greater conservation during evolution.

The routes leading to the discovery and identification of the different transferrins have given rise to several nomenclatures. The principle problem is the absence of a relationship between early names and those selected more recently to have better scientific or technical significance. The egg transferrin has been historically known as conalbumin and milk transferrin as "red protein." In the case of serum transferrins, however, a functional description was first applied. It was early described as the $\beta$ metal-binding protein because of its metal-binding properties known before the actual protein was observed. Two technically descriptive names for this protein were subsequently suggested. The one, transferrin, was selected on the basis of the protein's ability to transport iron in animal blood; the other, siderophilin, was selected on the basis of its ability to form an iron complex and not for any physiological function. In view of the fact that no biological functions for the egg-white and milk proteins have been demonstrated, the stem name of siderophilin might be preferable as a generic name for these proteins. McKenzie [205] has recently suggested that the name ferrin be used for this purpose. The name transferrin appears, however, to be too firmly established, and consequently it appears best to name all these proteins as transferrins: ovotransferrin for conalbumin, lactotransferrin for milk transferrins, and serum transferrin for the blood serum transferrins. Although captions and legends of reproduced figures and tables use other names, an attempt is made to follow this nomenclature in the text.

In this chapter we shall emphasize the biological, comparative, and genetic aspects of transferrins. The physical studies before 1965 have been discussed recently elsewhere [40] and therefore are covered here briefly.

## PREPARATION

Numerous different procedures have been developed for the separations and purifications of the different transferrins. Newer methods employ various modifications of older ones with the addition of fractionation by ion exchange and molecular exclusion as additional steps. The differential solubilities of the iron-free and iron complexes have been particularly useful in sequential procedures wherein the protein is fractionated first in one form and then refractionated in the other form.

## Serum Transferrin

Early fractionation procedures as well as most current ones have employed a particular Cohn serum fraction as a starting material for the preparation of serum transferrin. The first detailed method was described by Surgenor et al. [326] on the fractionation of the serum fraction IV-7 of Cohn [327]. Subsequent methods have used mainly fraction IV-4 or a related fraction, because these fractions can be obtained as by-products of commercial blood plasma fractionation procedures. The ethanol fractionation procedures for blood serum by which these starting fractions are obtained have been published in detail and are not described here. Two procedures applicable to the large scale preparation of serum transferrin have been published. One procedure [328] was designed for use with conventional large scale blood fractionation equipment and gave approximately 40% yields of transferrin from Cohn fraction IV-4. Another procedure [329] employs rivanol (6,9-diamino-2-ethoxyacridine lactate) as an agent for complexing the protein and has been reported to yield an electrophoretically almost pure preparation of transferrin with a relatively short procedure [329, 330]. A special method for preparing single crystals of serum transferrin has recently been described [331].

## Ovotransferrin

Earlier fractionation techniques for ovotransferrin employed precipitation by ammonium sulfate and acid [47]. Ethanol fractionation was also successfully employed [101], [332]. Warner and Weber [79] first crystallized iron-ovotransferrin from dilute ethanol solutions at low temperatures. More recently the ion exchangers CM-cellulose and DEAE-cellulose have been used to prepare large quantities of ovotransferrin in high purity [72, 78, 94, 333]. Azari and Baugh [334] have developed a procedure combining ion-exchange chromatography and crystallization from ammonium sulfate.

## Lactotransferrins

Procedures for the purification of lactotransferrins are similar to those for ovotransferrin and serum transferrin. Montreuil *et al.* [335] precipitated human lactotransferrin with ammonium sulfate at pH 7 and chromatographed it on Amberlite XE-64 ion-exchange resin. Purification of human lactotransferrin by chromatography on DEAE-cellulose has been employed in our laboratory [192, 336] and by Johanson [337] who further employed chromatography on calcium phosphate. Precipitation with rivanol has been used successfully [338]. Bovine lactotransferrin has also been purified by chromatography on DEAE-cellulose [339].

## Methods for Determination

Methods for the determination of transferrin are related to its chelation of iron [340, 341]. In serum the methods use the sum of the residual iron-binding (due to incomplete saturation of the transferrin) and the original total amount of iron to determine the total amount of serum transferrin. In egg white the problem is much easier because there is practically no iron present in egg white and a quite satisfactory method is the direct determination of the intensity of the salmon-pink color after addition of iron [342]

## CHEMICAL COMPOSITION

The chemical compositions of the various transferrins have many differences varying from minor to large ones. Tables 3-7, 4-6, and 6-1 compare data on amino acid and carbohydrate analysis. Large differences are in the content of histidines, prolines, tyrosines, and tryptophans. The reported absence of tryptophan in human lactotransferrin must be considered cautiously until confirmatory analyses are reported. A spectrophotometric analysis on a single preparation of human lactotransferrin made in our laboratory gave a value of 13 residues of tryptophan per 77,000 g [346].

TABLE 6-1. CARBOHYDRATE CONTENTS OF TRANSFERRINS

| | Reference | Human Serum Transferrin [343] | Chicken Ovo-transferrin [35] | Chicken Serum Transferrin [35] | Bovine Lacto-transferrin [344] | Human Lacto-transferrin [345] |
|---|---|---|---|---|---|---|
| 1. | Mol wt used | 90,000 | 76,000 | 90,000 | 86,100 | 95,000 |
| 2. | Carbohydrate content | 5.3% | 2.2% | – | 7.2% | 7.17% |
| 3. | Groups/mole | | | | | |
| a. | Sialic acid | 4 | 0 | 1 | 1 | 3 |
| b. | Mannose | 8 | | | | 7 |
| c. | Galactose | 4 | {3 | {12 | {20-21 | 10 |
| d. | Hexosamine | 8 | {6 | {6 | 10-11 | 13 |
| e. | Xylose | – | – | – | – | 1 |
| f. | Fucose | – | – | – | – | 5 |

Information on the terminal amino acids of the transferrins is incomplete. Of the ten, which have had their N-terminal amino acids determined, seven had alanine, one had valine, and in one case none could be found. No information on the C-terminal amino acids is available.

Rather extensive differences have been found in the carbohydrate residues of the different transferrins studied (Table 6-1). Chicken ovotransferrin has the lowest carbohydrate content and over half of these residues are glucosamine; it is also free of sialic acid [69]. The number of residues of sialic acid per mole vary greatly in the other transferrins. Jamieson [343] recently reported in detail on the carbohydrate of human serum transferrin. The carbohydrate was present on two glycopeptide chains, each with a molecular weight of 3675 g; approximately 2350 g of each was carbohydrate. The chains were linked to the protein moiety through an asparaginyl-glycosylamine linkage. The aspartic acid to sialic acid ratio in the glycopeptide fraction was approximately 1 to 2. Similar studies have not been done with ovotransferrin.

## PHYSICAL PROPERTIES

The metal-free transferrins have physical properties like other globular proteins of the same molecular size. Upon chelation of metal ions several extensive changes in the physical properties occur. Most of the other changes, however, are slight.

### SHAPE AND SIZE

Molecular weights of transferrins determined in several different ways are shown in Table 6-2. The $S_{20,w}$ values for all transferrins are clustered around 5.1. Most of the early values for the molecular weights of transferrins are similar and range from 86,000 to 93,000 g. Nearly all molecular weights obtained by iron-binding determinations are between 70,000 and 80,000 g. More recently obtained values by physical determinations are between 74,000 and 82,000 g. However, extreme values recently reported are 93,000 g by Bezkorovainy and Grohlich [348] and 68,000 g by Charlwood [349]. Most transferrins appear to have approximately similar molecular weights when examined by the same method in any one laboratory. In our laboratory, for the sake of convenience, we consider all transferrins to have molecular weights of 76,000 g. This obviously would not correct for differences in carbohydrate content.

Related to the question of molecular weights of transferrins is the question of the possible presence of subunits in the molecule. Tryptic peptide maps of chicken ovotransferrin [35], human serum transferrin [353], and rabbit serum transferrin [230] show fewer spots than would be predicted from their lysine and arginine contents. Figure 6-1 shows such a peptide map of rabbit serum transferrin. Supporting this idea is the finding of two similar oligosaccharides in human serum transferrin (HST) [343]. These data plus the presence of two iron-binding sites and the observed molecular weights that are higher than most single-chain proteins would a priori

TABLE 6-2  MOLECULAR WEIGHTS OF THE TRANSFERRINS

| Molecular Weight (g/Mole) | Method | $S_{20,w}$ | Reference |
|---|---|---|---|
| | | Human serum transferrin | |
| 74,000 | Sed.-diff. | 4.9 | [347] |
| 76,000 | Sed.-equil. | | [347] |
| 93,000 | Sed.-diff. | 5.0 | [348] |
| 68,000 | Sed.-equil. | 5.1 | [349] |
| 82,000 | Sed.-equil. | 5.2 | [331] |
| 76,000 | Sed.-equil. | | [21] |
| | | Swine serum transferrin | |
| 88,000 | Sed.-diff. | | [350] |
| | | Rat serum transferrin | |
| 67,000 | Sed.-equil. | 5.1 | [349] |
| | | Monkey serum transferrin | |
| 68,000 | Sed.-equil. | 5.2 | [349] |
| | | Rabbit serum transferrin | |
| 77,000 | Sed.-equil. | | [21] |
| | | Human lactotransferrin | |
| 95,000 | Sed.-equil. | | [351] |
| | | Bovine lactotransferrin | |
| 86,100 | Sed.-diff. | | [344] |
| | | Rabbit lactotransferrin | |
| 70,000 | Sed.-diff. | | [230] |
| | | Chicken ovotransferrin | |
| 86,000 | Sed.-diff. | 5.05 | [80] |
| 86,000 | Osmotic pres. | | [80] |
| 82,400 | Light scat. | | [80] |
| 76,000 | Light scat. | | [352] |
| 76,000 | Sed.-equil. | | [21] |

implicate transferrins as proteins composed of two subunits. On the basis of sedimentation velocity studies of the reduced-carboxy-methylated protein such subunits have been reported [354]. Determinations of N-terminal amino acids, however, indicate only one N-terminal amino acid for each molecule of human serum transferrin or ovotransferrin and none for human lactotransferrin. Other than for a single report [354] of two subunits all reports indicate a single polypeptide chain [21, 348].

Our laboratory has reported [21] that human serum transferrin, rabbit lactotransferrin, and chicken ovotransferrin all behave as a single polypeptide chain in the ultracentrifuge after reduction of the disulfide bonds and carboxymethylation of the exposed sulfhydryls. Figure 6-2 illustrates the relationships of sedimentation coefficients to molecular weights of

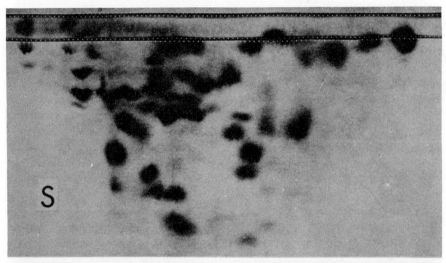

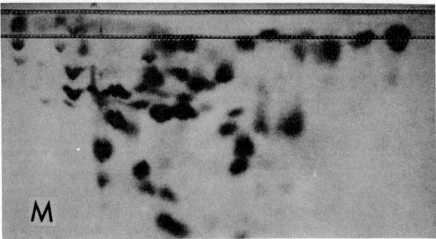

Fig. 6-1. Two dimensional peptide maps of soluble tryptic peptides from rabbit serum (S) and milk (M) transferrins, stained with ninhydrin. First dimension pH 4.7 (pyridine 25 ml, glacial acetic acid 25 ml to 1 liter with water) 40 V/cm, 1 hr; second dimension ascending chromatography *iso*-amyl alcohol : pyridine : water (35:35:30 by vol.). Reproduced from [230].

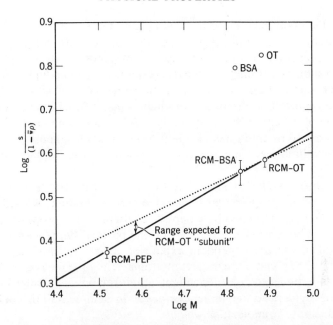

Fig. 6-2. Relation of sedimentation coefficients to molecular weights of RCM-proteins (reduced carboxymethylated) in 8 $M$ urea, and the behavior of nonreduced BSA (bovine serum albumin) and OT (chicken ovotransferrin) for comparison. The vertical bars indicate the range of $S_{20}$ values observed. BSA and PEP (pepsin) were included to provide reference points for proteins with a single polypeptide chain of the same size and slightly less than half the size of transferrin, respectively. The dashed line represents the "theoretical" slope with BSA as a reference point and the solid line our experimental slope. The double-headed arrow indicates the range of $s/(1-v\rho)$ values expected for a half-molecule of OT. Reproduced from [21].

reduced-carboxymethylated chicken ovotransferrin, bovine serum albumin, and porcine pepsin. It can be seen that the molecular weight of ovotransferrin is close to that of bovine serum albumin (single chain with molecular weight of 66,000 g) and much larger than pepsin (single chain with molecular weight of 32,700 g). The finding of a single-chain structure provides interesting material for speculation along genetic and physical lines [21]:

"On a genetic basis, it appears more convenient and more economical in terms of required DNA and m-RNA to construct a 76,000 Mw protein by the joining of two preformed 38,000 Mw halves than by uninterrupted synthesis of a single chain. Twice as much DNA (and m-RNA) would be required for the large single chain. The analyses of the composition of transferrins indicate that they consist of two similar polypeptide halves, each of which contains one iron-binding site (and in HST, one oligosaccharide each). Since the physical evidence indicates only one polypeptide chain per molecule, the apparent duplication in composition is probably not due to subunits. It might, however, result from a gene duplication process. Such duplications involving peptide segments of shorter lengths have been reported: γ-globulin light chains appear to have evolved from duplication of a 12,000 Mw polypeptide [289], and some haptoglobin α-chains appear to be multiples of 8,000 Mw peptide segments [355]. Perhaps the evolutionary antecedent of transferrins was 38,000 Mw with 1 $Fe^{+++}$."

It was also concluded [21] that the transferrin structure might be similar to the model prepared for bovine serum albumin. A model with two globular regions joined by a peptide chain has been proposed for bovine serum albumin [274, 275, 356] on the basis of fluorescence polarization studies and pH-dependent expansion below pH 4. Chicken ovotransferrin [357] and human serum transferrin [347] have also been reported to undergo pH-dependent conformational changes below pH 4.5 as evidenced by large decreases in sedimentation coefficient without decrease in molecular weight.

## Isolectric Points and Charge Relationships

Reported values for the isoelectric points (approaching isoionic pH values in most cases) also differ depending on the source of the transferrin, the conditions employed for the determination, and the laboratory in which the work was done. The effect of the formation of the iron complex on the isoelectric point will be discussed later. A pH of approximately 6 appears valid for the isoelectric point of chicken ovotransferrin and human transferrin. Higher isoelectric points of approximately pH 8 for bovine lactotransferrin and pH 9 for cassowary ovotransferrin appear to be well substantiated. Such large differences as found between chicken and cassowary ovotransferrins support the great evolutionary divergence known for these species. The acidity required for dissociation of iron from the iron complex of these two transferrins also differs. Electrophoretically this species difference manifests itself that on electrophoresis at pH 8.6 these two ovotransferrins migrate in opposite directions (Fig. 3-13).

An early observation of chicken ovotransferrin was that a new component formed with a slightly different electrophoretic mobility in acidic solutions [26]. Phelps and Cann [357] studied these effects in more detail and found that the $S_{20,w}$ values also changed in acidic solutions (Fig. 6-3).

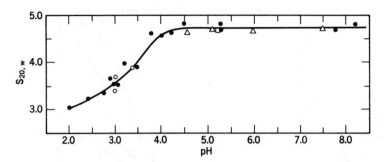

Fig. 6-3. Effect of pH on the corrected sedimentation constant of ovotransferrin at ionic strength 0.1, protein concentration 1.1-1.3 g/100ml: ●, ovotransferrin prepared by electrophoresis-convection; ○, ovotransferrin prepared by the acid precipitation method; □, ovotransferrin prepared by electrophoresis-convection, exposed for 1 hour to pH 2-3 before sedimentation at pH 5.2; △, ovotransferrin prepared by the acid precipitation method, exposed for 1 hour to pH 2.5-3.4 before sedimentation at pH 4.5-7.5. Reproduced from [357], Academic Press, 1956.

As mentioned earlier the molecular weight, however, did not change because there were corresponding changes in viscosity and in the observed diffusion coefficient. Other transferrins have since been observed to undergo similar changes in acidic solutions.

## PROPERTIES OF THE METAL-FREE AND METAL COMPLEXES OF TRANSFERRINS

The majority of definitive studies on the properties of the metal complexes have been done with human serum transferrin and chicken ovotransferrin.

### Absorption Spectra and Charge Relationships

Chicken ovotransferrin and its iron complex have more or less typical ultraviolet absorption spectra. The iron complex has a slightly higher extinction coefficient at 280 m$\mu$. Values for $E_{1\,cm}^{1\%}$ of 11.4 and 14.1 have been reported for human serum transferrin and its iron complex, respectively [331].

Robert Warner and his co-workers [358, 359] noted differences between the absorption of chicken ovotransferrin and those of its iron complex in the alkaline range in which tyrosines are assumed to show changes in ionization. Figure 6-4 is a reproduction of the plot for phenolic equilibria. In considering the roles of tyrosines in binding they concluded [359]:

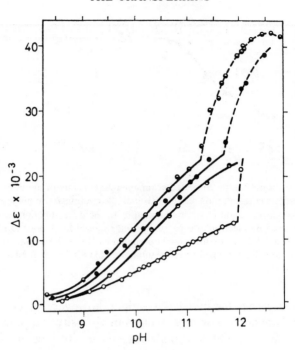

Fig. 6-4. Δ Phenolic equilibria; $\Delta_\epsilon$, was measured at 295 mμ and refers to the increment over the molar extinction at neutral pH:◐, 25°; ●, 15°; ◓, 5°; ○, iron ovotransferrin, 22°; all are at 0.1 ionic strength. Reproduced from [359].

"Any consistent choice shows a difference of six titratable phenol groups between conalbumin and its iron complex, establishing the nature and number of the chelating groups beyond question as three phenolate residues per ferric ion." The acceptance of a figure of six for the number of tyrosyl groups (from the spectrophotometric data) must await further studies because other studies have not substantiated the participation of six tyrosines (three per ferric ion) as is described in the next section.

Human serum transferrin and chicken ovotransferrin have been reported to bind cobalt, iron, copper, zinc, and manganese. The iron complex is salmon pink, and has an absorption maximum at 465 mμ. Complexes of copper or manganese are yellow (Fig. 6-5). Study of the displacement of hydrogen ions by metals during complex formation [361, 362] have shown the displacement of 2.9 equivalents per ferric ion and 1.9 equivalents per cupric ion at pH 8.5. The same value with iron was found in experiments at pH 7.5 and 9.5. Warner and Weber [79] concluded that three protons are displaced by iron and two by copper when the complex is formed and that the particular groups of the proteins involved in co-ordinating the metals must bear protons at least up to pH 9.5 and, therefore, must have pK values greater than 10.

The earlier studies [79] which show differences between electrophoretic mobilities of iron ovotransferrin and of metal-free ovotransferrin, have now been confirmed (Fig. 6-6) [363], and these later results have indicated the existence of three molecular species when insufficient iron is present to form the saturated 2:1, iron:transferrin complex. This can be interpreted as showing that such a mixture contains metal-free transferrin; 1:1, iron:transferrin, and 2:1, iron:transferrin. It also suggests that the two iron-binding sites have similar affinities for iron and that there is no cooperative binding [363, 364].

## Shape and Size

Chelation of metal causes little, if any, change in shape and size of transferrin. In a detailed study Fuller and Briggs [80] found small differences in the frictional ratios of iron-free and iron transferrin. A decreased antigenic reactivity of the metal ovotransferrin complexes as compared with metal-free ovotransferrin has been interpreted as reflecting an alteration of the protein structure as a result of chelation of the metal ion [365].

## Binding Affinities

All the transferrins have been found to bind two gram atoms of metal ions per mole of protein. The metal ions dissociate in acid solution but are stably bound in alkaline solutions (to pH 9 or 10). The relative strengths

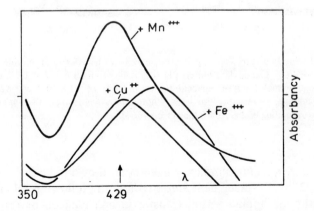

Fig. 6-5. Absorption spectra for colored complexes of human serum transferrins. Reproduced from [360].

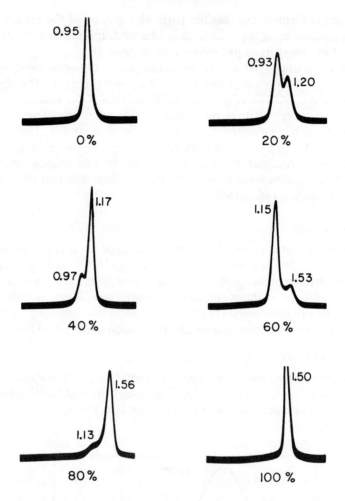

Fig. 6-6. Moving boundary electrophoresis diagrams of transferrin at varying degrees of saturation with $Fe^{3+}$. Cacodylate buffer, pH 6.7, ionic strength 0.1. $E = 13$ volts per cm. Numbers indicate mobilities of adjacent peaks, in units of $-1 \times 10^{-5}$ cm$^2$ volt$^{-1}$ sec$^{-1}$. Percentages indicate saturation with $Fe^{3+}$. Photographs were taken after approximately 20,000 sec, in ascending limbs. Reproduced from [363].

of ovotransferrin metal complexes have been shown to be $Fe^{3+} > Cu^{2+} > Zn^{2+}$ [342, 358]. Displacement studies show the following relative binding strengths for human serum transferrin and chicken ovotransferrin: $Co^{3+} > Fe^{3+} > Mn^{3+} > Cu^{2+}$ [366]. Probably the most accurate dissociation constants are for the iron complexes for which a value of $10^{-29}$ has been calculated [358].

The pH dependence of metal binding by different transferrins varies extensively. Iron complexes of human serum transferrins and all avian ovotransferrins have a pK (pH for half maximum dissociation of the metal) of approximately 6.5. Some, such as human lactotransferrin, are much more strongly associated at this pH and have a pK of approximately 5.0 [40]. The copper complexes have pK values approximately one unit higher (more alkaline) than the iron complexes.

## THE NATURE OF THE BINDING SITES AND THE STRUCTURE OF THE COMPLEXES

It is presently believed that the "active centers" of binding in the different transferrins are essentially identical and that the metal complex consists of the metal ion, the ligands contributed by the folding and orientation of the constituent amino acid residues in the protein, and a participating carbonate or bicarbonate ion [342]. No prosthetic groups or unusual constituents in the protein molecule have been found, and the problem has resolved itself into determining the nature of the protein groups involved and the actual physicochemical structure of the complex in the environment of the metal ion.

One approach to the study of the binding sites has involved the selective chemical modification of the side chain groups of the more chemically reactive constituent amino acids. In studies from our laboratory the iron-binding activity could be retained by careful treatment of metal-free transferrins with amino group reagents. Human serum transferrin, human lactotransferrin, and chicken ovotransferrin were modified by acetylation with acetic anhydride, carbamylation with KCNO, and succinylation with succinic anhydride [336]. It was concluded from this work that the amino groups on the periphery of the molecule were not critical for the maintenance of the tertiary structure and not involved in chelation of metal ions. The involvement of tyrosines was early indicated by the complete loss of chromogenic capacity (color on the addition of iron or copper) with iodination of tyrosines of metal-free ovotransferrin [367]. No losses in chromogenic capacity occurred when the iron complex was iodinated. This observation has been confirmed by further studies of the effects of iodination [97, 368, 369], acetylation of the phenolic hydroxyls [97], and nitration of the phenolic ring [370]. In the recent iodination studies [97] it was also found that iodination of the copper complex did not destroy the chromogenic capacity with either iron or copper. This would not seem to agree with earlier observations that the binding of iron involves an additional tyrosine over that required for the binding of copper. The "unchelated" third tyrosine in the copper complex ought to be iodinated as the essential tyrosines were apparently iodinated in the metal-free protein. However, the effects of this type of chemical modification can also be explained in many other ways. One explanation is that the third tyrosine is protected from iodination in the copper complex as well as in the iron complex. Table 6-3 shows the inactivation of chromogenic capacities upon acetylation with N-acetylimidazole and the reactivation (60-70%) upon deacetylation with hydroxylamine. These latter results point strongly to the involvement of tyrosyl hydroxyls in the chromogenic capacities for iron. Chromogenic capacities for copper were also lost upon acetylation.

Examinations of the physical properties of the metal complexes have also given insight into the nature of the binding sites and the structure of the complexes. Electron spin (paramagnetic) resonance studies indicated that two or more nitrogens were involved in the binding of copper, and that the metal ions do not interact with one another, that is, they must be greater than 9 or 10Å apart [192, 361, 371]. The ESR spectrum for copper ovotransferrin is given in Fig. 6-7. For the iron transferrin complexes the peak at g = 4.27 was characteristic of a highly asymetric electric field. A unique application of ESR studies was the recent proof by Aisen and co-workers [371] that copper and iron form uncolored complexes with transferrins in the absence of bicarbonate and that the formation of the colored complex with copper or iron requires bicarbonate or some other ligand (Figs. 6-8, 6-9). Studies of optical rotatory dispersion also have afforded penetrating insight into the possible groups involved in chelation of the metal and in the structure of the complex [360], [372]. Metal-free ovotransferrin had a plain negative rotatory dispersion between 300 and 675 m$\mu$ with a specific rotation, $[\alpha]_D^{10}$ of $-30°$. In the presence of iron the dispersion became anomalous due to the appearance of a negative Cotton effect (Fig. 6-10). The magnitude of the Cotton effect was a function of the amount of iron bound to the protein. Each mole of ovotransferrin was found by rotatory dispersion titration to bind two moles of iron. The Cotton effect had a trough at 505 m$\mu$, a peak at 420 m$\mu$, a breadth of 85 m$\mu$ and an amplitude of 16°. The inflection point near 470 m$\mu$ corresponded to the absorption maximum of the iron ovotransferrin complex. Nearly identical results were obtained with human serum transferrin. The amplitude of the Cotton effect of iron transferrin was less than that of ovotransferrin. $Mn^{3+}$ and $Cu^{2+}$ were also tested. $Mn^{3+}$ absorbed maximally at 429 m$\mu$ and gave rise to a positive Cotton effect. $Cu^{2+}$ absorbed maximally at 436 m$\mu$, but did not give rise to optical activity. Copper transferrins therefore do not have an asymmetric center as do iron and manganese transferrins. Circular dichroism spectra show still further differences between the complexes [373] but detailed interpretations have

TABLE 6-3. REACTIVATION OF N-ACETYLIMIDAZOLE-ACETYLATED TRANSFERRINS UPON DEACYLATION WITH HYDROXYLAMINE [97]

| Sample | OD (465 m$\mu$)[a] before Fe$^{3+}$ | OD (465 m$\mu$)[a] after Fe$^{3+}$ | Act. (%) |
|---|---|---|---|
| | Experiment A | | |
| Ovotransferrin | 0.007 | 0.242 | 100 |
| Ovotransferrin (50-fold excess reagent) | 0.027 | 0.067 | 20 |
| Ovotransferrin (50-fold excess reagent + hydroxylamine treatment)[b] | 0.038 | 0.211 | 70 |
| Ovotransferrin (100-fold excess reagent) | 0.024 | 0.030 | 3.4 |
| Ovotransferrin (100-fold excess reagent + hydroxylamine treatment)[b] | 0.035 | 0.185 | 66 |
| Ovotransferrin (100-fold excess reagent + imidazole added)[c] | 0.058 | 0.066 | <5 |
| Ovotransferrin (100-fold excess reagent + phenol added)[c] | 0.059 | 0.066 | <5 |
| | Experiment B | | |
| HS transferrin[d] | 0.0123 | 0.261 | 100 |
| HS transferrin (100-fold excess reagent) | 0.068 | 0.068 | 0 |
| HS transferrin (100-fold excess reagent + hydroxylamine treatment) | 0.057 | 0.192 | 60 |

[a]Reading of optical densities corrected to a sample concentration of 4.5 mg/ml. Readings made before and after additions of Fe$^{3+}$ as indicated.
[b]Treatment with hydroxylamine is described in text. Deacylated samples were found to contain no detectable O-acetyl groups.
[c]Treatment of the acetylated transferrin with 100-fold excess imidazole or phenol, as compared to the number of tyrosine residues, under the same conditions used for the original acetylation with N-acetylimidazole.
[d]HS transferrin is human serum transferrin.

not yet been reported. Further evidence that the two metal-binding sites are identical and do not interact was obtained from proton magnetic relaxation rate measurements [363].

In the case of the iron complex the binding site of transferrin has been pictured as consisting of three tyrosines, two histidines, and a bicarbonate in an octahedral arrangement around the iron [192]. The copper complex had two tyrosines and two histidines in a square planar arrangement. It was recognized that at least one of these formulas was wrong because they did not allow for bicarbonate in the copper complex. At least one of the ligands in the iron complex and copper complex might therefore be different. Now it has been shown that complexes can be formed without bicarbonate; there is even the possibility that the bicarbonate may not interact directly with the metal. Other studies showing that the binding site below pH 6.5 is different from that in more alkaline solution also suggest that we have a lot more to learn about the structures of these chelates [371].

## STABILITIES OF TRANSFERRINS AND THEIR METAL COMPLEXES

The main initial study of the relative stabilities of the transferrins and their metal complexes began as a result of a series of experimental failures. Dr. Paul Azari, then a graduate student in chemistry at the University of Nebraska, attempted to hydrolyze enzymatically the iron complex of chicken ovotransferrin in an effort to obtain fragments or a "peptide core" containing the colored iron complex. When it was discovered that no hydrolysis by either trypsin or chymotrypsin could be detected even after prolonged incubation, a general study was made of the stabilities of human serum transferrin and chicken ovotransferrin and their metal complexes [367], [374]. The iron complexes were found considerably more

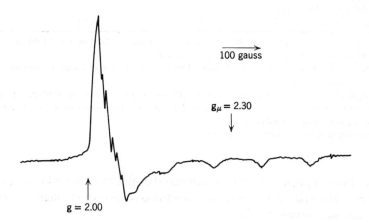

Fig. 6-7. The EPR spectrum at liquid nitrogen temperature for an aqueous solution of copper-ovotransferrin. The extra-hyperfine splitting on the peak near $g = 2.00$ is evidence for coupling to nitrogen atoms. Reproduced from [192].

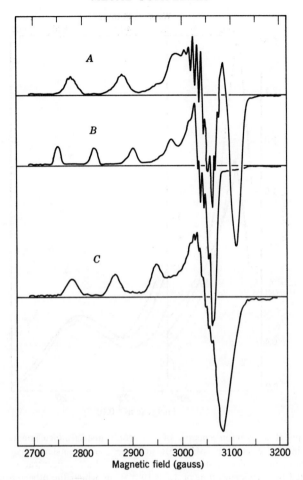

Fig. 6-8. EPR spectra obtained at 77° K of $^{65}$Cu-transferrin, pH 9.2 (*A*), $^{65}$Cu-transferrin-bicarbonate, pH 9.2 (*B*), and $Cu^{2+}$ in an excess of ammonia, pH 10.1 (*C*). The modulation amplitude was 5 gauss, the microwave frequency was 9149 MHz, and the microwave power was 1 milliwatt. Reproduced from [371].

stable to denaturation by heat, organic solvents, surface forces, high pressures, and exposure to high concentrations of urea or guanidine in aqueous solution. The iron complexes were also much more stable to chemical treatments that disrupted the structures of the metal-free protein, such as reductive splitting of disulfide linkages (Table 6-4). The much higher stability of the iron complexes has since been confirmed by more extensive studies on changes in viscosity, sedimentation, and optical rotation [348], [375] (Fig. 6-11).

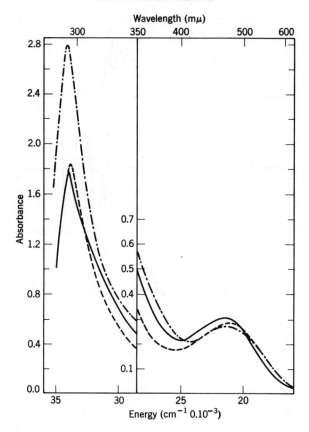

Fig. 6-9. Optical absorption spectra of —, iron-transferrin-bicarbonate (pH 9); ---, iron-transferrin-nitrilotriacetate (pH 9.5); and ·—·, iron-transferrin-oxalate (pH 9.4). The spectrum of the oxalate complex was taken at an $Fe^{3+}$ concentration of $8 \times 10^{-5}$ M, and normalized to a concentration of $1.2 \times 10^{-4}$ M, at which the other spectra were obtained. The reference cells contained apotransferrin in the same concentrations as the sample cells. Reproduced from [371].

Azari and Feeney [374] suggested that human serum transferrin and chicken ovotransferrin underwent structural changes on chelation of metal ions which stabilized the molecule to denaturation and proteolysis. This later was extended to include stabilization to chemical cleavage. A similar interpretation was also made by Glazer and McKenzie [375] who further suggested that the iron complex might provide two cross-links between widely separated sections of the peptide chains. However, the inactivation of chromogenic activity or binding activity by chemical modification of the metal-free transferrins could also be from effects changing the binding directly (active site modification) or indirectly, such as by

TABLE 6-4. COMPARISON OF THE EFFECT OF SULFITE AND UREA ON DISULFIDE BONDS CLEAVED OF CONALBUMIN AND IRON CONALBUMIN

| | Conalbumin | | Iron Conalbumin | |
|---|---|---|---|---|
| Urea Concentration[a] | Moles of $-S-S-$ Cleaved/ Mole Ovotransferrin | Optical Density at Ovotransferrin | Moles of $-S-S-$ Cleaved/ Mole Fe-Ovotransferrin | Optical Density at 470 m$\mu$[b] |
| M | | | | |
| 0 | 0.64 | 0.24 | 0 | 0.25 |
| 0.5 | 0.64 | 0.24 | 0 | 0.25 |
| 1 | 0.80 | 0.24 | 0 | 0.24 |
| 2 | 1.4 | 0.21 | 0 | 0.24 |
| 3 | 2.7 | 0.20 | 0 | 0.24 |
| 4 | 4.8 | 0[c] | 0.32 | 0.24 |
| 5.4 | 7.2 | 0[c] | 1.3 | 0.24 |
| 7.2 | 9.6 | 0[c] | 1.8 | 0.24 |
| Control | 0 | 0.25 | 0 | 0.25 |

[a]In addition to urea the reaction mixture contained 0.5 M sulfite and 5 mg/ml ovotransferrin or iron ovotransferrin. Solution was buffered with 0.2 M $(NH_4)NO_3$–0.1 N $NH_3$, pH 9.0. For the titration procedure see the text.
[b]Excess $Fe^{3+}$ was added to all mixtures and they were dialyzed for 18 hr. in 0.001 N $NH_3$ containing excess $Fe^{3+}$. The optical density readings were corrected for dilutions.
[c]These solutions became turbid after dialysis. No color of iron ovotransferrin was detected. (Table 3, of *Azari, P. R.* and *R. E. Feeney*: The resistances of ovotransferrin and its iron complex to physical and chemical treatments [367].

physical restriction of conformation or by redistribution of charges in the protein. Interpretations of the mechanisms that stabilize the molecule against modifications must consider all possible types of terminal effects. The following is a simplified visualization of stabilization as modified from Azari and Feeney [367].

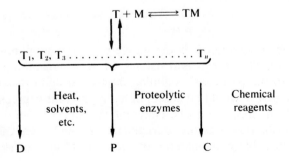

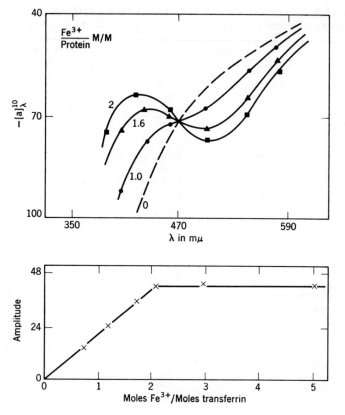

Fig. 6-10. Rotatory dispersion titration of transferrin with $Fe^{3+}$. In the upper portion of the figure, specific rotation at 10, $-[\alpha]^{10}_{\gamma}$, is plotted against wavelength. In the presence of $Fe^{3+}$, the rotatory dispersion of transferrin becomes anomalous due to a negative *Cotton* effect centered about the absorption maximum of the metal-protein complex at 470 m$\mu$. As with conalbumin (Fig. 1), the magnitude of the Cotton effect is a function of the amount of $Fe^{3+}$ bound to transferrin and becomes maximal at 2 moles of $Fe^{3+}$ per mole of protein (lower portion of figure). Conditions: identical to those for conalbumin. Reproduced from [360].

In the previous equation T is the native protein in equilibrium with the metal complex TM, which is resistant; $T_1$ to $T_n$ are the conformations susceptible to the particular condition impressed upon the protein. The various forms of T would be susceptible, in different degrees to changes caused by heat, solvent effects, enzymolysis, etc., to give the products: denatured forms (D), proteolytic fragments (P), and chemically modified forms (C). Possible combinations of these could also exist, such as C could rapidly undergo denaturation. There is also the possibility that different molecular forms of the metal complexes exist. These forms might have different susceptibilities, but the susceptibilities would not be the same as those of the metal-free proteins.

## BIOSYNTHETIC ASPECTS

The biosynthesis of the transferrins has received comparatively little study. It is commonly thought that serum transferrin is synthesized in the liver along with many other blood serum proteins. Although a decrease in the content of the major blood serum protein, the serum albumin, is usually accompanied by a decrease in the transferrin, a nutritional deficiency of iron does not cause a particular decrease in the serum transferrin.

The site of synthesis of chicken ovotransferrin is considered to be in the oviduct [149]. It is thought that all of the egg-white proteins are synthesized in the oviduct. On the other hand, Williams [35] has shown a close relationship of the chicken ovotransferrin and chicken serum transferrin; he reported that they differed only in their carbohydrate prosthetic groups. The apparent identity of the amino acid composition of the chicken serum transferrin and chicken ovotransferrin can be seen in Table 6-5. The biosynthetic origin of lactotransferrin is also unclear but identity or near identity with the blood serum transferrin has been proven in the case of the rabbit proteins [230].

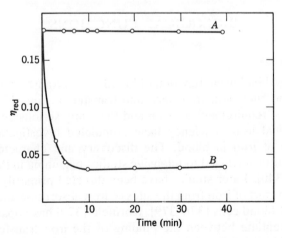

Fig. 6-11. Viscosity changes accompanying the denaturation and "renaturation" of conalbumin at pH 6.0. (A) Conalbumin (0.5 g/100 ml) in 6 M urea (0.05 M NaCl); (B) solution A diluted with an equal volume of 0.05 M NaCl. Reproduced from [375].

TABLE 6-5. AMINO ACID COMPOSITION OF RABBIT AND CHICKEN TRANSFERRINS

|  | Rabbit | | | | Chicken | | | |
|---|---|---|---|---|---|---|---|---|
|  | Serum[230] | | Milk[230] | | Serum[35] | | Egg White[a][35] | |
|  | Res/ 70,000g | Res/ 10,000g | Res/ 70,000g | Res/ 10,000g | Res/ 76,000g | Res/ 10,000g | Res/ 76,000g | Res/ 10,000g |
| Lys | 51.7 | 7.4 | 51.0 | 7.3 | 59.2 | 7.8 | 61.5 | 8.1 |
| His | 16.4 | 2.3 | 16.0 | 2.3 | 11.1 | 1.4 | 13.1 | 1.7 |
| Arg | 24.0 | 3.4 | 24.0 | 3.4 | 31.8 | 4.2 | 32.7 | 4.3 |
| Asp | 70.2 | 10.0 | 70.1 | 10.0 | 71.2 | 9.4 | 72.4 | 9.5 |
| Thr | 20.3 | 2.9 | 20.4 | 2.9 | 34.0 | 4.5 | 34.1 | 4.5 |
| Ser | 33.6 | 4.8 | 33.5 | 4.8 | 41.8 | 5.5 | 41.9 | 5.5 |
| Glu | 60.9 | 8.7 | 60.7 | 8.7 | 75.4 | 9.9 | 69.0 | 9.1 |
| Pro | 32.5 | 4.6 | 32.8 | 4.7 | 32.2 | 4.2 | 32.2 | 4.2 |
| Gly | 45.1 | 6.4 | 44.6 | 6.4 | – | – | 57.5 | 7.6 |
| Ala | 51.2 | 7.4 | 49.5 | 7.1 | 52.5 | 6.9 | 51.6 | 6.8 |
| Cys/2 | 32.2 | 4.6 | 31.4 | 4.5 | 22.8 | 3.0 | 24.3 | 3.2 |
| Val | 42.3 | 6.0 | 42.0 | 6.0 | 43.0 | 5.7 | 45.9 | 6.0 |
| Met | 5.6 | 0.8 | 5.4 | 0.8 | 11.2 | 1.5 | 10.5 | 1.4 |
| Ileu | 15.1 | 2.2 | 15.2 | 2.2 | 23.2 | 3.0 | 24.0 | 3.2 |
| Leu | 58.4 | 8.3 | 57.4 | 8.2 | 50.4 | 6.6 | 47.4 | 6.2 |
| Tyr | 23.2 | 3.3 | 22.8 | 3.3 | 19.6 | 2.6 | 20.2 | 2.6 |
| Phe | 24.7 | 3.5 | 24.2 | 3.5 | 23.8 | 3.1 | 25.4 | 3.3 |
| Try | 9.4 | 1.3 | 10.6 | 1.5 | 19.7 | 2.6 | 17.5 | 2.3 |
| NH$_3$ | 61.2 | 8.7 | 59.6 | 8.5 | – | – | – | – |

[a]Values for chicken ovotransferrin amino acid composition determined in our laboratory differ slightly from these calculated from the data of Williams [35]. See Table 2-4.

## BIOLOGICAL FUNCTIONS

### Iron Transport

The principal biological function of blood serum transferrins is thought to involve the chelation, transport, and transfer of iron. The relatively high biological requirement for iron and the many serious consequences of physiological iron deficiency have stimulated investigations on the biochemistry of iron in blood. The discovery and characterization of blood serum transferrin led to intensive studies beginning in the late nineteen forties [340]. Later studies have been directed primarily at the processes of transport of iron from the sites of iron absorption and storage to immature red blood cells [376–379]. Kornfeld [379] has apparently been able to differentiate between the binding of the iron transferrin to the immature red blood cell and the uptake of iron by its heme within the red cell. His studies employed transferrins that had various types of modifications of amino groups. It was mentioned earlier in this chapter that low level (25%) modifications of amino groups do not materially affect the iron-binding. In Fig. 6-12 are plotted values of $A$ binding of $^{59}$Fe transferrin to reticulocytes, $B$ binding capacity of carbamylated transferrins, and $C$ intracellular $^{59}$Fe absorption, all as a function of the percentage of carbamylation of amino groups. It is evident that as the extent of carbamylation was increased there were: (a) increases in the binding to reticulocytes, (b) no changes in binding capacity of the protein, and (c) a large decrease in the transfer of iron as measured by intracellular iron incorporation. Acetylation gave very similar results as trinitrophenylation at similar levels. Amidination, however, which in contrast to the other modifications results in only small changes in charge, caused a much higher amount (three-fold) of binding to the reticulocytes but not much impairment of release of iron to the cell. It is apparent that the binding and release of iron to the cell are related but independent mechanisms.

Related problems are the mechanism by which the iron is absorbed through the intestine and bound to the transferrin, and the mechanism by which the iron is released from the transferrin *in vivo*. Both dogs and hu-

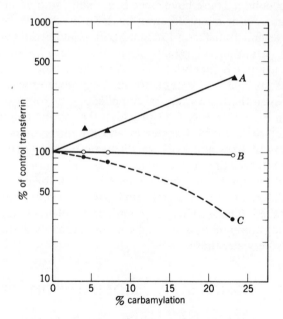

Fig. 6-12. Effect of increasing carbamylation on the biologic function of transferrin. The carbamylation procedure and the assay method are described in the text. ($A$) Binding of [$^{59}$Fe]transferrin to reticulocytes; ($B$) binding of $Fe^{3+}$ to transferrin as determined by the optical density method (see text); ($C$) intracellular $^{59}$Fe incorporation. All results are expressed as per cent of the untreated control values. Reproduced from [97].

mans deposit iron in the liver when the saturation level of iron in the transferrin exceeds 70% [380]. Paul Saltman's laboratory has been studying the effects of various buffers and other conditions on the association and dissociation of the iron complexes with the idea that these may be the key to understanding the way transferrin works in the body [381, 382].

No direct evidence for iron transport has been obtained for transferrins in milk or egg white. In fact no egg white has ever been found to contain more than one percent the amount of iron capable of being bound by the amount of ovotransferrin present. One possible relationship [230] has been noted in milk. The iron-binding capacity and the total iron of rabbit milk increase two to three weeks after the beginning of lactation.

ANTIMICROBIAL ACTIVITY

All the transferrins have potential antimicrobial activity due to their capacity to chelate iron and thereby inhibit microbial growth by iron deprivation. The discovery of the metal chelating properties of ovotransferrin followed the observation [54] of an iron-reversible inhibition of microbial growth in egg white. There have since been many studies showing the inhibition of growth of many different bacteria by chicken ovotransferrin and human serum transferrin [383–385]. It is more difficult to demonstrate an important antimicrobial role in blood serum than in egg white because of the much higher relative iron content in blood serum. The ovotransferrin is probably a major factor in the antibacterial activity of egg whites, but even there a variety of complicating factors exist. One such complication is seen in Fig. 6-13 in which the mutual counteractions of inhibitory activities of ovotransferrin and 8-hydroxyquinoline, and the further effects of $Fe^{3+}$ and $Co^{3+}$ are illustrated. Either ovotransferrin or oxine is inhibitory when tested alone. These inhibitions are specifically counteracted by iron for ovotransferrin and cobalt for oxine. The simultaneous addition of ovotransferrin and oxine prevents inhibition. It is evident that the inhibitory effects are intimately related to the presence of other chelating compounds and metal ions and probably to the equilibrium relationships between these.

## COMPARATIVE BIOCHEMISTRY

Transferrins that have been studied in detail, the blood serum transferrins and ovotransferrins, show a variety of multiple forms and significant differences in physical properties, such as isoelectric pH values and the acidity required for dissociation of the iron.

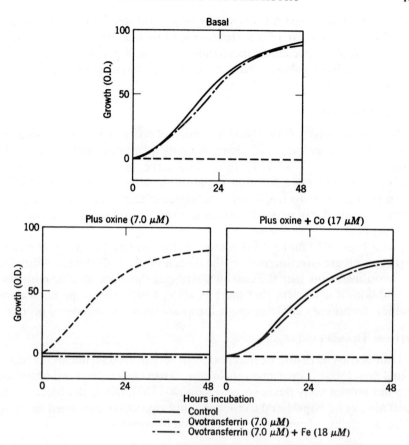

Fig. 6-13. Effect of ovotransferrin and 8-Hydroxyquinoline (oxine) on growth of *Staphylococcus albus*. Basal (medium) was nutrient broth with the additions of oxine or oxine plus cobalt (Co) as indicated. Growth curves are for basal with no further additions or with ovotransferrin or ovotransferrin plus iron (Fe) as indicated [163]. All conditions were as described for similar studies. Reproduced from data of [383], Academic Press, 1951.

## Serum Transferrins

The initial demonstration of genetically determined transferrin heterogeneity in human blood serum by Horsfall and Smithies [386] has since been followed by numerous reports on the sera of many other vertebrate animals as well as man. The serum transferrins appear to be rapidly assuming a position equivalent to the hemoglobins in the study of molecular genetics of the blood serum proteins of higher animals. These are discussed in Chapter 5.

The sera of all vertebrates studied contain transferrin. Indeed, it would be difficult to conceive of a vertebrate with its high requirement for iron not possessing a specialized mechanism for handling iron. Even Antarctic chaenichthyid fishes, which have no red blood cells, have serum transferrin [292].

## Ovotransferrins

All investigations on the ovotransferrins of different avian species have demonstrated they vary extensively not only quantitatively but also qualitatively [78, 82, 173, 174, 189]. The large quantitative differences (Table 3-1) represent a five-fold variation in some instances. In addition to the similarity in electrophoretic mobilities of some species, many properties were similar in most all species such as absorption spectra and heat stability [78].

From Figs. 3-13 and 3-14 it is evident that within this group of ratites there is a progressive increase in the direction of an alkaline mobility for the ovotransferrins with the cassowary having the most alkaline mobility. In addition, it is evident that there is also an increase in the number of multiple forms seen until five or six forms are found in cassowary protein.

## Other Transferrins

Transferrins have also been reported in human cord serum and cerebral spinal fluid [309]. Trace amounts of transferrins have also been observed in other animal body fluids such as seminal [387], but such observations must always be considered cautiously until rigorously examined because of the possibility of small amounts of blood in samples.

## Relationships of Blood Transferrins and Other Transferrins

A possible relationship between chicken serum transferrin and chicken ovotransferrin was observed many years ago by Hektoen and Cole [46] when they reported an immunological relationship between ovotransferrin and that serum fraction which is soluble in 64% saturated ammonium sulfate and in which most of the transferrin is found. Identity of ovotransferrin and chicken transferrin has also been reported by immunoelectrophoresis and agar diffusion [143]. Williams [35] has reported a close identity of chicken serum transferrin, ovotransferrin, and a transferrin in the livetin fraction of egg yolk. The egg yolk fraction was identified by starch-gel electrophoresis and was isolated as a pink protein from egg yolk. Ovotransferrin and the serum transferrin were compared by starch-gel electrophoresis, peptide patterns, immunoelectrophoresis, amino acid composition, and carbohydrate analyses. Williams interpreted his results that conalbumin and transferrin are glycoproteins which differ only in their carbohydrate prosthetic groups; they appeared to be identical in their peptide moiety (Table 6-5). Principal differences in carbohydrate content were in a much higher content of hexose and the presence of sialic acid in the serum transferrins. Treatment of the chicken serum transferrins with the enzyme neuraminadase (sialidase) converted three multiple forms into one. The close relationship between the chicken serum transferrin and ovotransferrin was also supported by comparison of ovotransferrin and serum transferrins from a hen bearing a mutant transferrin gene with the corresponding proteins from wild-type hens. The amino acid sequences of serum transferrin and ovotransferrins were considered as probably controlled by the same genetic locus.

Similar results have been obtained by Ogden et al. [388]. Three phenotypic types were considered: type A with two bands, and type B with two bands but with the slower and faster bands of each type coinciding with one another so that in another variant three bands are seen that correspond with the mobilities of the two bands of A and two bands of B. Examination of egg whites from hens with the three transferrin phenotypes gave results identical to those obtained with the serum transferrins in that the three corresponding ovotransferrins were found in the egg whites.

The recent data comparing rabbit serum transferrin and rabbit lactotransferrin show identity or nearly complete identity of the protein moiety (Table 6-5). This is not the case with human and bovine serum transferrin or human and bovine lactotransferrin (Table 4-6) whose amino acid contents are very different. Several fractions of milk proteins with iron-binding capacity but with differing amino acid contents have also been separated [389]. Thus it seems likely that gene duplication might have occurred early in these mammals and one or the other, or both, genes have undergone many mutations and diverged extensively.

The transferrins can be expected to receive more and more attention from biologists. Biologists have really only been concerned with the transferrins for about a decade, and yet there is now a large volume of biological literature on these proteins. Their unique property of combining reversibly with iron to form a red-colored complex and the ease of labeling them with radioactive iron afford the biologist with simple tests for their study. These properties together with their distributions in different fluids as genetically controlled multiple molecular forms should entice more and more investigators.

# 7
# Lysozyme

Lysozyme has been one of the more interesting enzymes since its discovery nearly 40 years ago by Sir Alexander Fleming [49]. During the intervening time there have probably been thousands of publications dealing with lysozyme from its medical aspects to the now highly detailed physics of the crystal of the molecules. This story is a fascinating one and the enzyme itself has also proven to be fascinating. The initial discovery of lysozyme was tinctured with an aura of excitement, and it was an example of excellence in scientific accomplishment. What more exciting thing could be found than an enzyme (in the days when information on enzymes was comparatively rare) which might cure some of mankind's ills by literally digesting the skins off infecting bacteria!

One of the best ways to introduce lysozyme is to quote directly this remarkable Fleming from his paper "On a Remarkable Bacteriolytic Element Found in Tissues and Secretions" published in the Proceedings of the Royal Society of Great Britain in 1922 [49]:

"In this communication I wish to draw attention to a substance present in the tissues and secretions of the body, which is capable of rapidly dissolving certain bacteria. As this substance has properties akin to those of ferments, I have called it a "Lysozyme" and shall refer to it by this name through the communication. The lysozyme was first noticed during some investigations made on a patient suffering from acute coryza. The nasal secretion of this patient was cultivated daily on blood agar plates. For the first three days of the infection there was no growth, with the exception of an occasional staphylococcus colony. The culture made from the nasal mucous on the fourth day showed in 24 hours a large number of small colonies which, on examination, proved to be large gram-positive cocci arranged irregularly but with a tendency to diplococcal and tetrad formation. It is necessary to give here a very brief description of this microbe as with it most of the experiments described below were done, and it was with it that the phenomena to be described were best manifested. The microbe has not been exactly identified, but for purposes of this communication it may be alluded to as the *Micrococcus lysodeikticus*."

This striking result led to further investigation, and it was noticed that one drop of diluted nasal mucus added to 1 cc of a thick suspension of the cocci caused a complete disappearance of the intact cells in a few minutes at 37°C. These two preliminary experiments clearly demonstrate the very powerful inhibitory and lytic action which the nasal mucus has upon *M. lysodeikticus*. Fleming demonstrated that this power is shared by most of the tissues and secretions of the human body, by the tissues of other animals, by some vegetable tissues, and to a very marked degree by chicken egg white. One of his experiments demonstrating lysozyme in human tears is shown in Fig. 7-1. Fleming considered lysozyme to be related to the "first line of defense" of the human body to many infections.

Sir Alexander Fleming went on to his extremely important humanitarian discovery of the first great mass produced antibiotic, penicillin. It was primarily for this latter work that he was knighted. Personal comments attributed to him throughout the years (hearsay) have indicated that he had a recurrent nostalgia for one of his first loves, lysozyme, and considered lysozyme and its mode of action one of the most important problems in medical bacteriology.

Long before Fleming reported a relationship between tears and egg white in their common lytic activity due to lysozyme, the Romans reputedly recommended egg white for the treatment of eye infections. Lysozyme is used as a pharmaceutical in some countries even today. At one time lysozyme was thought to be the cause of an intestinal inflammation (colitis) and sodium dodecyl sulfate was even recommended for oral therapeusis since it was known to inactivate lysozyme in solution.

Fleming's original enthusiasm and scientific perspicacity has been borne out by the extensive researches on lysozyme in the past decade and by its current importance in protein and enzyme chemistry. It is the third protein whose tertiary structure has been revealed by X-ray crystallography, being preceded only by myoglobin and hemoglobin. It is the first enzyme, however, that has had this distinction. This enzyme was also one of the first to yield information on the direction of synthesis of a peptide chain (from N-terminus toward C-terminus) [148] and was a key protein supporting the hypothesis that primary structure (amino acid sequence) dictates the conformation of a molecule [390]. The latter was demonstrated in experiments in which lysozyme was completely reduced and allowed to reoxidize to an active enzyme. It is also the basis of another

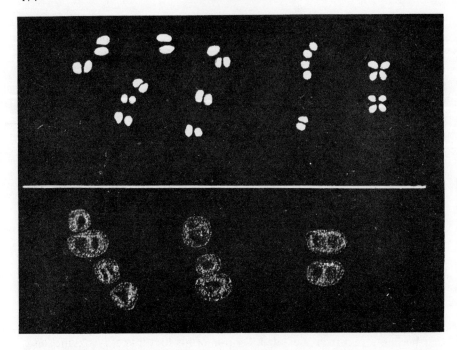

Fig. 7-1. Upper half: *Micrococcus lysodeikticus* before being acted on by tears; lower half: same partially dissolved by tears. Examined by Burri's method. Reproduced from [49].

milestone in enzyme chemistry. The virtual existence of the "enzyme—substrate" complex (an intermediate at the very core of our current theories of enzyme action) was shown by X-ray crystallographic studies on lysozyme-inhibitor complexes with analogous substrates acting as inhibitors. Lysozyme is an excellent example to discuss in this monograph because of the progress that has been made possible in a variety of fields through its study. Today the focusing of various of these findings is leading to a comparatively sophisticated understanding of the properties of lysozymes. These enzymes and the proteins related to them provide one of the most rewarding and best characterized series of homologous and analogous proteins now being studied.

## THE DEVELOPMENT OF OUR PRESENT KNOWLEDGE OF LYSOZYME

Historically, the initial discovery of lysozyme by Fleming was followed by a decade and a half of studies primarily biological in nature. During this time there were numerous reports on the bacteriostatic activity of various fluids and certain effects on the microorganisms concerned. Karl Meyer was one of the most active investigators of lysozyme during the mid-1940's. With his co-workers he developed a viscosimetric method for the assay of lysozyme [391] and described lysozyme activity in the latex of *Ficus* trees (due to a lysozyme chemically different from chicken egg-white lysozyme) [392]. They later described increases of the lysozyme contents in human gastric juice in cases of peptic ulcer [393] and in the human colon during ulcerative colitis [394]. Although lysozyme was crystallized in 1937 [395] it was not until 1946 that a simple, high yielding, and direct method for crystallizing lysozyme from egg white was developed by Alderton and Fevold of the U.S. Department of Agriculture, Research Laboratory, Albany, California [48]. For the next few years chicken egg-white lysozyme was apparently the most easily obtainable animal protein in a highly purified crystalline state. In this method, chicken egg-white lysozyme can be crystallized directly from egg white containing five per cent sodium chloride at pH 9.5. This simple method lead to a surge in studies on the chemistry and properties of chicken egg-white lysozyme. It rapidly became one of the standard proteins of protein chemistry. Enzyme chemists, however, did not show much interest in it because the substrate was either whole killed bacterial cells of *M. lysodeikticus* or particulate cell wall preparations. This absence of a simple substrate apparently impeded progress for many years.

Early studies on the chemistry of lysozyme were done in the laboratory of Dr. Heinz Fraenkel-Conrat also at the U.S. Department of Agriculture, Research Laboratory, Albany [396, 397]. In 1958 the application of the cellulose ion-exchange agents in one of our laboratories at the University of Nebraska [72] made possible the isolation of lysozymes from egg whites from many species of birds.

By this time studies on lysozymes were moving in a variety of directions, some of them seemingly unrelated. One direction was an increased interest in comparative and genetic aspects of lysozymes from different species with the realization of the utility of this approach. Another was a continuation of the chemical approach in attempts to elucidate the structure of lysozyme by chemical means. These studies are discussed in separate sections in this chapter. A third approach used a rapidly increasing understanding of the enzymatic action of lysozyme on carbohydrate substrates. The initial breakthrough in the latter was the observation that chicken lysozyme not only releases N-acetylamino sugars from cells of *M. lysodeikticus* but also degrades chitin, a linear polymer of N-acetylglucosamine (NAG) [398]. This led to the conclusion that the enzyme possess $\beta$-(1-4)-glucosaminidase activity. Shortly after this, a tetrasaccharide was isolated from lysozyme digests of *M. lysodeikticus* cell walls in Salton's laboratory [399]. This tetrasaccharide contained equal molar amounts of alternating units of $N$-acetylglucosamine ($NAG$) and $N$-acetylmuramic acid ($NAM$). It was shown to be hydrolyzed by lysozyme into a disaccharide, $NAG$-$NAM$, as illustrated in Fig. 7-2. At this stage the understanding of lysozyme activity was squarely in the middle of an understanding of cell membrane composition. Anyone familiar with the area of cell membrane structure knows that it has been rapidly developing since that time and the use of lysozyme in many of these studies has been a great benefit. Beginning to understand the substrate specificity of lysozyme in turn led to the use of many different model compounds in studying the activity and structure of lysozyme.

The next era saw the primary structure and connections of the disulfide bonds delineated. Many workers participated but two predominant groups were in the laboratories of Jolles and Canfield [86, 401]. This very quickly led into plans for concerted attack on the three dimensional structure by X-ray diffraction [402]. The finale, as of now, is the actual demonstration by physical means of an enzyme-inhibitor complex that is apparently closely related to the productive enzyme-substrate complex. This was done by the X-ray diffraction analysis of lysozyme crystals containing tri-$NAG$, which is bound in a crevice of the lysozyme molecule [403]. From these observations the protein side chains involved in the catalytic action of lysozyme have been identified and a detailed mechanism of action has been proposed [404, 405].

General Distribution

Lysozyme, an enzyme with rather restricted specificities for substrates of restricted distribution, may prove to be distributed more diversely than

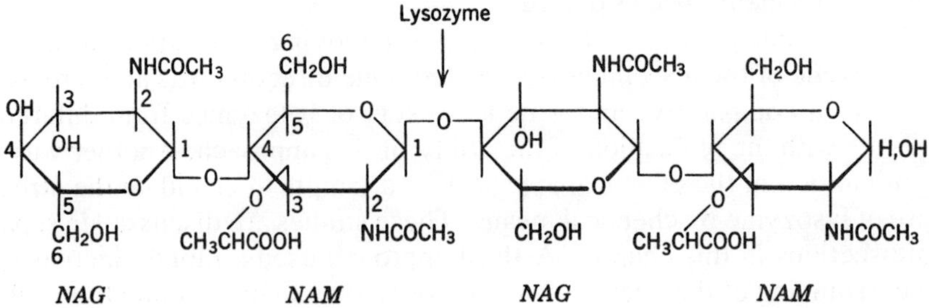

Figure 7-2. The cell-wall tetrasaccharide with the $\beta$-(1-4) glycosidic linkage hydrolysed by lysozyme shown by an arrow. The formula is drawn in a way that is now unconventional to resemble more closely the actual atomic arrangement. Redrawn from [400] (*Proc. Roy. Soc. 167B*, 378 (1967)).

any other enzyme. It is enough to bring man down from his lofty attitude to observe that at least analogous enzymes exist in Milady's tears and also in one of the lowest of all organisms, the bacteriophage!

Lysozymes from many sources have been isolated and studied. Some of the sources from which lysozymes have been isolated and the approximate concentrations of the enzyme are given in Table 7-1. Other studied egg-white lysozymes are listed in Tables 3-1 and 3-5.

## ISOLATION

Chicken egg-white lysozyme is the most extensively studied lysozyme and as a result the best characterized. Unless stated otherwise, the properties of lysozyme described in the following sections refer to this chicken enzyme. Isolation procedures usually take advantage of three properties of lysozymes: low molecular weight, very alkaline isoelectric point, and the general stability of most lysozymes. Ion-exchange chromatography on a cation exchange agent such as CM-cellulose is the most common method for purification [72], [416]. Goose egg-white lysozyme has been isolated using CM-cellulose [176]. The low molecular weight of these enzymes has permitted them to be obtained from avian egg whites by gel filtration [185].

Isolation of lysozymes from other sources such as animal tissues or plants requires an initial extraction followed by differential precipitation and in most instances ion-exchange chromatography. In comparative study of lysozyme, chromatographic behavior has been used as an indication of similarities and differences between these lysozymes. Thus chicken lung and chicken egg-white lysozymes chromatograph differently on Amberlite CG-50 whereas lysozymes from human leucocytes, saliva, and placenta chromatograph similarly to each other [198]. All lysozymes from human tissues and secretions seem to be identical to each other. Even from the same source, several forms of lysozyme may be obtained as was found in duck egg white [199] and snail digestive juice [406]. Also, it is reported that lysozyme can be isolated by absorption from an extract onto chitin at pH 5.5 in 1% NaCl followed by elution with 0.5–1.0% acetic acid [417]. A second absorption is carried out in 0.1 $M$ phosphate buffer at pH 7.9–8.0 and the chitin is regenerated with 2–5% acetic acid.

A lysozyme has been crystallized from papaya latex as a mercury derivative [412]. Relatively homogeneous preparations by several criteria have been prepared by this method employing repeated crystallizations [413]. An enzyme that resembles lysozyme by cleaving the glycosidic bond between NAM and NAG in a bacterial cell wall preparation has

TABLE 7-1   SOURCES AND CHARACTERISTIC PROPERTIES OF LYSOZYMES

| | Enzyme Content (mg/kg) | Distinguishing Property | References[a] |
|---|---|---|---|
| *Animal and Bird Lysozymes* | | | |
| Dog spleen | 50 | | 176 |
| Dog kidney | 15 | | |
| Rabbit spleen I | 400 | | |
| Rabbit spleen II | 400 | | |
| Hen lung I | 10 | | |
| Hen lung II | 10 | | |
| Hen egg white | 4,500±20% | | |
| Duck egg white I | 800–1,500 | | |
| Duck egg white II | 800–1,500 | | |
| Duck egg white III | 800–1,500 | | |
| Goose egg white | 1,200±20% | | |
| Snail digestive juice (*Helix pomatia*) | 5,000–10,000 | 2 Muramidases | 406 |
| *Nephthys hombergi* (annelid) | 20 | | |
| *Human Lysozymes* | | All human lysozymes appear similar | |
| Milk | 30–3,000 | | 248 |
| Colostrum | 90–1,020 | | 248 |
| Saliva | 150–200 | | |
| Placenta | 15–20 | | |
| Urine (normal) | traces | | |
| Plasma | 10±5 | | |
| Normal leucocytes | 10,000±30% | | |
| Leucocytes LMC | | | |
| Tears | 7,000±20% | | |
| *Bacteriophage* | | | |
| T4 | | 2 Cysteine residues | 407 |
| T2 | | | 408 |
| λ | | | 409 |
| *Streptomyces griseus* | | 2 Muramidases | 410,411 |
| *Papaya* | | Apparently lacks involvement of tryptophan in binding | 412,413 |
| *Streptococcus faecalis* | | | 414 |
| *Staphylococcus aureus* | | | 415 |

[a]Data summarized in Jolles [198] except where indicated.

been partially purified from *S. faecalis* cell walls. This enzyme has been referred to as autolysin and appears to be an integral part of the cell wall since it is activated by the action of trypsin on the disrupted walls. The concentration of the enzyme was greatest in cells near the end of the log growth phase. A partial purification was obtained by gel filtration [414]. Cold ethanol fractionation and gel filtration were used to purify a lysozyme from *Staphylococcus aureus* strain 524 [415]. It is this type of relative ease of isolation and wide distribution in such divergent species that is rapidly making lysozyme one of the favorite proteins for comparative and genetic biochemical studies.

## GENERAL PHYSICAL AND CHEMICAL PROPERTIES

One of the early observations on chicken lysozyme was its comparatively high stability in neutral or slightly acidic, aqueous solutions exposed to normal lighting conditions in a laboratory. Solutions containing only 100 μg/ml stored for six years at pH values ranging from 3 to 9 showed considerable residual activity [418]. At pH 4 to 5, 77% of the activity was retained after the six year period. Even at pH 9.5 nearly all the activity was retained after storage at 37°C for seven days. The enzyme is highly unstable to the presence of small amounts of cupric ion (a condition to which many proteins are stable) (Fig. 7.3). At pH 5.4 lysozyme retains its native properties when boiled (100°C) for one minute. This remarkable stability to "denaturation," although still not completely understood, is

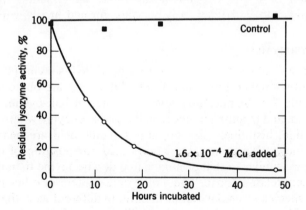

Fig. 7-3. The inactivation of lysozyme by copper at 37°C. The concentration of lysozyme was $3.3 \times 10^{-5}$ $M$; the buffer was borate adjusted to pH 9.3 at 37°C. Reproduced from [418], Academic Press, 1956.

partially explainable now in terms of the protein's size, shape, and distribution of disulfide bonds.

## Chicken Egg-White Lysozyme

The complete amino acid sequence of this lysozyme was independently published by Jolles and co-workers [401] and by Canfield [86] in 1963. The same year, Canfield attempted to identify the positions of the four disulfide bonds formed by the eight half-cystine residues of the molecule by incubating hen oviduct preparations with cystine-$^{35}$S and isolating lysozyme-$^{35}$S. Peptides containing the radioactive label were isolated from a pepsin digest of lysozyme and the cystine oxidized to cysteic acid. Amino acid analysis of these small fragments indicated cystine bridges between half-cystine residues 1 and 8, and 2 and 7 [416]. The positions of the four disulfide bonds were soon established between half-cystine residues 1–8, 2–7, 3–5, and 4–6 (numbered from amino terminal) [419]. The amino acid sequence including the disulfide bonds is shown in Fig. 7-4. X-ray analysis of the three-dimensional structure of the crystalline enzyme has been established at 2 Å resolution by Blake and co-workers [402]. The molecular model projected from the X-ray results had a cleft in which substrate-like inhibitors were bound. A schematic representation of the molecule is given in Figure 7-5 showing the conformation of the molecule. Table 7-2 summarizes many of the physical constants determined for chicken and other lysozymes. Other physical studies on chicken lysozyme included an investigation of a pH dependent reversible dimerization that occurs between pH 5 to 9. Higher polymers are formed at pH values above 10 [422].

## Other Lysozymes

The amino acid composition of goose egg-white lysozyme [176] is quite different from chicken lysozyme. Large differences are obvious from the data in Table 7-3. Significantly fewer residues of half-cystine, arginine, aspartic acid, and tryptophan occur in the goose enzyme whereas the values for lysine, histidine, tyrosine, and glutamic acid are nearly double those of chicken lysozyme. The amino acid composition of duck lysozyme is more similar to that of human and dog lysozyme than it is to that of the closely related goose. The amino acid content of other proteins of these two species is similar rather than quite different as is the case with lysozyme. However, biological properties may differ as is seen by the enzyme inhibitory properties of duck and goose ovomucoids. As is described in Chapter 8, goose ovomucoid is a good inhibitor of bovine trypsin and has essentially no activity against α-chymotrypsin whereas duck ovomucoid is a good inhibitor for both enzymes. Goose lysozyme is three

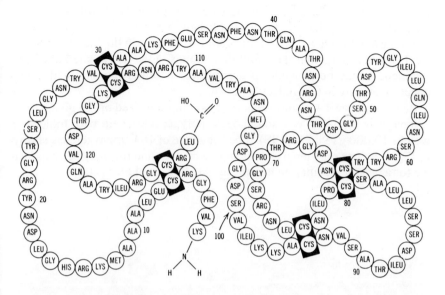

Fig. 7-4. The structure of egg white lysozyme indicating the positions of the four disulfide bonds. ASN and GLN denote asparagine and glutamine, respectively. Reproduced from [420].

TABLE 7-2  PHYSICAL CONSTANTS OF LYSOZYMES

| Lysozyme | Molecular Weight (g) | $\bar{v}$ (cm³/g) | $S_{20,w}$ (Svedberg units) | $D_{20,w}$ (cm²/sec × 10⁷) | $E_{1cm}^{1\%}$ ($\lambda = 280$) | Reference |
|---|---|---|---|---|---|---|
| Chicken egg white | 14,307 | 0.703 | 1.91 | 11.2 | 26.35±0.18 | 87 |
|  |  |  |  |  |  | 86 |
| Human milk | 16,000 | 0.721 | 2.19 | 11.9 |  | 421 |
| Papaya | 25,000 |  | 2.57 |  |  | 412 |
| Papaya "E" | 24,600±1,500 |  | 2.70 |  | 23.8±0.3 | 413 |
| Bacteriophage |  |  |  |  |  |  |
| T2 |  |  | 1.9 |  |  | 408 |
| T4 | 19,000±1,000 | 0.741 | 1.9 |  | 12.8 | 407 |
| λ | 17,900 |  | 2.06 | 10.5 |  | 409 |
| Snail |  |  |  |  |  |  |
| (Helix pomatix) |  |  |  |  |  | 406 |
| I | 21,000 | 0.72 |  |  |  |  |
| II | 24,000 | 0.72 |  |  |  |  |

times as active as the chicken enzyme in lysis of *M. lysodeikticus* cell walls at pH 6.2. It has not been established that the two lysozymes act at the same site in cell walls, and this could be a factor in the increased activity. If it is assumed that the catalytic residues of the active site have been retained, it may be possible to account for extensive amino acid differences elsewhere in the molecule [424].

Bacteriophage T4 lysozyme is the best characterized of the phage lysozymes; it is reported as a single basic polypeptide chain with molecular weight 19,000 g ($S_{20,w}$ = 1.9S). The sequence of 164 amino acids is shown in Fig. 7-6. The enzyme has no disulfide bridges and this may be responsible for a lack of stability on heating above 40°C at pH 6.5 [407].

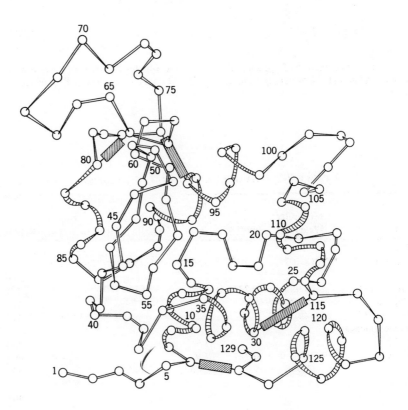

Fig. 7-5. Schematic drawing of the main chain conformation of lysozyme (by W. L. Bragg). Reproduced from [402].

TABLE 7-3  AMINO ACID COMPOSITIONS OF LYSOZYMES AND AVIDINS

| | Lysozymes | | | | | | | | | Avidins | |
|---|---|---|---|---|---|---|---|---|---|---|---|
| Source | Chicken[a] Egg White | Duck[b] Egg White II | Goose[c] Egg White | Dog[b] Spleen | Rabbit[b] Spleen | Human[d] Tear | T4[e] Phage | Papaya[f] | Streptomyces[g] | Streptomyces[g] Subunit (15,000 g) | Chicken[h] Subunit (17,000 g) |
| Ala | 12 | 11–12 | 10 | 15 | 12±1 | 12–13 | 15 | 21 | 16 | 17 | 5 |
| Arg | 11 | 12–14 | 6–7 | 8 | 6 | 11–12 | 13 | 13 | 7 | 4 | 8 |
| Asp | 21 | 18–20 | 13–14 | 18 | 20 | 17 | 22 | 22 | 19 | 12 | 15 |
| CySH | 8 | 8 | 3–4 | 8 | 8 | 6 | 2 | 8 | — | — | 7 |
| Glu | 5 | 5 | 10 | 10 | 11 | 9–10 | 13 | 11 | 4.5 | 9 | 12 |
| Gly | 12 | 12 | 14 | 10 | 12±1 | 12 | 11 | 26 | 15 | 17 | 12 |
| His | 1 | 0 | 3–4 | 1 | 1 | 1 | 1 | 3 | 3 | 2 | 1 |
| Ile | 6 | 5–6 | 9 | 5 | 7 | 5 | 10 | 11 | 3 | 3 | 8 |
| Leu | 8 | 8 | 4–5 | 9 | 10 | 8 | 16 | 12 | 4 | 8 | 7 |
| Lys | 6 | 6 | 11 | 8 | 6 | 5 | 13 | 10 | 3 | 4 | 10 |
| Met | 2 | 2 | 2 | 2 | 2 | 2 | 5 | 4 | 0 | 0 | 2 |
| Phe | 3 | 1 | 2 | 3 | 3 | 2 | 5 | 12 | 5 | 2.5 | 8 |
| Pro | 2 | 2 | 2–3 | 8 | 5 | 3 | 3 | 18 | 2 | 4 | 2.5 |
| Ser | 10 | 10–11 | 6–7 | 9 | 9 | 6–7 | 6–7 | 16 | 15 | 10 | 10 |
| Thr | 7 | 7 | 8–9 | 7 | 7 | 5–6 | 11 | 13 | 12 | 19 | 22 |
| Trp | 6 | 5–6 | 2–3 | 6 | 2 | 5 | 2–3 | 7 | 7 | 8 | 4.5 |
| Tyr | 3 | 4–5 | 6 | 3 | 3 | 5 | 6 | 13 | 7 | 6 | 1 |
| Val | 6 | 6 | 7 | 9 | 6 | 7–8 | 9 | 8 | 5 | 7 | 8 |

Asparagine and glutamine reported with aspartic acid and glutamic acid, respectively. Data from [a][86]; [b][198]; [c][176]; [d][423]; [e][407]; [f][412]; [g][410,411]; [h][411].

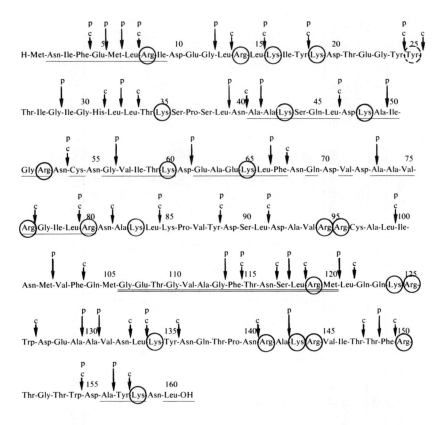

Circles show the amino acids which were hydrolysed by trypsin at their carboxy terminal side and broken circle shows the amino acids hydrolysed by chymotryptic action of trypsin preparation used. Arrows show the points which were hydrolysed by chymotrypsin (c) and pepsin (p). The amino acid sequence of some parts of the structure was confirmed by peptides which could be isolated after cyanogen bromide cleavage or dilute acid hydrolysis of the protein (pH 2·0, 105 to 106°C, for 70 hr). These portions are underlined; single underlines (————) indicate the peptides obtained from dilute acid hydrolysis and double underlines (======) indicate the peptide from cyanogen bromide cleavage.

The sequence has been corrected in the following positions:

           45       45
  (a) Gln ⟶ Glu
        133   135     133   135         139
  (b) Leu-Lys-Tyr ⟶ Leu-<u>Ala</u>-Lys-<u>Ser</u>-Arg-<u>Trp</u>-Tyr,

where the underlined amino acid residues are newly inserted. The total amino acid residues of T4 phage lysozyme is 164 (personal communication from Dr. M. Inouye). [425]

Figure 7-6. The amino acid sequence of T4 phage lysozyme

Other lysozymes of phage origin have been partially purified from bacteriophage T2 [408, 426, 427] and bacteriophage λ [409]. The T2 enzyme had a pH optimum similar to that of the T4 enzyme and a sedimentation coefficient of 1.75. λ-lysozyme consists of a single polypeptide chain containing 159 to 160 residues; it has methionine as its N-terminal and valine as its C-terminal residues. This enzyme contains no disulfide bridges and thus appears quite similar to T4-lysozyme in several respects. A weak immunological cross-reaction between these enzymes also indicates a similarity between the two enzymes. Few similarities can be found between the bacteriophage lysozymes and the lysozymes of plant origin or from higher vertebrates. Future developments may show a link between these groups of lysozymes but the lack of shared physical and chemical traits is now usually interpreted as evidence for convergent evolution.

## ENZYMATIC ACTIVITY AND ACTIVE SITE STUDIES

One of the first direct probes into the active site of lysozyme was the chemical modification studies of Fraenkel-Conrat [396, 397]. Many of the methods used in these earlier studies have now been modified so that they are less rigorous and cause less side reactions. Nevertheless, these early studies showed a requirement for a variety of functional groups in lysozyme for the maintenance of enzymatic activity. These studies showed the requirement for the tertiary structure, the disulfide bonds, amino groups (lysines), and carboxyls (aspartic acids or glutamic acids). The early study had the distinction of preparing one of the first chemical derivatives of an enzyme which retained an undenatured structure as evidenced by the capacity to be crystallized; the derivative was acetyllysozyme. This study was particularly important at the time in that it provided one of the first examples of a very stable enzyme (stable to physical conditions normally causing denaturation) whose enzymatic properties could be changed by chemical modification.

### Enzymatic Characteristics

Salton [428] has stated the requirements of an enzyme to be a lysozyme as follows: "The enzymatic reaction should result in (a) liberation of glucosamine and muramic acid from bacterial cell walls, (b) liberation of reducing groups, and (c) lysis of bacteria under proper conditions." Other workers have expanded these requirements to include: (d) a basic protein, (e) low molecular weight (approx. 15,000 g), and (f) stability at acidic pH (100°C and pH 4.5 for 1 min) [198]. In light of recent developments we feel that the latter three requirements are too restrictive to encompass established or proposed mechanisms accounting for the evolutionary development of lysozymes. It must be stressed that the most important property for identification of a protein is its biological activity! Following this consideration, lysozymes' genetic and chemical relation to other proteins must be considered as is currently being done using amino acid sequences and essential amino acid residues.

To gain an understanding of the biological function of lysozyme it is necessary to understand the nature of its substrates and the mechanisms by which the catalytic activity is expressed. The structure of the cell wall of *M. lysodeikticus* and the linkage susceptible to lysozyme hydrolysis both in whole cells and the cell wall tetrasaccharide is indicated in Fig. 7-2. It has been previously mentioned that lysozyme will degrade chitin, a linear polymer of NAG. Lysozyme also hydrolyzes derivatives of chitin such as hydroxyethyl or glycol chitin but will not hydrolyze polyglucosamine or cellulose (both containing $\beta$-(1-4) linkage) indicating the importance of the N-acetyl group. The rate of hydrolysis of these substrates varies with the method used for detecting the reaction products and with the conditions of the assay. The pH of optimum activity is substrate dependent; lysozyme action on whole cells of *M. lysodeikticus* is optimum at pH 6 to 6.5 [429], whereas on the cell wall tetrasaccharide or oligosaccharides derived from chitin the optimum pH is close to 5 [430–432].

A transglycosylation reaction has been described for lysozyme [433, 434]. Lysozyme digestion of the cell wall tetrasaccharide NAG(-NAM-NAG-)$_1$NAM produces in addition to the disaccharide (NAG-NAM) a number of compounds corresponding to the tri, tetra, and penta oligomers of the disaccharide NAG(-NAM-NAG-)$_{2-4}$NAM. The lag period may permit synthesis of larger oligosaccharides that serve as catalysts for the hydrolysis by the following scheme proposed by Sharon [432] for the production of disaccharides:

$$D_2 \xrightarrow{\text{very slow}} D, \quad \text{where } D = \text{NAG-NAM},$$

$$2\,D_2 \xrightarrow{\text{slow}} D_3 + D,$$

$$D_3 + D_2 \xrightarrow{\text{fast}} D_4 + D,$$

$$2\,D_3 \xrightarrow{\text{fast}} D_5 + D, \text{ etc.}$$

A similar scheme may be written for the hydrolysis of chitin oligosaccharides. Transglycosylation is also catalyzed by lysozyme in the presence of chitin oligosaccharides. Under the proper conditions insoluble chitin-like polymers are formed from tri- or tetra-NAG and a lag period is noted in the hydrolysis of the lower oligosaccharides [435]. With the cell wall tetrasaccharide it is convenient to use an assay system that detects the appearance of disaccharide in the reaction mixture. Sharon [432] reported that disaccharide was not released at the beginning of incubation of lysozyme with tetrasaccharide. That is, a lag period occurs the length of which is inversely proportional to enzyme concentration. These observations would be expected if the hydrolysis of the tetrasaccharide involves a transfer reaction rather than a direct cleavage to disaccharide.

## Chemical Modification

The effects of various chemical treatments on chicken egg-white lysozyme activity are summarized in Table 7-4. Much attention has been given to the importance of the tryptophan residues. Optical absorbance changes observed when lysozyme complexed with some polymers of NAG indicated the transfer of exposed indole groups to a hydrophobic environment [436]. The effects of chemical modification of tryptophan residues agree with the involvement of one or more of these residues in the activity of lysozyme. Appropriately controlled, modification of only one of the six tryptophan residues can be accomplished with N-bromosuccinimide (NBS) [436] or iodine [437] to produce an inactive enzyme. The residues modified by NBS and iodine are tryptophan-62 [438] and residue 108 [439], respectively. X-ray data indicate both of these residues present in the cleft in which substrate molecules are presumed to be bound.

The importance of intact disulfide bonds to the activity of chicken lysozyme is indicated by the inactivation of lysozyme following their modification using thioglycolate or mercaptoethanol in urea [397], [440]. Sulfitolysis of lysozyme modified 2.5 of the possible 4 disulfide bridges of native lysozyme, and a progressive inactivation of the enzyme was observed. The highest inactivation occurred at a low pH (pH 4.7). The bridges formed by half cystine residues 1–8 and 2–7 were proposed as the most likely ones to be affected [441]. Other studies have also indicated that the disulfide bridge formed by half-cystine residues 1–8 is not essential for enzymatic activity [442, 443]. Recently a derivative of chicken lysozyme containing 16 half-cystine residues (rather than the normal 8) was prepared by reaction with cystine at pH 9.5. This modified enzyme was inactive. Under the proper conditions the enzyme could be reactivated to approximately 80% of the original activity. Two components formed on reoxidation, one indistinguishable from the native, and the second having only 50 to 60% of the original specific activity [444]. From the evidence currently available it appears that the disulfide bridges of lyso-

TABLE 7-4  ENZYMIC ACTIVITY OF HEN EGG-WHITE LYSOZYME AFTER VARIOUS TREATMENTS

| Treatment | Amino Acids or Groups Modified | Activity | Reference |
|---|---|---|---|
| Oxidation | Cys, Met, Try | Inactivation | 396 |
| Photo-oxidation | His, aromatic amino acids | Reduced activity | 456 |
| Ozonization (slow) | Try, Met | Inactive | 457 |
| in formic acid | 2 Try modified: | Active | 458, 459 |
|  | 6 Try modified: | Inactive |  |
| Reduction | Cys | Reversible inactivation | 460, 461 |
| Esterification | Free carboxyl groups | Depends on the reagent | 396 |
| Acetylation | Free NH$_2$ groups | Progressive inactivation | 396, 462 |
|  |  | Shift in pH optimum | 447 |
| Guanidination | Lys | Active | 462 |
| Xanthination | Try | Inactivation | 463 |
| Iodination | His, aromatic amino acids | Inactivation | 396, 437 |
|  |  |  | 400, 464 |
| Iodoacetic acid (pH 5.5) | Partially Met, Lys | Formation of closely related active substances | 465 |
| Urea |  | Formation of closely related active substances | 466 |
| NBS | Mainly Try | Progressive inactivation | 436, 437 |
|  |  |  | 449, 467 |
| Incorporation of radioisotopes |  | Active up to 20 mc/mmole | 468 |
| Carboxypeptidase | Release of Arg-LeuOH | Active | 198 |
| Aminopeptidase | Release of H.Lys-Val | Active | 469 |
| Cupric ion | Unknown | Inactive | 418 |
| Sulfitolysis | Disulfide bonds | Progressive inactivation | 441 |
| Intramolecular cross-linking; | Lys | Active | 470 |
|  |  |  | 471 |

zyme may be partially but not completely reduced and still maintain biological activity. And under the proper condition reduced lysozyme may be reoxidized to an active form [390].

Many attempts have been made to demonstrate the involvement of the unique histidine residue of chicken lysozyme in its mechanism of action. Treatment of lysozyme with diazo-1-H-tetrazole in attempts to modify this single histidine produced an inactive enzyme. However, the interpretation of this result was difficult because of a possible interfering modification of tyrosine residues [445]. In another report carboxymethylation of histidine and approximately three of the six $\epsilon$-amino groups of lysine resulted in an unexplained activation of lysozyme rather than an inactivation. More extensive modification was reported to yield an inactive enzyme. In this report it was concluded that neither histidine nor lysine participated directly in the active site of lysozyme [446]. Several lysozymes from duck egg white are reported to be active even though no histidine is present in the molecule [199].

Conclusions mentioned at the beginning of this section that amino groups (lysines) were necessary for the activity of chicken lysozyme have recently been reinvestigated [447]. Acetylated lysozyme retained full activity when assayed with glycol chitin. But, against cells of *M. lysodeikticus* in neutral media, the activity decreased progressively with the extent of acetylation. A pH profile of the reaction showed that the pH optimum of the modified enzyme was shifted to lower values. This is illustrated in Fig. 7-7. At the lower pH values the lysis of *M. lysodeikticus* does not proceed to completion with the modified or native lysozyme. However, when the pH is raised to near neutrality a spontaneous lysis occurs indicating that both enzymes were hydrolyzing susceptible bonds even though a large change in turbidity was not observed. These results were interpreted as indicating the first step in lysis of cells by lysozyme involves an interaction between the positive charge of lysozyme and the negative charges on the surface of the bacterial cells. Modification of the positive amino groups causes a decrease in the interaction near neutral pH values. The remaining steps involve hydrolysis of the susceptible bond followed by lysis of the cell.

INHIBITORS OF ENZYMATIC ACTIVITY

The competitive inhibitors of lysozyme are found in a cleft in the surface of the molecule first seen in low-resolution models made from X-ray studies. A substrate of lysozyme is presumably bound in the same site as the inhibitors shown in Fig. 7-8. Some of the residues of this portion of the molecule are indicated in Fig. 7-9. Three tryptophans, 62, 63, and 108, are located in this area in general agreement with chemical evi-

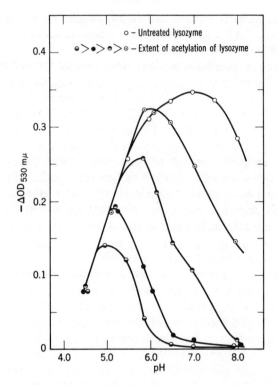

Fig. 7-7. pH-Dependence of lytic activity of acetylated lysozyme toward bacterial cells, *M. lysodeikticus*. Redrawn from [447].

dence involving tryptophans in the binding of substrate [437], [449]. The residues implicated in the catalytic activity from X-ray studies were 52 (aspartic acid) and 35 (glutamic acid) [400], [404], [450].

The use of the crystal structure of lysozyme-NAG and lysozyme with the β-(1,4)-linked dimer of NAG for interpreting the properties of the enzyme in solution has been supported by reports indicating these inhibitors bind with equal strength to the enzyme in both states [451]. A stoichiometry of 1 to 1 was reported for the association of lysozyme with both the monomer and dimer of NAG [451].

Oligosaccharides prepared from the cell walls of bacteria susceptible to the enzymic activity of lysozyme consist of alternating units of β-(1-4)-linked NAG and its 3-0-lactyl ether, NAM. Chitin and oligosaccharides derived from it contains β-(1-4)-linked NAG units. Binding constants of oligomers of NAG to lysozyme have been determined by fluorescence spectroscopy [452, 453], ultraviolet difference spectra [454], and ultraviolet spectroscopy and inhibition experiments [455]. Such studies indicate that the binding constant increases from the monomer to the trimer of NAG. These and other saccharides that inhibit the activity of lysozyme on *M. lysodeikticus* are used to estimate the structural requirements of the substrate and to study the binding site of the enzyme. Such studies support the predictions made of the binding specificity of the sub-sites within the cleft. Saccharides such as cellobiose that lack the 2-acetamido group do not affect the activity of lysozyme. Association constants of various saccharides with lysozyme are shown in Table 7-5 [453]. The fluorescense maxima of each of the enzyme-saccharide complexes are shifted to shorter wavelengths by about 10 m$\mu$. This phenomenon is generally inter-

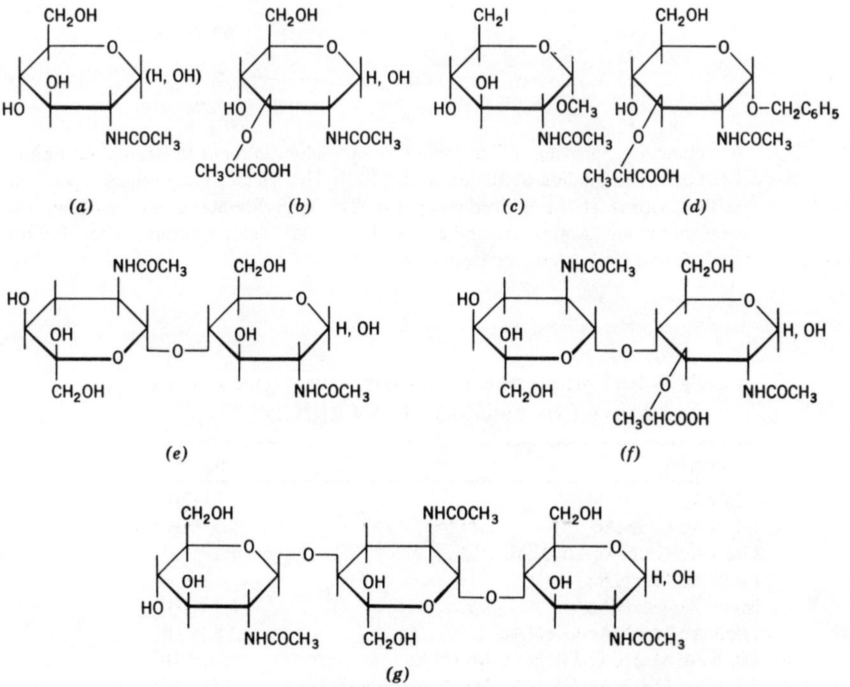

Figure 7-8. Lysozyme inhibitors: (a) N-acetylglucosamine (NAG); (b) N-acetylmuramic acid (NAM); (c) 6-iodo-$\alpha$-methyl-N-acetylglucosaminide; (d) $\alpha$-benzyl-N-acetylmuramic acid; (e) di-N-acetylchitibiose (di-NAG); (f) N-acetylglucosaminyl-N-acetylmuramic acid (NAG-NAM); (g) tri-N-acetylchitotriose (tri-NAG). Redrawn from [400]. (*Proc. Roy. Soc. 167B*, 378 (1967)).

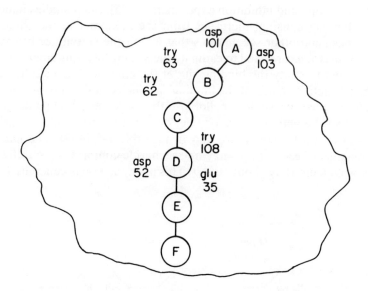

Fig. 7-9. A schematic drawing of a hexa-(N-acetylglucosamine)-lysozyme complex, from the crystallographic studies of Blake et al. [400]. The saccharide units are specified by letters starting with A at the nonreducing end. The approximate locations of certain critical protein side-chain groups are indicated. Tri-(N-acetylglucosamine) when bound occupies sites A, B, and C. Reproduced from [448].

TABLE 7-5  ASSOCIATION CONSTANTS WITH HEN'S EGG WHITE LYSOZYME AT pH 5.4 AND 25°

| Saccharide | $K_a$, $M^{-1}$ |
|---|---|
| GlcNAc | 15–20 |
| GlcNAc-GlcNAc | $5.5 \times 10^3$ |
| GlcNAc-GlcNAc-GlcNAc | $1.1 \times 10^5$ |
| GlcNAc-MurNAc | 20 |
| MurNAc-GlcNAc | $1.1 \times 10^4$ |
| GlcNAc-MurNAc-GlcNAc | $2.8 \times 10^5$ |
| GlcNAc-MurNAc-GlcNAc-MurNAc | $2.1 \times 10^3$ |
| GlcNAc-MurNAc-GlcNAc-MurNAc dimethyl ester | $1.9 \times 10^3$ |
| GlcNAc-MurNAc-GlcNAc-MurNAc pentapeptide | $7 \times 10^3$ |

All saccharides are $\beta(1\rightarrow 4)$ linked. All constants, except for that of GlcNAc-MurNAc were determined by fluorescence emission measurements. The constant for GlcNAc-MurNAc was derived from equilibrium dialysis experiments [453].

preted as a shift of one or more tryptophans to a more hydrophobic environment. The variation of the intensity of fluorescence that occurs with complex formation has been analyzed by a difference fluorescence technique in an attempt to analyze the contributions of individual tryptophan residues [452]. A conformational shift has been observed with X-ray analysis by Phillips [405] in which tryptophan-62 is shifted by 0.75 Å concommitant with a slight closing of the cleft when tri-NAG is bound.

The active site of lysozyme is now often discussed in terms of sub-sites designated by letters, A, B, C, D, E, and F. Each sub-site interacts with a single sugar residue (Fig. 7-9). The contributions of each sub-site to the total binding of substrate is studied. The residues involved in each of these sub-sites have been identified from structural studies by Phillips [405]. These studies suggest the site of bond breaking is between sub-sites D and E. Also from structural studies the lactyl residue of NAM appears to be pointed out of the enzyme. However, binding studies utilizing fluorescence measurements for determining association constants indicate that the lactyl group also interacts with the enzyme.

## Comparative Enzymatic Activities

Lysozymes from chicken egg white, duck egg white, goose egg white, and human milk were compared for their activities on two oligosaccharides from chitin, chitotetraose (tetra-$NAG$), and chitopentose (penta-$NAG$) [472]. All lysozymes except the goose enzyme digested the oligosaccharides with formation of transglycosylation products. The goose enzyme had very low activity at pH 4.7 and no activity at pH 6.7. The duck enzymes had the highest activity and acted more completely than the remaining lysozymes. When investigating the enzymatic activity of the egg-white lysozymes of the ratite birds, Osuga and Feeney [82] noted that kiwi lysozyme acted incompletely in reducing the turbidity of a suspension of *M. lysodeikticus* cells (Fig. 7-10).

Many workers have noted an increased resistance of *M. lysodeikticus* walls to chicken lysozyme when the walls are O-acylated [473, 474]. An autolysin has been isolated from *S. faecalis* and found to have $N$-acetylmuramidase activity [414]. O-acylation of *S. faecalis* walls has been found to greatly increase their susceptibility to chicken lysozyme rather than decrease it. If *M. lysodeikticus* and *S. faecalis* walls are acetylated and then de-O-acylated both walls are more sensitive to chicken lysozyme. However, the activity of the autolysin was affected differently against similarly treated substrates. *S. faecalis* walls became more susceptible and *M. lysodeikticus* walls became much less sensitive. This autolysin did

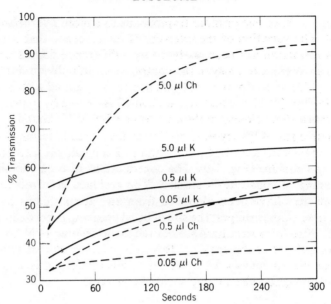

Fig. 7-10. The lysis of *Micrococcus lysodeikticus* cells by kiwi and chicken egg white. The lysis of 300 μg of cells in pH 7 phosphate buffer as measured by a decrease in turbidity at 540 mμ. The changes in the percentage transmission (%T) in seconds are plotted for kiwi (K) and chicken (Ch) egg white. The initial turbidity of the cell suspension was such as to give a transmission of 30%. Reproduced from [82] Academic Press 1968.

not release soluble products from *M. lysodeikticus* cell walls but it was able to hydrolyze over 90% of the glycosidic bonds between N-acetylmuramic acid and N-acetylglucosamine in *S. faecalis* walls. The activity of *S. faecalis* autolysin was shown to be maximal toward the end of the exponential growth phase. It has been postulated that autolysis plays a role in wall growth and division by opening bonds in the cell wall structure to permit enlargement [475].

A lysozyme isolated from *S. aureus* was similar to chicken egg-white lysozyme in its optimal temperature for reaction, pH optimum, activation by NaCl and $Ca^{2+}$ ions, inhibition by sodium citrate and EDTA, and inactivation by cuprous ions ($Cu^+$) and sodium dodecyl sulfate [415]. Since it is inactivated at 56°C, it differs in temperature stability from chicken lysozyme. Though the mechanism of action was similar (mucopeptide N-acetylmuramyl hydrolase), the specificity appeared different. The *S. aureus* lysozyme would act on the cell walls of *S. epidermidis* in contrast to chicken egg-white lysozyme. A preliminary report of lysozyme occurrence in *S. aureus* indicates all *S. aureus* strains probably produce a lysozyme (503 strains tested) and at least 13 of 35 strains tested of *S. epidermidis* produce an enzyme capable of lysing *M. lysodeikticus*. The activity of *S. aureus* lysozyme was 40 times less than chicken lysozyme against *M. lysodeikticus* cells but only 7 times less active against purified cell walls. Staphylococcal lysozyme has no lytic effect on living and heat killed *S. aureus* 524 cells or cell walls but will act on the cell walls if the phosphodiester teichoic acid-murien linkage is disrupted. This was interpreted to mean that the enzyme is inhibited by teichoic acid linked by an ester bond in the cell wall [415].

Papaya lysozyme lyses *M. lysodeikticus* cell walls at one-third the rate of chicken lysozyme and has an amino acid composition that is quite different [412, 413]. This lysozyme has a molecular weight of 25,000 g and a chitinase activity 400 times greater than chicken lysozyme when tested against tetra-NAG. The binding site of papaya lysozyme apparently does not involve tryptophans [413].

Two enzymes have been isolated from the digestive juice of the snail *Helix pomatia*, both of which hydrolyze mucopeptide from the cell wall of *M. lysodeikticus* and have been shown to be muramidases. One, like chicken lysozyme, hydrolyzes hydroxyethyl chitin whereas the other does not.

A comparison of its enzymic activity based on initial rates with that of chicken egg-white lysozyme showed T4 phage lysozyme to be over 6 times more active against *M. lysodeikticus* and 250 times more active against *E. coli* (Table 7-6). However, the egg-white lysozyme activity was more extensive; it was able to hydrolyze 64% of the N-acetylmuramide linkages in *M. lysodeikticus* cell walls whereas the phage lysozyme hydrolyzes only 20%. As a result, T4 phage lysozyme decreased the original turbidity of the cell preparation 30% whereas chicken egg-white lysozyme decreased it by 83%. This situation appears similar to that observed for the kiwi lysozyme. The phage enzyme seems best adapted to its specific substrate, *E. coli*.

## EVOLUTION AND GENETICS

Avian egg white may prove to be the best source of lysozymes for the purposes of studying comparative biochemistry and genetics. Lysozyme can be obtained in large amounts in pure form from species whose relationship has been established by other means.

TABLE 7-6  SUBSTRATE SPECIFICITY AND COMPARISON WITH EGG WHITE LYSOZYME

| Substrate and Enzyme | Amount of Enzyme in Reaction Mixture μg | Activity[a] sec | Ratio of T4 to Egg White Activity[b] |
|---|---|---|---|
| *M. lysodeikticus* | | | |
| Egg white lysozyme | 20 | 46.8 | 6.3 |
| T4 phage lysozyme | 23.8 | 6.1 | |
| *E. coli* | | | |
| Egg white lysozyme | 10 | 26.9 | 250 |
| T4 phage lysozyme | 0.119 | 9.0 | |

[a] Time required for an optical density decrease of 0.05 unit at 350 mμ. The reciprocal of the values is a linear function of the activity.
[b] Ratios are calculated on the basis of unit weight of the enzymes [407].

Much work of this type has already employed avian egg white lysozymes [82, 176]. One study, an immunological comparison of the egg-white lysozymes of members of the avian order Galliformes, now indicates close relationship between the chicken, partridge, and American quails. Taxonomically these birds had been considered to be more distantly related (Table 3-5) [185].

## Genetics

Extensive genetic studies have been carried out employing sequence studies of lysozymes from strains of T4 phage [476]. The lysozyme of one T4 phage carrying two frame-shift mutants was found to differ from the wild-type by a sequence of four amino acids resulting from mutations causing in one case the addition of two bases, and in the other the deletion of two bases [477]. The low molecular weight of lysozyme is advantageous for sequence analysis in genetic studies.

Lysozyme is playing a key role in elucidating the mechanism of immunity observed with temperate bacteriophages. The activity of lysozyme is used to detect the expression of the prophage gene R (the structural gene for lysozyme) following heter-immune superinfection. In this way evidence supporting various theories concerning the nature of the immunity phenomena is obtained [478–480].

## Relationship to Other Proteins

In discussion of bovine α-lactalbumin the observation by Brew and co-workers [224] of extensive homology in its amino acid sequence with that of chicken lysozyme was described. They proposed that the genes for these two proteins arose from a common ancestral gene. The homology between these two proteins is considerable since the initial report showed 40 residues of the aligned sequences to be identical and it is probable that this could be extended.

At the time of this writing the amino acid sequence of human lysozyme has not been completed, but it is well underway in Canfield's laboratory. Human lysozyme obtained from the urine of a patient with monocytic leukemia is being used in the sequence analysis [481]. Approximately 80 g of lysozyme was excreted by one patient over an eight month period. From initial results it appears that the homology between bovine $\alpha$-lactalbumin and human lysozyme is greater than that reported between bovine $\alpha$-lactalbumin and chicken lysozyme [482]. Further studies should provide evidence for determining the point of divergence of lysozyme and $\alpha$-lactalbumin in the development of mammals.

Other investigators have postulated a number of other proteins to be related to lysozyme. The amino acid sequence of RNase has been compared to that of lysozyme by statistical analysis of the genetic codes that could code for these proteins, and there is a possibility that they were related through an ancestral gene [424]. Utilizing a time scale similar to that based on the evolution of cytochrome c as summarized by Nolan and Margoliash [483] the divergence of these two proteins would be placed near the beginning of life itself (between 1400 and 2200 billion years). The value of extending theories on homologues to such distant relationships may be questioned by many investigators, but it is through such extensions that we will discover the genetic factors which dictate the existence of homologues.

Another alkaline protein, avidin, discussed in previous chapters has been the subject of a recent report by N. M. Green [411] in which he proposed a relationship existing between lysozyme and avidin. He compared physicochemical characteristics of chicken egg-white avidin and lysozyme with an avidin (Streptavidin) and a lysozyme from *Streptomyces griseus*. The only general similarity was the involvement of tryptophan residues in the binding of substrate. He noted, however, that the amino acid compositions of corresponding lysozymes and avidins (from the same species) had striking resemblances. Comparison of the amino acid composition of a subunit of avidin (Table 7-3) with that of lysozyme shows more similarities than found between trypsinogen and chymotrypsinogen or hemoglobin and myoglobin which have been related by sequence studies. Similarities of composition is not proof of sequence homology, which still remains to be investigated. The similarities in composition may only reflect the fact that both of these proteins have an unusually high isoelectric point. However, the degree of similarity between the contents of amino acids suggests further investigations may support the much earlier proposal of Meyer [484] that the difficulty of separating avidin and lysozyme was an indication of their relationship. Other proteins should eventually be shown to have diverged like lysozyme into several different homologous forms with very different functional properties.

# 8
# Inhibitors of Proteolytic Enzymes

The naturally occurring protein inhibitors of proteolytic enzymes are unique materials in biological science. They have the capacity of combining with and inactivating those proteins which have the capacity of hydrolyzing and degrading one of their own species. The inhibitor, proteolytic enzyme, and protein substrate have very different roles in the inhibitory process!

Within any given individual many homologous and analogous forms of these inhibitors may exist in different tissues and secretions. The homology and analogy is carried further, and indeed is thereby more interesting scientifically, in that the different inhibitors may react with different analogous and homologous forms of proteolytic enzymes. These homologous and analogous forms of the inhibitors and the enzymes provide a unique opportunity for studying the relationship of the structures of the proteins to the properties required for their interactions. The reaction between enzyme and inhibitor results in the formation of a complex of the two proteins that is usually stable over a wide pH range. Since one of the partners of the complex is an active enzyme whereas the other has an enzyme-inhibiting activity, it is frequently possible to use these biological activities to measure the dissociation of the complex by determining the amount of one or both of the interacting proteins that are free. Figure 8-1 is the illustration of some of the different ways that bovine trypsin and bovine $\alpha$-chymotrypsin combine with four different avian ovomucoids. As is described later, when an inhibitor is capable of combining with more than one enzyme, the combining sites on the inhibitor may be overlapping or nonoverlapping. Such systems have many characteristics that lend themselves well to physical-chemical studies and thus provide a useful tool for the fundamental study of protein-protein interactions.

## DIAGRAMMATIC REPRESENTATION OF THE ENZYME-INHIBITOR COMPLEXES

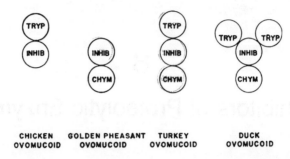

Fig. 8-1. Diagrammatic representation of the enzyme-inhibitor complexes. Reproduced from [81].

Inhibitors are widely distributed in both animals and plants. In plants they are particularly plentiful in the legumes. The principal plant inhibitors that have been studied are from soybeans, lima beans, pea, and potatoes; the principal animal inhibitors that have been studied are from pancreas, colostrum, blood plasma, and avian egg white. Inhibitors have also been found in seminal fluid, urine, and certain intestinal parasites. Some of these inhibitors may have an important-function in biological control mechanisms and may even have practical significance in animal and human nutrition. All of these subjects are under investigation in many laboratories.

The discovery of the inhibitors of proteolytic enzymes has been closely oriented to the history of the proteolytic enzymes that they inhibit. Inhibitors of bovine trypsin and $\alpha$-chymotrypsin were reported as early as 1900. However, most of the early reports are difficult to interpret because of the questionable purity and identification of enzymes used in these studies. The first real understanding of the fundamental nature of the problem began with the crystallization of a trypsin inhibitor from bovine pancreas and the crystallization of its complex with trypsin by the Rockefeller Institute team of Kunitz and Northrup in the mid-1930's [485]. Approximately a decade later the trypsin inhibitory activity was noted in soybeans by Ham and Sandstedt [486] and Bowman [487]. This was followed several years later by the crystallization of a soybean inhibitor by Kunitz [488]. Ovomucoid from chicken egg white had been recognized as a separate protein for many years but it was not until 1947 that Lineweaver and Murray [53], possibly stimulated by the other discoveries of inhibitors, identified chicken ovomucoid as an inhibitor of trypsin. During the next few years inhibitory activities were found in many biological materials and an excellent general review of inhibitors of proteolytic enzymes was published in 1954 by the eminent father and son team of Michael Laskowski, Sr. and Michael Laskowski, Jr. [489]. Methods for the assay and isolation of a number of inhibitors was published in 1955 by M. Laskowski, Sr. [490]. During the last decade there have been many investigations on these inhibitors. A recent conference has considered their pharmacological applications [491] and their general properties have been reviewed in 1966 [492].

In this chapter we attempt to describe and relate the properties of several of these inhibitors of proteolytic enzymes. In keeping with the principal aim of this book, the emphasis is on the comparative properties. Space does not permit a detailed discussion of many of the extensive chemical and physical-chemical studies on these interesting proteins.

## GENERAL PROPERTIES OF INHIBITORS

### PANCREATIC INHIBITORS

The inhibitors from bovine pancreas have probably received the most attention and are the best characterized inhibitors of proteolytic enzymes. Two inhibitors from bovine pancreas have been intensively studied (Table 8-1, Fig. 8-2). The first inhibitor studied is now called the Kunitz or basic inhibitor [485]. It is the only inhibitor whose primary amino acid sequence is known at this time. Its physical-chemical properties and the characteristics of its interaction with trypsin have been intensively studied. The second inhibitor is popularly called the Kazal inhibitor [519]. It differs from the basic inhibitor in its physical and chemical properties as well as in certain of its inhibitory characteristic.

The basic pancreatic inhibitor is remarkably stable in acid solution. It can be heated to 80°C in 2.5% trichloroacetic acid without inactivation. Even at pH 12 only 50% of the activity is lost after 24 hours at room temperature. One most striking thing about the structure is its low molecular weight of approximately 6300 g. It consists of a single polypeptide chain of only 58 amino acids. The presence of only three disulfide bonds and the absence of tryptophan, histidine, and cysteine have helped to expedite its sequence determination. The protein is reported to inhibit several trypsin-like proteolytic enzymes including Kallikrein and plasmin. It apparently also inhibits bovine $\alpha$-chymotrypsin. The basic trypsin inhibitor has now been well characterized (Table 8-2, Fig. 8-2) and is under study in many laboratories [521–525]. Another inhibitor found in bovine lung and parotid gland has been described as an inhibitor of the trypsin-like en-

TABLE 8-1 NATURALLY OCCURRING INHIBITORS OF PROTEOLYTIC ENZYMES[c]

| Source | Type | pI | Molecular Weight g/mole | Enzymes Inhibited | Remarks | References |
|---|---|---|---|---|---|---|
| Egg white | | | | | | |
| Chicken | Ovomucoid | 3.8–4.4 | 28,000 | Tryp. | Contains approx. 20% carbohydrate | 493,494,83 |
| | Ovoinhibitor | — | 46,500 | Tryp., Chym., subtilisin | Tryp. and Chym. inhib. simultaneously. Chym. and subtilisin compete | 88 |
| | Papain Inh. | — | 12,700 | Papain, Ficin | Papain & Ficin compete | 64 |
| Tinamou | Ovomucoid | 4–5 | 28,000 | Chym., subtilisin | | |
| Turkey | Ovomucoid | 4–5 | 28,000 | Tryp., Chym., subtilisin | Tryp. and Chym. inhib. simultaneously | 81 |
| Penguin | Ovomucoid | 4–5 | 28,000 | Tryp., Chym. subtilisin | Subtilisin strongly inhibited | 193 |
| Quail (Jap) | Ovomucoid | 4–5 | 28,000 | Tryp. | Human trypsin strongly inhibited[c] | 196a |
| Pancreas | | | | | | |
| Bovine | Tissue (basic) | 10.1 | 6,513 | Tryp., Chym. | — | 495,496 |
| | Juice (Kazal) | — | 6,155 | Tryp. | — | 497,498 |
| Porcine | Juice (Kazal) | — | 6,024 | Tryp. | — | 499 |
| Blood | | | | | | |
| Human | $\alpha_1$ | 4.0 | 45,000 | Tryp., Chym. | Greater affinity for Chym. | 500,501 |
| | 1 | 2.8 | 16,400 | Tryp., Chym. | Weak inhib. of Chym. | 502 |
| Bovine | | 3.8 | 39,000–71,000[a] | Tryp., Chym. | Weak inhib. of Chym. | 503 |
| Ovine | | 4.3 | 40,600 | Tryp., Chym., Plasmin, Proteinase A | — | 504 |

202

TABLE 8-1 CONTINUED

| Source | Type | pI | Molecular Weight g/mole | Enzymes Inhibited | Remarks | References |
|---|---|---|---|---|---|---|
| Colostrum Bovine | | 4.2 | 10,500 | Tryp. | Inactivated by pepsin[c] | 194 |
| | | | | | Resistant to pepsin | 505 |
| Porcine | | — | — | Tryp. | Has tryptophan | 247 |
| Soybean | Crystalline (Kunitz) | 4.5 | 21,000 | Tryp. | | 506 |
| | "Acetone insol." | 4.2 | 24,000 | Tryp., Chym. | No tryptophan | 507 |
| | STI $A_1$ | — | 14,300 | Tryp. | — | 508 |
| | STI $A_2$[b] | 4.5 | 21,600 | Tryp. | — | 508 |
| | 1.9 S inhibitor | 4.0 | 16,400 | Tryp., Chym. | High cystine, no glycine | 509 |
| | $F_1$ | | | Tryp., Chym | Has tryptophan | 510 |
| | $F_2$ | | | Tryp., Chym. | Has tryptophan | 510 |
| | $F_3$ | | | Tryp. | Has trypt., no tyrosine | 510 |
| Lima bean | | 3.6 | 10,000 | Tryp. | Resolved into 4 | 511,512,513 |
| Navy bean | | — | 23,000 | Tryp. | No data for Chym. | 514 |
| Pea (Black-eyed) | | — | 17,000 | Tryp., Chym. | — | 515 |
| Potato | 1 | — | 38,000 | Tryp., Chym. $B.$ $subtilis$ proteinase, $S.$ $griseus$ proteinase, Carboxipeptidase B | Tryp. only inhib. with casein as substrate | 516,517,518 |

[a]39,000 g/mole is the molecular weight obtained from trypsin inhibition studies and 71,000 g/mole is the molecular weight obtained by sedimentation–diffusion studies. It has been suggested that the inhibitor may exist in solution as a dimer.

[b]The acetone soluble inhibitor and the $A_2$ ($STI_2$) inhibitor of Rackis et al. [508] are both believed to be similar to Kunitz's soybean inhibitor.

[c]Tryp. and Chym. refer to bovine trypsin and bovine chymotrypsin, respectively. Chicken ovomucoid and nine other ovomucoids tested do not inhibit human trypsin. Japanese quail ovomucoid, lima bean inhibitor, bovine colostrum inhibitor, and bovine Kunitz pancreatic inhibitor are good inhibitors of human trypsin [196a].

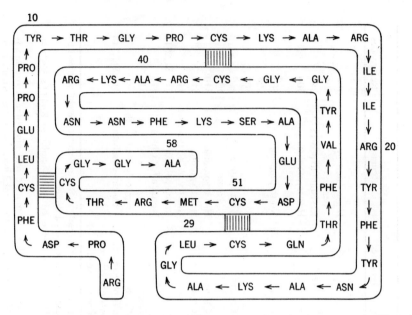

Fig. 8-2. A schematic representation of the structure of the bovine pancreatic (Kunitz) trypsin inhibitor. The molecular weight is 6,513. Reproduced from [496].

zyme Kallikrein. These inhibitors have more recently been found to be identical to one other and very similar to the basic pancreatic inhibitor [526, 527].

In contrast to the basic inhibitor, the bovine Kazal pancreatic inhibitor is restricted in its specificity inhibiting apparently only trypsin. Its amino acid content is known (Table 8-2) and its primary structure is partially characterized [497]. A primary difference between the Kunitz and Kazal inhibitor, is the much greater susceptibility of the Kazal inhibitor to proteolysis by trypsin.

SOYBEAN INHIBITORS

As a consequence of the observation that soybeans had a growth retarding effect on experimental animals, an inhibitor was sought in the beans [486, 487]. A potent trypsin inhibitor from soybeans was reported in 1944. Crystallization of a trypsin inhibitor was achieved a few years later [488], but it was not until 15 years later that Birk [528] found a second inhibitor in soybean which was also an inhibitor of bovine chymotrypsin. Since that time a number of different preparations of inhibitors have been isolated from soybeans and this has resulted in a rather extensive and confusing nomenclature. There are apparently at least four or more inhibitors in soybeans. In addition there are probably multiple forms of several of the inhibitors. In Table 8-1 and Table 8-3 we have listed four different soybean inhibitors. We have attempted to use the several nomenclatures used by the authors; until further characterization is achieved, however, we would prefer to keep some designation such as $STIA_1$, $A_2$, etc. rather than to impose a new nomenclature. The "classical soybean trypsin inhibitor," SBTI, has received considerable study. Older commercial preparations of SBTI contained impurities of the other proteins that could be removed by ion-exchange chromatography [508].

All these different inhibitor preparations have been found to show different specificities against trypsin and chymotrypsin, and they can probably be as well compared on this basis as on any other with our present understanding of their properties. Frattali and Steiner [510] have listed their order of inhibitory capacity for trypsin as follows: STI ($F_2$) or $SBTIA_2 = SBTIA_1 =$ purified inhibitor $AA = 1.9$ S inhibitor $> F_1 > F_3$.

TABLE 8-2  AMINO ACID COMPOSITION OF SOME INHIBITORS FROM ANIMAL ORIGIN

|  | Porcine | | Bovine | | Ascaris suis |
|---|---|---|---|---|---|
|  | I | II | (Kazal type) | Basic "Kunitz" |  |
| Reference: | 499 | 499 | 497 | 495 | 520 |
| Ala | 1 | 1 | 1 | 6 | 4 |
| Arg | 2 | 2 | 3 | 6 | 4 |
| Asp | 4 | 4 | 7 | 5 | 4 |
| CySH | 6 | 6 | 6 | 6 | 8 |
| Glu | 7 | 6 | 7 | 3 | 10 |
| Gly | 4 | 4 | 5 | 6 | 5 |
| His | 0 | 0 | 0 | 0 | 0 |
| Ile | 3 | 3 | 3 | 2 | 3 |
| Leu | 2 | 2 | 4 | 2 | 0 |
| Lys | 4 | 4 | 3 | 4 | 8 |
| Met | 0 | 0 | 1 | 1 | 0 |
| Phe | 0 | 0 | 0 | 4 | 0 |
| Pro | 5 | 4 | 4 | 4 | 5 |
| Ser | 6 | 5 | 2 | 1 | 1 |
| Thr | 6 | 5 | 4 | 3 | 3 |
| Try | 0 | 0 | 0 | 0 | 1 |
| Tyr | 2 | 2 | 2 | 4 | 2 |
| Val | 4 | 4 | 4 | 1 | 2 |
| Molecular weight | 6024 | 5609 | 6155 | 6513 | 7600 |

TABLE 8-3  AMINO ACID COMPOSITION OF SOME INHIBITORS OF PLANT ORIGIN[a]

| | Soy Bean | | | | | Lima Bean | | | | | |
|---|---|---|---|---|---|---|---|---|---|---|---|
| | $F_1$ | $F_3$ | (Kunitz) STI | Low Mol. Wt. Inhibitor | 1.9S Inhibitor | Fr. 4 | Fr. 6 | Component 2 | Potato Inhibitor 1 | Navy Bean | Black-eyed Pea |
| Reference: | 510 | 510 | 506 | 529 | 509 | 513 | 513 | 512 | 518 | 514 | 515 |
| Ala | 5 | 5 | 9 | 4 | 8 | 3 | 3 | 3 | 12 | 8 | 10 |
| Arg | 9 | 8 | 9 | 2 | 4 | 2 | 2 | 2 | 14 | 7 | 5 |
| Asp | 21 | 30 | 29 | 12 | 23 | 14 | 14 | 14 | 32 | 30 | 21 |
| CySH | 14 | 12 | 4 | 14 | 26 | 12 | 14 | 14 | 8 | 30 | 13 |
| Glu | 23±1 | 49±2 | 21 | 7 | 14 | 7 | 8 | 5 | 24 | 17 | 14 |
| Gly | 10 | 10 | 18 | 0 | 0 | 1 | 1 | 0 | 42 | 5 | 4 |
| His | 3 | 5 | 2 | 1 | 2 | 6 | 6 | 3 | 2 | 10 | 7 |
| Ile | 7 | 7 | 14 | 2 | 4 | 4 | 4 | 4 | 25 | 9 | 5 |
| Leu | 11 | 13 | 16 | 2 | 4 | 4 | 4 | 3 | 31 | 6 | 2 |
| Lys | 11±1 | 21 | 11 | 5 | 10 | 4 | 4 | 4 | 16 | 11 | 8 |
| Met | 6 | 10 | 3 | 1 | 2 | 0 | 0 | 0 | 2 | 1 | 0 |
| Phe | 4 | 2 | 9 | 2 | 4 | 1 | 2 | 1 | 5 | 4 | 6 |
| Pro | 8 | 6 | 10 | 6 | 12 | 7 | 7 | 6 | 13 | 16 | 8 |
| Ser | 15 | 11 | 13 | 9 | 18 | 12 | 13 | 12 | 13 | 35 | 28 |
| Thr | 6 | 5 | 8 | 2 | 4 | 5 | 5 | 3 | 11 | 14 | 5 |
| Try | 1 | 3 | 2 | 0 | 1 | 0 | 0 | 0 | — | 0 | 2 |
| Tyr | 5 | 0 | 4 | 2 | 4 | 1 | 1 | 1 | 3 | 4 | 10 |
| Val | 4 | 2 | 12 | 1 | 2 | 1 | 1 | 1 | 32 | 2 | 2 |
| Molecular weight | 18,300 | 23,400 | 21,500 | 7,974 | 16,400 | 9,100 | 9,667 | 8,291 | 37,700 | 23,000 | 16,923 |

[a] Calculated whole number of residues.

The approximate order for chymotrypsin is: 1.9 S inhibitor ⩾ AA ⩾ SBTIA$_1$ > STI (F$_2$) > F$_1$ ⪢ F$_3$. In addition to these differences, there are obviously outstanding variations in the amino acid contents. Component F$_3$ is distinctive in lacking tyrosine but having tryptophan, whereas F$_1$ and F$_3$ have large numbers of disulfide bonds. Although several of these different inhibitors appear pure by the criteria used for their examination, it is thus evident that in the soybean alone inhibitors are found with very different compositions and very different inhibitory activities!

## LIMA BEAN INHIBITORS

The first lima bean inhibitor separated by Fraenkel-Conrat and co-workers [511] was quite different from the first isolated from soybeans. Its molecular weight was less than half that of the soybean inhibitor. In addition, its inhibitory activity was inactivated by acetylation of its amino groups whereas the soybean inhibitor was unaffected by acetylation. It differed also from the first soybean inhibitor in not having any tryptophan. More recently lima bean inhibitors have been separated and divided into four components by workers at Rockefeller Institute [512], and six components in our laboratory [513]. In the latter study, six chromatographically distinct inhibitors were found in a highly inbred genetic type of lima beans. In addition, one of the chromatographically homogeneous fractions was a "double-headed" inhibitor inhibiting both trypsin and chymotrypsin. The trypsin inhibitory activity was approximately twice that of the chymotrypsin inhibitory activity and thus had some resemblance to one or more of the fractions from soybeans. All the fractions studied have been found to be devoid of tryptophan and to contain many disulfide bonds.

## POTATO INHIBITORS

Two crystalline proteins strongly inhibiting widely diverse proteolytic enzymes have been isolated from potato tubers [516, 517], [530, 531].

One of the potato inhibitors (inhibitor 1) was found stable over a wide pH and temperature range. One of the most interesting aspects of this inhibitor is its combining capacities with trypsin and chymotrypsin. It is a comparatively weak inhibitor of trypsin but a very strong inhibitor of chymotrypsin. This "strength" is not only a matter of a low dissociation constant of the complex, but also the number of molecules of chymotrypsin bound to one molecule of inhibitor. The inhibitor apparently is capable of combining with as many as four molecules of chymotrypsin.

## Avian Egg-White Inhibitors

Chicken ovomucoid is one of the inhibitors that has received a great deal of study. This attention arose from its mucoid nature and its inhibitory properties. In common with the other inhibitors described earlier, ovomucoid was eventually found to be contaminated with one or more other proteins with inhibitory activity. The principal protein contaminating most chicken ovomucoid preparations proved to be a second inhibitor called ovoinhibitor [93]. This contaminant was responsible for the inhibitory activity of commercial preparations of ovomucoids against bovine $\alpha$-chymotrypsin. Purified chicken ovomucoid was found to have no detectable activities against $\alpha$-chymotrypsin. Even these more purified preparations of ovomucoid appear very heterogeneous because of the presence of multiple molecular forms. Longsworth, Cannan, and MacInnes in 1940 [26] first described the heterogeneity of ovomucoid in terms which remain true today: "ovomucoid, though not electrically separable into more than one component, shows complexity as indicated by 'reversible boundary spreading.'" After nearly thirty years the problem has progressed only slightly further, in that a partial separation into fractions of different electrophoretic mobilities has been achieved [43], [97]. Because differences in content of sialic acid were found, an obvious possibility was that the heterogeneities were due to this variation but this has not proven to be the primary cause. A more likely possibility is that heterogeneity is due in part to more subtle differences such as differences in amide content or perhaps differences in content of carbohydrate other than sialic acid residues.

The amino acid content of chicken ovomucoid is given in Table 8-4. One of the most important characteristics of the protein is that nearly one quarter of its composition is carbohydrate. The other characteristic, in common with most other inhibitors, is the absence of tryptophan. It is a moderate sized protein (molecular weight 28,000–29,000 g) available in comparatively large quantities since it is easily purified from the highly concentrated natural source. These factors together with the variation in specificities of the ovomucoids from different species (Table 3-8) make these particular inhibitors excellent materials for research on structure-function relationship. There is a large amount of information concerning the physical and chemical properties of chicken ovomucoid but much of this information does not seem to be appropriate for discussion here. The reader is rather referred to a review by Melamed [533]. This information is also currently under preparation for publication elsewhere [534].

The other main inhibitor in chicken egg white is the ovoinhibitor originally described by Matsushima [56], [535]. He observed that chicken egg

TABLE 8-4  CHEMICAL COMPOSITION OF EGG-WHITE INHIBITORS

| Constituents: | Ovomucoids[a] [residues/mole (28,000 g)][b] | | | | | | | | | Ovoinhibitor [residues/mole (44,000 g)][c] |
|---|---|---|---|---|---|---|---|---|---|---|
| Species: | Chicken | Turkey | Cassowary | Emu | Ostrich | Rhea | Tinamou | Duck | Penguin | Chicken |
| Alanine | 11.7 | 8.6 | 7.3 | 7.0 | 5.6 | 5.2 | 8.5 | 8.4 | 5.4 | 14.8 |
| Arginine | 6.3 | 5.8 | 1.0 | 0.0 | 3.4 | 0.0 | 3.5 | 1.2 | 3.1 | 14.6 |
| Aspartic acid | 31.9 | 27.2 | 25.8 | 27.1 | 27.3 | 28.7 | 30.8 | 33.0 | 28.0 | 33.8 |
| CySH | 17.5 | 16.7 | 19.9 | 16.7 | 19.9 | 17.4 | 19.6 | 20.2 | 18.8 | 18.2 |
| Glutamic acid | 14.9 | 19.0 | 17.7 | 16.6 | 18.0 | 20.1 | 18.1 | 20.1 | 15.7 | 27.5 |
| Glycine | 16.1 | 17.3 | 14.3 | 14.9 | 17.0 | 14.6 | 17.6 | 19.2 | 14.7 | 24.0 |
| Histidine | 4.3 | 5.2 | 3.1 | 3.0 | 2.2 | 4.1 | 3.4 | 3.5 | 2.2 | 10.1 |
| Isoleucine | 3.2 | 4.4 | 6.0 | 6.0 | 4.5 | 3.7 | 4.7 | 2.6 | 4.1 | 12.6 |
| Leucine | 12.2 | 13.5 | 12.5 | 13.4 | 15.3 | 12.6 | 9.4 | 13.5 | 12.6 | 16.3 |
| Lysine | 13.6 | 11.2 | 16.5 | 16.2 | 14.8 | 15.8 | 17.1 | 17.2 | 13.5 | 17.2 |
| Methionine | 1.9 | 1.8 | 1.0 | 0.9 | 1.0 | 0.9 | 0.0 | 7.9 | 2.0 | 2.4 |
| Phenylalanine | 5.3 | 3.2 | 3.1 | 3.1 | 3.3 | 5.0 | 4.7 | 4.7 | 2.5 | 7.0 |
| Proline | 7.7 | 8.8 | 10.4 | 8.9 | 11.4 | 9.7 | 14.0 | 10.4 | 9.5 | 12.4 |
| Serine | 12.5 | 10.0 | 14.2 | 13.4 | 18.3 | 16.7 | 12.5 | 13.3 | 13.4 | 18.9 |
| Threonine | 14.6 | 14.2 | 11.8 | 12.2 | 17.0 | 16.2 | 14.0 | 19.9 | 16.0 | 22.9 |
| Tryptophan | 0.0 | 0.0 | 0.0 | 0.0 | 0.0 | 0.0 | 0.0 | 0.0 | 0.0 | 0.0 |
| Tyrosine | 6.7 | 6.8 | 10.6 | 6.4 | 11.1 | 8.2 | 10.4 | 11.5 | 10.1 | 9.5 |
| Valine | 16.0 | 15.7 | 15.4 | 15.1 | 16.9 | 17.9 | 17.6 | 17.4 | 19.4 | 18.9 |
| Sialic acid | 0.3 | 2.2 | 4.7 | 8.9 | 2.9 | 5.3 | 0.1 | 0.0 | 6.8 | — |
| Hexose | 16.7 | 18.0 | 16.4 | 15.6 | 15.2 | 16.9 | 15.9 | 12.8 | 14.8 | 9.5 |
| Glucosamine | 21.0 | 18.5 | 16.2 | 15.1 | 16.0 | 14.5 | 16.4 | 10.2 | 12.8 | 7.4 |
| N-Terminal | Ala | Val | Val | Val | Val | Val | Val | Val | Val | — |
| Recovery (g) | 27,400 | 27,100 | 27,300 | 26,900 | 28,200 | 27,200 | 27,500 | 28,000 | 27,100 | — |

[a] [82], [193], [532].
[b] The value of 28,000 g was used for the molecular weight of chicken ovomucoid and assumed for the molecular weights of the other proteins.
[c] [88]

white contained an inhibitor of the fungal proteinase from *Aspergillus oryzyae* and the bacterial proteinase from *Bacillus subtilis*. This bacterial enzyme we shall call subtilisin in this discussion. Workers in our laboratories at the University of Nebraska [81], reasoning from the substrate specificity of subtilisin concluded that the ovoinhibitor might also inhibit bovine chymotrypsin and this proved to be the case. This led to the observation that the weak inhibitory activities of commercial preparations of chicken ovomucoid were due to the presence of contaminating amounts of ovoinhibitor [93]. Chicken ovoinhibitor has a molecular weight of approximately 46,500 g [88] and is a "double-headed inhibitor" inhibiting both trypsin and chymotrypsin at the ratio of two molecules of each enzyme to one molecule of inhibitor [81], [88]. Thus this gives a total of 2 + 2 or 4 total molecules of enzymes per molecule of inhibitor.

Fossum and Whitaker [64] have recently partially characterized another inhibitor, existing only in trace amounts. It constituted approximately 0.1% of the solids of chicken egg white as based on a reported purification factor of 110. This inhibitor is the first protein to have inhibitory activity against ficin and papain (Table 8-1).

INHIBITORS FROM BLOOD

Trypsin inhibitors from blood plasma were known before the beginning of this century. Landsteiner [536] referred to over half a dozen individuals who reported inhibitors in blood before 1900. There has been a particular interest in blood inhibitors because of the possible relationship to diseases involving proteolysis and to the blood-clotting mechanism. There is a long list of reports of blood inhibitors since 1900 and much credit should be given to the many investigators responsible for this work. We shall necessarily have to restrict ourselves to the results of the last few years and, in addition, must necessarily discuss the blood plasma inhibitors in only a general and descriptive approach. The blood inhibitors exist in comparatively small amounts and there is not so much definitive physical-chemical information on their properties as there is for the inhibitors from other sources.

We now know that there are several inhibitors of proteolytic enzymes in human blood plasma. Various of these inhibit the blood-clotting enzymes, plasmin and fibrinolysin, as well as either trypsin or $\alpha$-chymotrypsin or both. Two inhibitors have definitely been identified in human serum and demonstrated by means of paper electrophoresis [537]. An inhibitor has been prepared from the $\alpha_1$-globulin fraction [500, 501] and an inhibitor has also been prepared with properties of an $\alpha_2$-globulin [502]. The inhibitor from the $\alpha_1$-globulin fraction was both an inhibitor of trypsin and $\alpha$-chymotrypsin (Fig. 8-3 and 8-4); and it also had a greater affinity for $\alpha$-

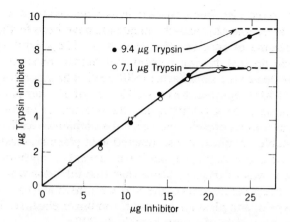

Fig. 8-3. Inhibition of trypsin, at two concentrations, with increasing amounts of $\alpha_1$-serum inhibitor. Although the reaction is stoichiometric, some dissociation of the trypsin-inhibitor complex is indicated by the curvature as the equivalence point is approached. Redrawn from [501].

chymotrypsin than trypsin. Unfortunately, no data were given as to whether it was "double headed" and would inhibit trypsin and $\alpha$-chymotrypsin simultaneously. The molecular weight was estimated at 45,000 g. The complex of trypsin and the inhibitor was stable to dissociation over a wide pH range.

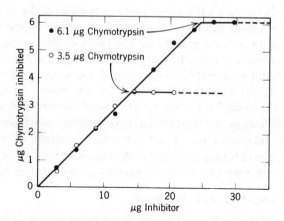

Fig. 8-4. Inhibition of chymotrypsin, at two concentrations, with increasing amounts of $\alpha_1$-serum inhibitor. There is little or no indication of dissociation of the chymotrypsin-inhibitor complex. Redrawn from [501].

Inhibitors have also been studied in bovine, ovine, and avian blood plasma. In fact, more physical-chemical data have been forthcoming from the blood plasma of species other than man. The inhibitory activity in bovine blood against bovine trypsin and human plasmin was demonstrated to be due to the same factor [538, 539]. The laboratory of M. Laskowski, Sr., [503] reported the crystallization of another plasma inhibitor when they described the isolation and characterization of one from bovine blood. It was a mucoprotein and reacted stoichiometrically with trypsin forming a stable complex; it also reacted with plasmin, elastase preparations and with $\alpha$-chymotrypsin but it was a weaker inhibitor of these enzymes than it was of trypsin. More than one inhibitor was also demonstrated in sheep serum [504]; one was demonstrated to inhibit trypsin, $\alpha$-chymotrypsin, and plasmin, but not thrombin or elastase. This inhibitor had a molecular weight of approximately 46,000 g.

## Inhibitors from Milk

The inhibitors from milk are found in quantity only in the colostrum, the milk of the first few days after parturition. The laboratory of the Laskowskis [540] gave us another first in "inhibitor science" when they reported the presence of trypsin inhibitors in bovine and human colostrum. The same laboratory then reported the crystallization of the bovine inhibitor [229] and several years later the crystallization of the swine colostrum inhibitor [247]. In the swine colostrum the highest concentration was found in the first day after parturition when 1 ml of milk inhibited 2 mg of trypsin. In common with many other inhibitors, the swine colostrum inhibitor was also devoid of tryptophan.

There has been very little information reported on the general properties of the crystalline inhibitors from colostrum. The swine inhibitor was considered to be a small protein on the basis of its combining capacity with enzymes but no molecular weight has been given. More recent studies in our laboratory [194] have also calculated a comparatively low molecular weight for bovine colostrum inhibitor based on its combining capacity with trypsin. This inhibitor was found not to inhibit $\alpha$-chymotrypsin. One observation, perhaps trivial, was that the colostrum and the lima bean trypsin inhibitors had nearly identical contents of amino groups based on their combining capacity with trypsin. As are described later in both of these inhibitors lysine is a necessary group for their functions.

## Inhibitors from Other Sources

Inhibitors of proteolytic enzymes have been found in many sources other than those described before. Most of these observations are doubtlessly valid but some of them must await the application of more definitive

## GENERAL PROPERTIES OF INHIBITORS

methods before the actual presence of inhibitors can be considered proven. The assay for inhibition of proteolytic enzymes is simple [93] and yet it can easily give uninterpretable or erroneous results. It seems quite clear, however, that there are inhibitors of proteolytic enzymes existing in small concentrations in many biological fluids.

Three other inhibitors from animal sources include those from urine, seminal fluid, and from the parasitic round worms, *Ascaris lumbricoides* and *Ascans suis*. Preparations of inhibitors from human urine inhibit α-chymotrypsin and plasmin as well as trypsin [502], [541]. Comparative studies of seminal fluid inhibitors were done in our laboratory. Inhibitory activities against trypsin and chymotrypsin were found in the chicken, turkey, bovine, ovine, and human species. Partial purification was done only with inhibitors from human seminal fluid. The comparative activities of these preparations against trypsin and α-chymotrypsin were essentially identical, indicating that the inhibitors were "double-headed." However, caution is necessary in interpreting these observations, because it is known that seminal fluid may contain some components from blood plasma.

The presence of an inhibitor of a proteolytic enzyme in an intestinal parasite is an obvious possibility and, indeed, one was early observed in parasitic round worms by Mendel and Blood [542]. N. M. Green [543] described in detail a trypsin inhibitor from *A. lumbricoides*. One inhibitor from *A. lumbricoides* has been crystallized and reported to inhibit only α-chymotrypsin [544]. Thus, there may be mixtures of inhibitors, some specific for trypsin and some for α-chymotrypsin. It has now been reported that there are at least three trypsin inhibitors which can be obtained from *A. suis* [545] and that some of these also inhibit α-chymotrypsin. One of the trypsin inhibitors has recently been studied in more detail [520]. Probably the most interesting of all the observations is that here we find another small inhibitor—with a molecular weight between 7600 and 8200 g and the absence of some amino acids—histidine, methionine, leucine, and phenylalanine.

The many different inhibitors observed in plants deserve a lengthy review in some journal devoted to plant chemistry or physiology. Liener [546] has included some treatment of the subject in a general discussion of toxic substances in plants. One inhibitor which appears to have been characterized in some detail is the "double-headed" inhibitor for trypsin and α-chymotrypsin from the black-eyed pea [515]. It inhibits two moles of chymotrypsin and one more of trypsin. It is thus similar to the potato inhibitor. The molecular weight of approximately 17,000 places it also within the general realm of the lower molecular weight inhibitors. Although the inhibitor contained a small amount of tryptophan, two per mole, here again we find that one of the amino acids, methionine, usually found in proteins is lacking (Table 8-2). Another bean inhibitor has also recently been described in navy beans (Table 8-3) [514]. Doubtlessly there will be many of these inhibitors isolated from plants.

## COMPARATIVE BIOCHEMISTRY

The most striking aspect of the comparative and genetic biochemistry of the inhibitors is the large number of different types of proteins that possess this activity and their many different sources. There is a multitude of analogous and homologous proteins exhibiting this activity. One thing they have in common, other than their inhibitory activity against proteolytic enzymes, is that many of them are relatively small and do not seem to require the usual complement of amino acids for their function. The group has generally been called "trypsin inhibitors" despite the fact that very early in the study of inhibitors, Kunitz and Northrup [485] reported that the pancreatic basic trypsin inhibitor also inhibited $\alpha$-chymotrypsin. It is now being found that inhibitors sometimes inhibit more than one molecule of enzyme per molecule of inhibitor and frequently inhibit several different enzymes, such as trypsin and $\alpha$-chymotrypsin. These binding sites usually do not overlap. Some of the trypsin inhibitors will also inhibit kallikrein and perhaps other proteolytic enzymes. Some of the $\alpha$-chymotrypsin inhibitors will also inhibit bacterial proteinases (subtilisin). Enzymes such as thrombin, plasmin, fibrinolysin, and fungal proteinase, may or may not combine with the inhibitors at similar sites or overlapping sites with trypsin or $\alpha$-chymotrypsin. A very few of the inhibitors have broad specificity such as the ovoinhibitor from chicken egg white that inhibits trypsin, chymotrypsin, subtilisin, and fungal proteinase.

### The Presence of Different Forms and Types of Inhibitors in the Same Biological Material

Every inhibitor that has been studied in detail has been found to exist in multiple molecular forms or to occur with different inhibitors. Why this happens is completely not understood as of this time. There is always the possibility that the multiple molecular ("isozymic") forms may be due to breakdown of the original inhibitor either in the tissue or because of the procedures used to purify it for study. This doesn't seem to be the general cause of these multiple forms of the inhibitors because they have been obtained from such very different types of tissues by many different techniques and yet show many multiple forms. Chicken ovomucoid is a "broadly heterogeneous" family of proteins [99] and the ovoinhibitor may be even more heterogeneous [547]. The legume inhibitors show an equivalent or even greater heterogeneity than the egg-white inhibitors. The reasons for these heterogeneities may be more interesting for their biosynthetic aspects than they are for the aspects as functional inhibitors.

The existences of inhibitors with apparently quite different structures and functions in the same materials are of direct interest to our subject of homology and analogy. Thus chicken egg white contains three inhibitors very different in their structure and in their inhibitory mechanism. Why should egg white contain a collection of such materials? Here we are dealing with some of the fundamental problems related to biological constancy and variation. It would appear likely that when we understand why such different forms of inhibitors are present we will understand the reason why inhibitors exist and what function they may have.

It is even more intriguing to ask why the egg white of the chicken should have an ovomucoid that is the inhibitor of trypsin, why the turkey should have an ovomucoid that is a "double-headed" inhibitor inhibiting trypsin and $\alpha$-chymotrypsin, and why the golden pheasant should have an inhibitor that is primarily a $\alpha$-chymotrypsin inhibitor. One obvious explanation here is that we simply have not tested all the enzymes which might be inhibited by the different ovomucoids. Perhaps a more realistic explanation is that there is no relative advantage to the kind of inhibitor they have in their eggs. This would harken back to the idea that the ovomucoids were present simply because they are convenient proteins to put in the egg white rather than have a function as inhibitors of proteolytic enzymes. Any teleogical reasoning at this stage is dangerous. Much more data are needed to understand these functions.

Finally, what is the evolutionary origin of these inhibitors? Are they possibly proteins that developed resistance to proteolysis and the capacity for forming tight complexes for some reason other than to function as inhibitors?

## Homologous Inhibitors Currently Under Study

The pancreatic inhibitors of different species are currently receiving much attention in several laboratories. The acidic, or Kazal, inhibitors from bovine and porcine pancreatic juice have been highly purified [497–499]. Two different inhibitors were obtained in the ratio of 4:1 from porcine juice (Table 8-1) [499].

A trypsin inhibitor has been isolated from human pancreatic juice with basic properties resembling the bovine (Kunitz) basic pancreatic inhibitor, but many of the properties were more similar to the bovine acidic (Kazal) inhibitor [548]. For example, the human inhibitor did not inhibit bovine $\alpha$-chymotrypsin in contrast to the bovine basic inhibitor which inhibits α-chymotrypsin. Other pancreatic inhibitors studied include ones from rat and dog [549, 550].

The homologous avian egg-white ovomucoids have not been studied extensively as yet except in our laboratory. The current successes of the use of the homologous egg-white lysozymes (see Chapter 7) would suggest that the homologous ovomucoids will also receive general attention.

## The Use of Homologous and Analogous Structures in Studying Function

The use of homology and analogy to study protein function is one of the main interests in our laboratories and probably the main reason why this book has been prepared. The inhibitors of the proteolytic enzymes have provided an area for many interesting studies on this point. The interest and the utility is considerably increased in these studies by the use of homologous enzymes like trypsin and α-chymotrypsin. In other words, small differences can be at a sophisticated level insofar as the interaction and fitting of the two proteins with one another. Small differences in both interacting proteins might therefore result in larger differences in the observed interaction.

It has thus been possible to obtain homologous proteins such as the chicken, turkey, and cassowary ovomucoids, all of which are trypsin inhibitors with different inhibitory properties. The chicken and cassowary ovomucoid differ by the fact that chicken ovomucoid is inactivated by arginine modification and cassowary ovomucoid, by lysine modification. The turkey ovomucoid differs by the fact that it is a trypsin inhibitor inactivated by lysine modification but an inhibitor of α-chymotrypsin as well. The application of an analogous protein in this comparison is the use of the chicken ovoinhibitor. The chicken ovoinhibitor has a very different amino acid composition and molecular weight than the chicken ovomucoid (Table 2-4), and although sequence data are not yet available to prove the absence of homology these two proteins appear to be analogs rather than homologs. The ovoinhibitor is an inhibitor whose anti-trypsin activity is inactivated by arginine reagents like chicken ovomucoid, rather than by lysine reagents as turkey and cassowary ovomucoids. Yet it is a "double-headed" or "multi-headed" inhibitor and also inhibits α-chymotrypsin-like enzymes. When working with the double-headed inhibitors it is possible to use the "other head" or other inhibitory activity as a control in chemical modification. A big problem to the biochemist who uses chemical modification as a probe for structure and function is the possibility of side reactions or unidentified reactions causing the effects that he believes may be due to the primary modification. This is much easier when the investigator can work with a protein which has two activities and he can use the other activity as a control for the first activity. Thus, in the case of turkey ovomucoid and ovoinhibitor, modification of the first with lysine reagents and the second with arginine reagents without affecting their α-chymotrypsin inhibitory activities is a good control for nonspecific peptide-bond splitting, etc.

A different area wherein these homologous proteins can be very useful in protein chemistry is in the development of methods for modifying specific amino acid residues or, indeed, in studying the function of such residues in the general physical chemistry of proteins. Thus, in the case of the ovomucoids there are homologous proteins which have arginines varying in number from 0 to 6 and methionines varying from 0 to 6. By the use of these homologous proteins containing different amounts of arginine it was possible to identify a side reaction between the ε-amino groups of lysines and the reagent for modifying arginines, 1,2-cyclohexanedione [196].

# MECHANISMS OF INHIBITORY ACTIVITY

The interaction of proteolytic enzymes and their protein inhibitors have been studied by many different laboratories utilizing a variety of approaches. Much is known about the characteristics of the interactions and certain of the mechanisms involved, although there is not general agreement as to a common mechanism for the inhibition. The formation of the enzyme inhibitor complex appears to have characteristics found in nonenzymatic associations such as the antigen antibody reaction or the polymerization of subunits as well as characteristics found in an enzyme-substrate interaction. Various of the studies are described later.

## STUDIES WITH MODEL SYNTHETIC COMPOUNDS

The studies with model synthetic compounds have shown that proteolytic enzymes may be inhibited by a variety of compounds and probably through a variety of mechanisms. The easiest studies to understand are those with substrate-type inhibitors, such as benzamidine or tryptophan, which inhibit trypsin or α-chymotrypsin, respectively. This type of study with substrate-like inhibitors has a direct bearing on some of the mechanisms currently postulated for the inhibitory interactions of the protein inhibitors.

Another type of model compound is the "nonsubstrate" polyelectrolyte. Poly-α-L-glutamic acid, polycysteic acid, and heparin were found to inhibit trypsin [551]. The specificity of the inhibition was shown by the fact that poly-α-D-glutamic acid did not inhibit trypsin. One of the interesting findings was a certain minimum size of the polyanion required for inhibition and that the quantity of polyanion used was critical in the inhibitory mechanism. In the presence of excess polyanion the enzyme was not inhibited. The suggested mechanism was that the polyanions formed a bridge over the active sites of the enzyme, as shown in Fig. 8-5. An excess of the polyanion would then form multiple bindings with the enzyme. This would leave the active site exposed, and the mixture would be active.

## Studies of the Inhibitory Complex by Enzymatic Assays

Since the inhibitory process involves the formation of a physical complex, the interactions may be studied either by, (a) examining for free enzymatic activity, or (b) by studying the degree of association by physical means (as done with nonenzymically functioning proteins). In general, these methods have given similar results, but the nature of the interactions are sufficiently complicated so that they are as yet incompletely understood. One of the easiest ways to determine the formation of the complex is by method (a), that is, the direct titration of the enzyme by the inhibitor and the determination of the residual enzymatic activities at the various steps of the titration. Figure 8-6 shows the decreasing activity of trypsin with increasing concentrations of three different inhibitors: the basic pancreatic inhibitor of Kunitz, the soybean inhibitor ($STIA_2$), and chicken ovomucoid. It is evident from this figure that the inhibitors differ as to their extents of association with trypsin. This is easily seen by the departure from stoichiometric inhibition in the neighborhood of the equivalence point. From such plots Green [552] calculated dissociation constants for the complexes with trypsin as follows: pancreatic inhibitor, $2 \times 10^{-10} M$; soybean inhibitor, $2 \times 10^{-10} M$; and ovomucoid, $5 \times 10^{-9} M$. Figure 8-6 also shows that in these data varying the substrate concentration over a tenfold range did not materially affect the dissociations. In spite of numerous attempts by others to calculate dissociation constants for these inhibitors, these older values of Green are still as good as the more recent ones.

Similar titrations have also been made with $\alpha$-chymotrypsin and with either trypsin or $\alpha$-chymotrypsin in the presence of an excess of the other enzyme for the "double-headed" inhibitors [81], [193]. A direct titration of trypsin and $\alpha$-chymotrypsin by chicken or duck ovomucoid is seen in Fig. 8-7. A repeat of this with duck ovomucoid in the presence of four times the amount of the other enzyme as needed to saturate the combining site or "head" for that enzyme on the inhibitor is given in Fig. 8-8. Here it is seen that a direct titration is obtained with trypsin or $\alpha$-chymotrypsin in the presence of the appropriate substrate, even though the other enzyme is present in four times the amount required to saturate the ovomucoid.

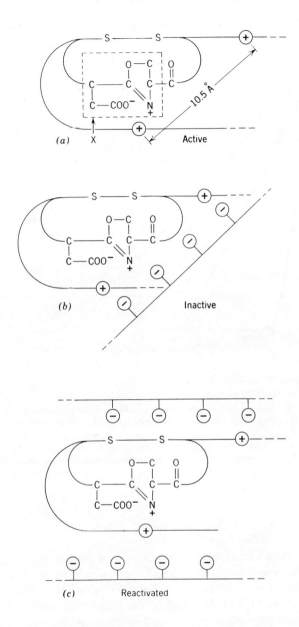

Fig. 8-5. Proposed mechanism for the inhibition and reactivation of trypsin by polyanions: (a) Schematic drawing of the active site (X) and location of two cationic groups (+) on opposite sides of the active site; (b) Inactive complex of a single polyanion molecule with the two cationic groups blocking the active site; (c) Reactivated complex of two polyanion molecules with the cationic groups. Redrawn from [551] Academic Press, 1960.

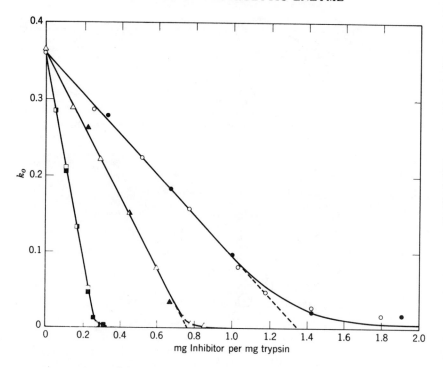

Fig. 8-6. Effect of inhibitor concentration on trypsin activity at different substrate concentrations. Each point was determined from the slope of an esterase run in 0.01 M borate buffer + 0.01 M CaCl$_2$. [Trypsin] = 0.72 μM; □, ■, pancreatic inhibitor; △, ▲, soy bean inhibitor; ○, ●, ovomucoid. Open symbols, 0.005 M BAEE; filled symbols, 0.05 M BAEE. The continuous curves were calculated from the expression, $k_o = a + \sqrt{a^2 - K_i T}$, where $a = \frac{1}{2}(T - I - K_i)$. I and T represent total inhibitor and trypsin concentrations. The values of I, T, and $K_i$ have been multiplied by a factor to bring them to the same units as $k_o$. Reproduced from [552].

There was no trace of any cooperative interactions or interferences between the two binding sites on duck ovomucoid.

The rates of association and dissociation of the complexes can give insight into the actual mechanism itself and can also be used in calculation of the dissociation constants. N. Michael Green [552] first showed the value of this approach by determining the times required for displacement of protein inhibitors from trypsin by the substrate, benzoylarginine ethyl ether (Fig. 8-9). No displacement of the pancreatic inhibitor by the synthetic substrate was observed during 20 minutes. This is in agreement with its low rate of reaction with trypsin determined by other methods and also with the low dissociation constants of the complex. The rate of dis-

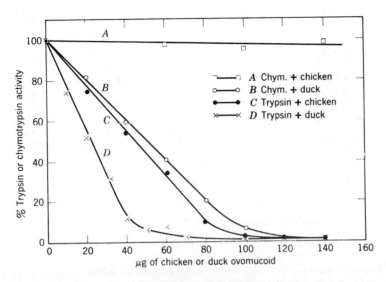

Fig. 8-7. Activity of 100 μg of trypsin or chymotrypsin in the presence of increasing amounts of chicken or duck ovomucoid. Conditions for assay are given under methods. Reproduced from [81] Academic Press, 1968.

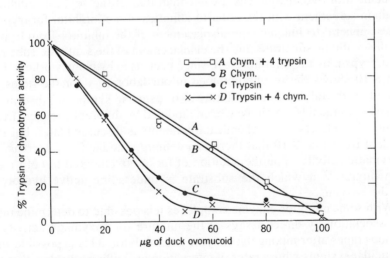

Fig. 8-8. Activity of 100 μg of trypsin or chymotrypsin in the presence of increasing amounts of duck ovomucoid. Four hundred micrograms of trypsin were added in the reaction mixture for $A$, and 400 μg of chymotrypsin were added in the reaction mixture for $D$. Reproduced from [81] Academic Press, 1968.

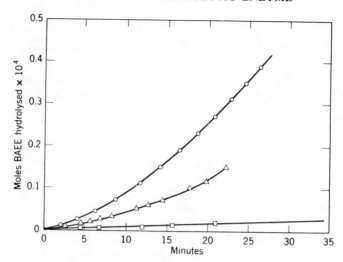

Fig. 8-9. Displacement of protein inhibitors from trypsin by BAEE, as shown by increase of the reaction velocity with time. The substrate was added after inhibitor and trypsin had combined. [BAEE] = 0.05 M; [CaCl$_2$] = 0.01 M; [T] = 0.78 μM. □, pancreatic inhibitor; △, soy bean inhibitor; ○, ovomucoid. Reproduced from [552].

placement of ovomucoid was greater than that of the soybean inhibitor, again in agreement with the order of affinities of the inhibitor for trypsin. These general techniques, the displacement of the inhibitor from trypsin by the synthetic substrate, and the displacement of the synthetic substrate from trypsin by the inhibitor have since been used by many workers. Figure 8-10 shows the results obtained in our laboratory for the times required for the inhibition of a mixture of trypsin and synthetic substrate by turkey ovomucoids which have been modified by different chemical treatments. The effects of the chemical treatments is discussed later, but it is evident from Fig. 8-10 that there is a definite time lag for a decrease of enzymatic activity upon the addition of turkey ovomucoid to both reaction mixtures in which the substrate is undergoing active hydrolysis by the enzyme.

With some inhibitor-enzyme complexes it is possible to determine rates of association by direct assays of the mixture for enzymatic activity at various times after mixing the enzyme and inhibitor. This is possible with particular systems whose rates of interaction are sufficiently slow that the times required for decreases of activity are easily measured by conventional equipment. This is the case for the interaction of turkey ovomucoid

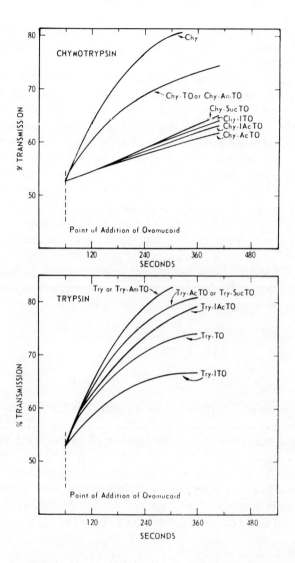

Fig. 8-10. Effect of chemical modification on delay time assays. Abbreviations used: Chy, chymotrypsin; TO, turkey ovomucoid; AcTO, acetylated turkey ovomucoid; Try, trypsin; AmTO, amidinated turkey ovomucoid; SucTO, succinylated turkey ovomucoid; ITO, iodinated turkey ovomucoid; IAcTO, iodinated and acetylated turkey ovomucoid. To a mixture of enzyme-buffer and substrate, was added turkey ovomucoid solution within 18–25 seconds and the inhibition recorded on a chart at 395 m$\mu$. The weight ratios were: (a) for chymotrypsin and turkey ovomucoids 22:15; (b) for trypsin and turkey ovomucoids 18:15. Reproduced from [553] Academic Press, 1966.

with bovine α-chymotrypsin, which requires approximately three-quarters of a minute for 50% interaction in those concentrations usually used for assay conditions [553].

As we might expect, some imperfections in the inhibitory mechanisms do occur. One of these is the actual enzymatic hydrolysis of the inhibitor by the enzyme. Gorini and Audrain [554, 555] observed the hydrolysis of ovomucoid by trypsin in the complex of trypsin and ovomucoid (Fig. 8-11). This was followed by the observation that the Kazal pancreatic inhibitor was also susceptible to hydrolysis by trypsin in the complex [556]. When the enzyme-inhibitor complexes were incubated, the appearance of free trypsin and the loss of inhibitor occurred at the same rate. This phenomenon was termed "temporary inhibition." Because the stoichiometric complexes of either of these inhibitors with trypsin show 5 to 10% the enzymatic activity of the enzyme, it has been assumed that this enzymatic activity is due to a dissociation of the enzyme from the complex to an extent of 5 to 10% and that the active free enzyme hydrolyzes either the free inhibitor or the inhibitor in the enzyme-inhibitor complex. This reaction has been represented as follows [489]:

$$T + I \underset{}{\overset{fast}{\rightleftharpoons}} TI$$

$$TI + T \underset{}{\overset{slow}{\rightleftharpoons}} TIT$$

$$TIT \underset{}{\overset{slow}{\rightleftharpoons}} 2T + products$$

In these equations T is the enzyme trypsin; I is the inhibitor; TI is the enzyme-inhibitor complex; and TIT is the enzyme-inhibitor complex with an additional molecule of trypsin capable of hydrolyzing the inhibitor. This representation thus indicates that the free trypsin acts enzymatically on the trypsin inhibitor complex to give proteolyzed products of the inhibitor. No proof has ever been presented, however, for this particular representation. As is discussed later, there is also the possibility that the enzymatic reaction could occur in the complex, TI, itself.

### Study of the Associations by Physical Methods

Physical methods have proved to be a valuable adjunctive procedure in studying the associations and dissociations of the complexes. Because of the relatively small dissociation constants of most of the complexes, several of the more commonly employed methods have been useful primarily to show the existence of complexes and their size rather than to study equilibria. Some of the early definitive examinations of the complexes by analytical centrifugation, free boundary electrophoresis, and viscosity

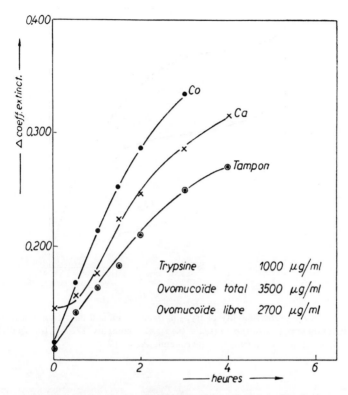

Fig. 8-11. Active proteolysis of ovomucoid in the ovomucoid-trypsin complex. Note the increased proteolysis when cobalt (Co) or calcium (Ca) is added to the buffer (Tampon). T = 36°C. Reproduced from [555].

measurements were done in Prof. F. F. Nord's laboratory at Fordham University [557]. The complexes were seen both ultracentrifugally (increase in size) and electrophoretically (change in charge). The large differences in electrophoretic mobility of trypsin, ovomucoid, and their complex are seen in Fig. 8-12. It has also been possible to show the formation of heteromolecular trimers with the double-headed inhibitor, turkey ovomucoid, when mixtures of trypsin, $\alpha$-chymotrypsin, and turkey ovomucoid are examined in the ultracentrifuge [81]. More recently, the very simple techniques of gel electrophoresis have made possible the observation of complexes of different inhibitors with different enzymes (Fig. 8-13).

Formation of complexes has also been observed by a fluorescence quenching technique [84] and by utilizing the spectral shifts of the acridine dye, proflavin, when it is bound or displaced from $\alpha$-chymotrypsin

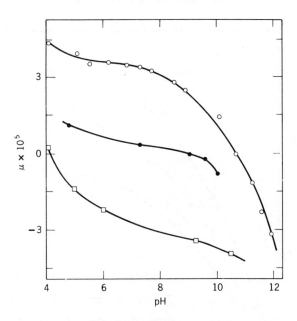

Fig. 8-12. Electrophoretic mobility of trypsin, ovomucoid, and their equimolecular complex as a function of pH. ○ = trypsin ($\Gamma/2 = 0.13$); ● = complex ($\Gamma/2 = 0.10$); □ = ovomucoid ($\Gamma/2 = 0.10$). Reproduced from [557] Academic Press, 1954.

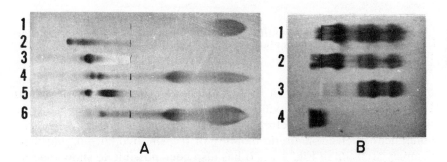

Fig. 8-13. Starch-gel electrophoretic patterns of inhibitors, enzymes, and their complexes. (A) 2 M urea starch-gel electrophoretic patterns with the Tris-citric acid in the gel buffer and succinic acid-sodium hydroxide in the bridge buffer at pH 6.0 and 4°. Staining was with nigrosine. The samples were all 10% solution, except the potato inhibitor which was 2.8%. The patterns are as follows (ratios are volumes of initial solutions): (1) turkey ovomucoid; (2) potato inhibitor; (3) α-chymotrypsin; (4) turkey ovomucoid: α-chymotrypsin (1:1); (5) potato inhibitor: α-chymotrypsin (1:1); (6) turkey ovomucoid: α-chymotrypsin (2:1). (B) Polyacrylamide electrophoretic patterns with a Tris-citric acid buffer system at pH 8.6. and 4°. Staining was with aniline blue-black. Solutions were 5%. The patterns are as follows (1) soybean trypsin inhibitor: trypsin (2:1); (2) soybean trypsin inhibitor: trypsin (1:1); (3) soybean trypsin inhibitor; (4) trypsin. Reproduced from (43) Academic Press, 1967.

and trypsin by the protein inhibitor [558]. The proflavin displacement technique also lent itself well to determining the rates of interaction of turkey ovomucoid with trypsin and chymotrypsin using rapid mixing, stop-flow techniques. By this procedure a second-order rate constant, $K_2$, was calculated for the association of turkey ovomucoid with bovine α-chymotrypsin (Fig. 8-14). This value for $K_2$ was $1.62 \times 10^4$ mole/liter/sec. A corresponding reaction with trypsin was more difficult to measure due to autolysis, but $K_2$ was approximately 5 times as fast as that for α-chymotrypsin at pH 7 in phosphate buffer.

The laboratory of M. Laskowski, Jr. has recently pioneered in the development of studies of protein-protein associations by potentiometric techniques [559]. The potentiometric technique compared well with other procedures as a method of titrating the amount of inhibitor bound to an enzyme and of determining dissociation constants.

STUDIES ON INTERACTION WITH CHEMICALLY MODIFIED INHIBITORS

The beginning of work on the chemical modification of inhibitors was in the laboratory of Fraenkel-Conrat at the Western Regional Research

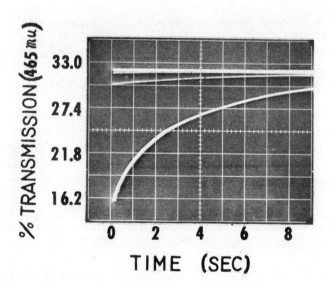

Fig. 8-14. Stopped-flow determinations of the reaction between turkey ovomucoid and proflavine-chymotrypsin mixture. Final concentrations in reaction mixture were 0.01, 0.1, and 0.1 mM for proflavine, α-chymotrypsin, and turkey ovomucoid, respectively. Reproduced from [558].

Laboratory of the U.S. Dept. of Agriculture. Here it was shown that acetylation of lima bean trypsin inhibitor deprives it of trypsin inhibitory activity while acetylation of chicken ovomucoid has little or no effect on activity [511]. This observation on the differential inactivation of these two trypsin inhibitors was one of the earlier definitive examples of the differential effect of chemical modification on the biological activity of proteins. During the last decade this approach has led to knowledge of the groups involved in trypsin inhibition by many different trypsin inhibitors. One of the early successes in this area was the demonstration by Stevens and Feeney [95] that the modification of amino groups of turkey ovomucoid destroyed its inhibitory activity for trypsin but not for $\alpha$-chymotrypsin. At the same time the activities of chicken ovomucoid against trypsin and golden pheasant ovomucoid against $\alpha$-chymotrypsin were not decreased, while a very weak activity of the golden pheasant ovomucoid against trypsin was destroyed by modification of amino groups. This successful differential modification of turkey ovomucoid strongly indicated that the amino groups were highly specific for inhibition of trypsin and that the epsilon amino groups of lysine were probably in "the reactive sites or active sites" of combination with trypsin. This work was further extended to the determination of the effect of modification, particularly charge changes, on the rates of reaction of turkey ovomucoid with trypsin and $\alpha$-chymotrypsin [553], and to the determination of the numbers of amino groups involved in the reactive site of a number of different trypsin inhibitors [194].

From a comparison of the effects of modification on the rates of inhibition of trypsin and $\alpha$-chymotrypsin by chemically modified turkey ovomucoids (Fig. 8-10), it is easily seen that iodination greatly increased the rate of reaction with trypsin and with $\alpha$-chymotrypsin in this type of assay. This was substantiated in the case of $\alpha$-chymotrypsin by direct observations of the time required for inhibition of $\alpha$-chymotrypsin when allowed to stand with inhibitor for different periods of time before the addition of substrate. These effects of iodination were confirmed in the case of $\alpha$-chymotrypsin inhibitory activity which is not destroyed by amino group reagents. Modification by reagents that caused little or no change in charge (amidination), caused no apparent change in rate of inhibition of $\alpha$-chymotrypsin, whereas all those modifications which made a more acidic protein (succinylation, iodination, or acetylation) greatly increased the rate.

The quantitation and modification of free amino groups in several inhibitors of trypsin and chymotrypsin was done with trinitrobenzenesulfonic acid (TNBS). This method was also used to do a kinetic analysis of both the relative rates of modification of amino groups and the loss of trypsin-inhibitory activity. This comparison made possible the estimation of the numbers of amino groups required for inhibitory activity. The trypsin-inhibitory activity of the following inhibitors was destroyed by modification with TNBS: lima bean trypsin inhibitor, a bovine colostrum inhibitor, and the ovomucoids of cassowary, duck, penguin, and turkey. It was found that either one or two "fast" reacting amino groups (depending on the inhibitor) were essential for the activity of six trypsin inhibitors [194], [513]. With three of these inhibitors, one fast reacting amino group was found essential for inhibitory activity. The semilogarithmic plot for loss of amino groups and loss of trypsin inhibitory activity of bovine colostrum inhibitor on modification with TNBS is shown in Fig. 8-15. The rate constant for inactivation is approximately double that for modification of all the amino groups of a total of three amino groups per mole (which all react at the same rate). This result was considered to fit a model that implicated two amino groups at the active site, the modification of either of which caused inactivation. In contrast, the rates for inactivation and modification of fast reacting amino groups of turkey ovomucoid were identical (Fig. 8-16) fitting the model for one amino group at the active site. The summary of these studies on amino groups in trypsin inhibitors is shown in Table 8-5.

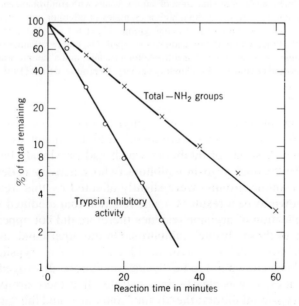

Fig. 8-15. Semilogarithmic plot for loss of amino groups and loss of trypsin-inhibitory activity in bovine colostrum inhibitor by modification with trinitrobenzenesulfonic acid. Reproduced from [194].

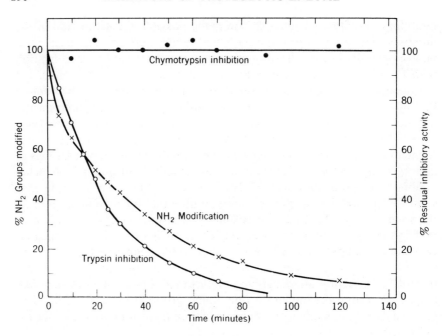

Fig. 8-16. Time course of modification of amino groups with trinitrobenzenesulfonic acid and loss of inhibitory activity in the inhibitor. Samples of inhibitor were incubated with a 10-fold molar excess of TBS to free amino groups at pH 8.5 and 35° in the dark. After various periods of incubation, the reaction was stopped. The number of amino groups which had been modified was determined spectrophotometrically, and a sample was assayed for residual trypsin- and chymotrypsin-inhibitory activity. Reproduced from [194].

Studies on the modification of arginyl residues have also been conducted [196]. The arginines were modified by treatment with 1,2-cyclohexanedione. A series of inhibitors which had previously been demonstrated to be "lysine" trypsin inhibitors (whose activity is destroyed by reagents for amino groups) were slightly affected by this treatment and this was shown to be a result of a side reaction that modified lysine residues. Modification of arginine residues therefore did not appear to cause inactivation of these "lysine" inhibitors. On the other hand, modification of two inhibitors, chicken ovomucoid and soybean trypsin inhibitor, which had been shown not to be "lysine" inhibitors, destroyed their inhibitory activities. It was clearly demonstrated that there are two distinct classes of trypsin inhibitors: the "lysine" inhibitors and the "arginine" inhibitors. A summary of the classification of inhibitors is given in Table 8-6.

TABLE 8-5  FIRST-ORDER RATE CONSTANTS FOR
TREATMENT OF INHIBITORS WITH
TRINITROBENZENESULFONIC ACID [194]

| Inhibitor | First-Order Rate Constant ($k$ min$^{-1}$) | |
|---|---|---|
| | Modification of "Fast" Amino Groups | Loss of Trypsin-Inhibitory Act. |
| Turkey ovomucoid | 0.087 | 0.081 |
| Penguin ovomucoid | 0.094 | 0.092 |
| Cassowary ovomucoid | 0.043 | 0.059 |
| Colostrum inhibitor | 0.058 | 0.122 |
| Duck ovomucoid | 0.089 | 0.051 |
| Lima bean inhibitor | a | 0.092 |
| Chicken ovomucoid | 0.041 | <0.006 |
| Tinamou ovomucoid | 0.042 | b |

[a] Difficulties encountered in distinguishing between rates of three differently reacting amino groups.
[b] Tinamou ovomucoid inhibits chymotrypsin, not trypsin. No loss in activity against chymotrypsin was observed.


TABLE 8-6  COMBINING SITE AMINO ACIDS OF
INHIBITORS OF TRYPSIN

| Inhibitor | Essential Amino Acid | Reference |
|---|---|---|
| Soybean trypsin inhibitor (Kunitz) | Arginine | 196, 565 |
| Chicken ovoinhibitor | Arginine | 532 |
| Chicken ovomucoid | Arginine | 196, 565 |
| Bovine Pancreatic Inhibitor (basic)[a] | Arginine | 532 |
| Soybean inhibitor (STI$_2$; acetone insoluble) | Lysine | 532 |
| Lima bean inhibitors | Lysine | 194 |
| Turkey ovomucoid | Lysine | 95, 194 |
| Cassowary ovomucoid | Lysine | 194 |
| Penguin ovomucoid | Lysine | 194 |
| Duck ovomucoid | Lysine | 194 |
| Colostrum inhibitor | Lysine | 194 |

[a] There is some doubt about the bovine pancreatic inhibitor.

Kassell and Chow [496] found that modification of amino groups of the bovine pancreatic inhibitor might, or might not, cause loss of inhibitory activity depending upon the type of modification. Previous studies had shown that modification by acetylation or succinylation caused inactivation. They now found that amidination or guanidination did not inactivate the inhibitor and concluded that "attachment of the $\epsilon$-amino groups of lysine to trypsin is not necessary for the activity of the inhibitor." Haynes and Feeney [560] obtained similar results with several other inhibitors. Guanidination of four "lysine inhibitors" did not abolish their activity. These homoarginine inhibitors were no longer "lysine inhibitors." They were inactivated by 1,2-cyclohexanedione but not amino group reagents. Thus, it appeared that the inhibitor had been transformed from a "lysine inhibitor" into an "homoarginine inhibitor," (Table 8-7). Since the peptide bond involving homoarginine is not considered to be split by trypsin, these results indicated that proteolysis was unessential for the inhibitory mechanism (This is discussed later). A further note of caution is necessary, however, because all chemical modifications may affect properties other than the functions of a particular residue. One of the more evident properties is that of charge. Sometimes charge change may cause inactivation, while other times it may increase the apparent activity.

TABLE 8-7 EFFECTS OF TREATMENT WITH 1,2-CYCLOHEXANEDIONE ON THE TRYPSIN-INHIBITORY ACTIVITY OF NATIVE AND GUANIDINATED TRYPSIN INHIBITORS [560]

| Inhibitor | % Residual Trypsin-Inhibitory Activity[a] |
|---|---|
| Native chicken ovomucoid | 37 |
| Guanidinated chicken ovomucoid | 35 |
| Native lima bean inhibitor | 73 |
| Guanidinated lima bean inhibitor | 13 |
| Native turkey ovomucoid | 83 |
| Guanidinated turkey ovomucoid | 12 |

[a] Values for chicken ovomucoid and lima bean inhibitor were obtained from assays with ester substrate; those of turkey ovomucoid were from caseinolytic assays.

Various other chemical modifications have been applied to inhibitors. One of the simplest observations was that maintenance of the tertiary structure was required for inhibitory activity. This was shown by a number of workers who found that reduction of disulfide bond caused loss of activity [98], [561, 562]. Other amino acid side chains such as methionines, have also been found to be unessential [558], [563].

## Enzymatic Modifications of Inhibitors

Workers in the laboratory of Michael J. Laskowski, Jr., of Purdue University have reported evidence suggesting that the "driving force" in the inhibitory reaction is actually a proteolysis of a peptide bond in the inhibitor. This work arose from the observation of "an overshoot" in the number of protons released when the inhibitor and enzymes are mixed. The "overshoot" was due to the release of three protons per molecule of soybean trypsin inhibitor on mixing with trypsin, followed by a relatively slow uptake of one proton. The uptake of this relatively slow proton was considered to be evidence for the proteolytic splitting of a peptide bond in the inhibitor. This reaction is shown in Fig. 8-17 in which $S$ is the original native (virgin) inhibitor. However, when the inhibitor was incubated with catalytic amounts of trypsin in acid solution (pH 3.75), the overshoot no longer occurred, as shown by $S^*$. Finally, when the inhibitor was first treated with trypsin in this strong acid solution and then treated with carboxypeptidase B at neutral pH the entire proton uptake mechanism was lost, as shown by $S_c^*$. $S^*$, the inhibitor formed by treatment of the native inhibitor with trypsin, was still active, while after sequential trypsin treatment and carboxypeptidase treatment the inhibitor, $S_c^*$, had lost all activ-

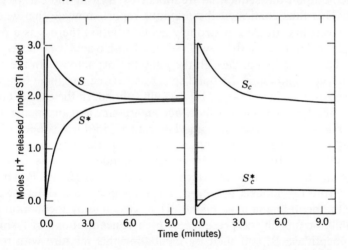

Fig. 8-17. Hydrogen ion release upon the addition of soybean trypsin inhibitor and of its derivatives to trypsin at pH 3.75. In all cases, 2 ml of $1.1 \times 10^{-4}$ M inhibitor (Worthington SI 5495) or derivative were added to 7 ml of $8.3 \times 10^{-5}$ M trypsin (Worthington TRLSF 6105). All solutions were 0.50 M in KCl and 0.05 M in $CaCl_2$; temperature was 20°. $S$, Virgin inhibitor; $S^*$, modified inhibitor (preincubated with 0.05 mg per ml of trypsin for 24 hours at pH 3.75); $S_c$, virgin inhibitor treated with 0.1 mg per ml of carboxypeptidase B (Worthington COB DFP 28), at pH 7.5 for 8 hours; $S_c^*$, modified inhibitor treated with carboxypeptidase B. At pH 3.75, stoichiometric complex formation releases 3.35 moles of hydrogen ion per mole. Reproduced from [564].

ity. The following representation was given to show the sequence of events in this mechanism:

$$T + S \underset{}{\overset{\text{fast}}{\rightleftharpoons}} TS \underset{}{\overset{\text{slow}}{\rightleftharpoons}} T + S^*$$

The fast reaction was considered to be the formation of the initial complex and the slow reaction was the proteolytic reaction. These authors also found similar effects with chicken ovomucoid and considered the mechanisms to be identical. They made the following statement:

"On the basis of these experiments and of several other studies now in progress in our laboratory, we propose that the trypsin-inhibitory reaction consist of a cleavage of one especially sensitive bond in the inhibitor by trypsin, and of subsequent formation of a *covalent* bond between trypsin and inhibitor (probably an ester bond between the active site of trypsin and the newly formed COOH-terminal of the inhibitor)."

In further studies from the same laboratory [565] it was reported that the free arginine was released from soybean trypsin inhibitor and chicken ovomucoid upon the sequential treatment by trypsin and carboxypeptidase B. This thus confirmed the previous report showing that when the inhibitor was first treated in strongly acidic solution there was a peptide bond split and that when the new C-terminal amino acid was subsequently removed by carboxypeptidase B, the inhibitor lost activity. In the case of soybean trypsin inhibitor the bond split was between arginine 64 and isoleucine 65 and this was bridged by a disulfide bond. In the case of chicken ovomucoid, the bond split was between an arginine and an alanine. In still further publications from the same laboratory [566] it has been claimed that they have achieved a resynthesis of the peptide bond in the soybean trypsin inhibitor. In addition, the original peptide bond was split at the arginine and the arginine was removed by carboxypeptidase B. Then not only was the resynthesis achieved, but they inserted a different amino acid in the peptide chain—a lysine. This was done by incubation of the split inhibitor first in the presence of high concentrations of lysine and carboxypeptidase B, and then by incubating this mixture with trypsin. Unfortunately, at the time of this writing, the only information available is in abstract form which states the resynthesized product was obtained in very small amounts. Nevertheless, it would seem highly probable that some resynthesis of the peptide bond has been achieved.

Parts of this work of Laskowski's laboratory has apparently been confirmed in two other laboratories, that of Greene [567] and Birk [568].

## MECHANISMS OF ACTION

Both the Kazal pancreatic inhibitor and the soybean trypsin inhibitor-chymotrypsin inhibitor lost their trypsin inhibitory activity when incubated in acid solution with trypsin. On the other hand, other laboratories have not obtained results indicating the splitting of the peptide bond in the complex under conditions in which inhibition occurs (neutral or slightly alkaline pH). In studies with the basic pancreatic inhibitor there was no evidence that the inhibitor was split in the process of complex formation [569]. A series of different experimental approaches to the requirement of the splitting of a peptide bond in the inhibitory mechanism in our laboratory have all indicated that this is not required for inhibition. On the other hand, it was possible to obtain splitting of the peptide bonds in acid solution of a number of different ovomucoids and to subsequently remove newly exposed carboxy terminal amino acids using carboxypeptidase B without any significant loss of inhibitory activity [570]. In addition, it has been possible to take chicken ovomucoid, which had the peptide bond split by trypsin (the $S^*$, which can be inactivated when subsequently incubated with carboxypeptidase B) and treat any newly exposed N-terminal amino acid with trinitrobenzenesulfonic acid to make the TNBS derivative without any loss in inhibitory activities [571]. This showed that resynthesis of the peptide bond could not be responsible for the activity of the trypsin treated protein. There was loss, however, with the Kunitz soybean trypsin inhibitor upon such treatment. The latter loss is best explained on the basis of a prevention of a resynthesis of the bond as required by Laskowski's mechanisms.

### INTERACTIONS WITH INACTIVE ENZYMES

Studies from our laboratories have shown that strong complexes are formed between inhibitors and inactive enzymes [572]. The enzymatically inactive derivative of $\alpha$-chymotrypsin with L-1-tosylamido-2-phenylethyl chloromethyl ketone (TPCK-chymotrypsin) form a complex with turkey ovomucoid. Also, the enzymatically inactive derivative of 1-chloro-3-tosylamido-7-amino-2-heptanone with trypsin (TLCK-trypsin) forms a complex with chicken ovomucoid. These inactive enzymes competed with the active ones in the formation of complexes and these competitions could be demonstrated enzymatically and physically. In addition, the anhydro-derivative of chymotrypsin (produced by elimination of a tosyl group from tosyl chymotrypsin) was completely devoid of any enzymatic activity and yet formed a good complex with turkey ovomucoid. This work was actually stimulated by studies of Foster and Ryan from the Washington State University [573] who showed that the potato chymotrypsin inhibitor formed an enzymatically competitive complex with anhydrochymotrypsin. Figure 8-18 shows the interaction of different amounts of anhydrochymotrypsin with turkey ovomucoid. In this procedure a competitive enzymatic assay was employed in which different amounts of anhydrochymotrypsin were mixed with turkey ovomucoid and incubated for 37° for 10 minutes. Then $\alpha$-chymotrypsin was added to this mixture and after 3 minutes of incubation the enzymatic activity of the mixture was determined. Thus, when chymotrypsin activity was found, it was an indication that this amount was not bound which was the result of competition of the inactive enzyme for the inhibitor. In the actual case of the material used in Fig. 8-18, it is evident that approximately three times as much of their particular preparation of anhydrochymotrypsin was required as active $\alpha$-chymotrypsin. Part of this requirement for a larger amount of anhydrochymotrypsin is undoubtedly due to the presence of "denatured" anhydrochymotrypsin (i.e., anhydrochymotrypsin further modified or changed in ways other than merely the active site serine). The competitive properties of anhydrochymotrypsin were therefore better than the three to one ratio indicated in Fig. 8-18. Gel electrophoretic experiments also confirmed the formation of complexes with the inactive enzymes.

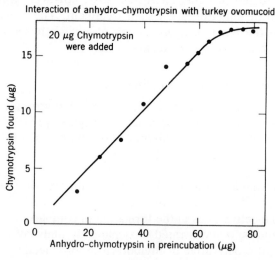

Fig. 8-18. The interaction of different amounts of anhydro-chymotrypsin with turkey ovomucoid. Competitive enzymatic assays (Method B). Various amounts of anhydro-chymotrypsin were preincubated with 20 $\mu$g of turkey ovomucoid at pH 8.2 and 37° for 3 min. Then 20 $\mu$g of native $\alpha$-chymotrypsin were added and the incubation was carried for 3 min under the same conditions. After the substrate-indicator solution was added, the rate of the hydrolysis of substrate was determined spectrophotometrically. Reproduced from [572].

## Comparisons of Theories for Mechanisms of Action

It is generally agreed that the inhibition of proteolytic enzymes by these protein inhibitors requires the formation of a physically demonstrable complex which is strongly associated. There is no general agreement, however, as to what the "driving force" is for this interaction. Relatively early in the consideration of these interacting forces, Nord's laboratory discussed the interaction between chicken ovomucoid and trypsin as follows [557]:

"We would therefore rather consider the inhibiting action of ovomucoid as another instance of competitive substrate inhibition so well known in the case of other enzymes, for example, lipases. The rate of action of an enzyme upon different substrates does not have to parallel their Michaelis-Menten constants, and a substrate, which is acted upon only very slowly but forms a stable enzyme-substrate complex will, therefore, act as an inhibitor."

Since this earlier observation the question of "competitive inhibition" has been argued many times.

The more recent theories from M. J. Laskowski, Jr.'s laboratory on the requirement of proteolysis as part of the mechanism has been discussed before. This is a very popular theory as of the time of writing this book. The observations from Laskowski's laboratory have been concerned with only two inhibitors and no studies have been reported with chymotrypsin inhibitors. There is no question that proteolysis may occur under conditions in which the complex exists. In fact, proteolysis under such conditions was observed with both the Kazal pancreatic inhibitor and chicken ovomucoid and termed "temporary inhibition" as described above. It is, of course, possible that this type of proteolysis may be unrelated to that which is involved in an actual inhibitory mechanism. It must also be realized that there are many different types of inhibitors and that there may be different types of mechanisms. It is probable, however, that a general mechanism involving a particular lysine or arginine for or combination with trypsin may be the requirement for most trypsin inhibitors and that, correspondingly, a particular amino acid residue such as tyrosine or phenylalanine may be required for combination with the binding site of enzymes like chymotrypsin.

Of direct bearing on the problem are recent studies showing that, in complexes of the basic pancreatic trypsin inhibitor (PTI) and trypsin, part of the trypsin can be removed by proteolysis without proteolysis of the inhibitor itself [569]. In this work, 48% of the original molecule of trypsin was still bound to the inhibitor. It was postulated that there was more than one bond between the two partners involved in the complex. In a current article from our laboratory an attempt has been made to discuss these various theories involved in the interaction between the enzyme and inhibitor [560]. Admittedly, the discussion in this article is prejudiced strongly in favor of the observations of this laboratory. Based on the observations cited previously in this chapter, as well as on other observations, the following conclusions were made:

The rate limiting step is most likely involved in conformational changes. The proposed general mechanism of action assumes a specific residue for which the enzyme has high affinity. This residue would be a particular lysine or arginine in the case of trypsin inhibitors and a chymotrypsin susceptible residue such as tyrosine, tryptophan, alanine or methionine in chymotrypsin inhibitors. This residue serves as "the recognition site," or binding site of the inhibitor to a binding site of the enzyme. In addition, other noncovalent bonds or forces strengthen the association, possibly as a result of a conformational change causing a better fitting. If the peptide bond of a particular residue serving as the binding site on the inhibitor can be cleaved only very slowly, or if the equilibrium is very strongly in favor of the intact bond, then the protein will be an inhibitor for the enzyme.

In many respects this latter proposed mechanism is very similar to the one described previously by Nord's laboratory. It differs in that it requires a single particular residue which is a "substrate type" residue and it also evokes the necessity for close fitting and binding of other sites and probably the requirement for conformational changes. It differs from the theories of Laskowski, Jr. in that it does not require the actual proteolytic step. Actually, all of these mechanisms are very similar!

It may be necessary to test many models before a mechanism is finally established. And, indeed, there may be different mechanisms for different inhibitors. Haynes and Feeney [560] attempted to summarize the characteristics required for a possible general mechanism in the following series of reactions:

$$E + I \stackrel{1}{\rightleftharpoons} (EI) \stackrel{2}{\rightleftharpoons} (EI)^+ \stackrel{3}{\rightleftharpoons}$$
$$(EI^*) \stackrel{4}{\rightleftharpoons} E + I^*$$

where E is enzyme; I is inhibitor; (EI) is the initial Michaelis-type complex; $(EI)^+$ is the stable, inhibited complex; $(EI)^*$ is the complex containing the inhibitor with the peptide bond cleaved; and $I^*$ is the free inhibitor with the peptide bond of the active-site residue cleaved. As indicated, step 2, which probably includes a rate-limiting conformational change, may be comprised of more than one step, and step 4 may include more than one step. The equilibrium between $(EI)^+$ and $(EI^*)$ is very strongly in favor of $(EI)^+$. The conversion of $(EI)^+$ to $(EI^*)$ by bond cleavage may be slow or nonexistent, depending on the inhibitor, and, if it does occur, it may or may not affect the inhibition.

## POSSIBLE FUNCTIONS AND ROLES OF PROTEIN INHIBITORS OF PROTEOLYTIC ENZYMES

The most attractive idea for a general role of naturally occurring protein inhibitors is that they control the action of proteolytic enzymes or esterases in the many different tissues and fluids in which both the inhibitors and these enzymes occur. This may indeed be the case, and certainly circumstantial evidence indicates that they may even be of critical importance in controlling such processes. In many instances, however, not only is there no proof for such a mechanism but they may even have some other function.

### CONTROL OF ACTIVATION OF ZYMOGENES OR PRECURSORS TO OTHER BIOLOGICALLY ACTIVE SUBSTANCES

One of the most obvious places where one of these inhibitors might have a function is in the pancreas where it could control activation of the zymogens, trypsinogen and chymotrypsinogen by trypsin. Both the zymogens and the inhibitors are present in the pancreas, and the inhibitors may serve as a control to prevent the rapid activation of the zymogens if a small amount of trypsin should be formed. Once in the intestine, activation of the trypsinogen to trypsin by the enterokinase would then allow the rapid activation of both trypsinogen and chymotrypsinogen.

Although there is no evidence for the role of pancreatic inhibitors in controlling the activation of the recently discovered trypsin-like enzyme responsible for conversion of proinsulin to insulin [574], these inhibitors might possibly play a role in controlling this conversion. The present evidence indicates that the enzyme which does the conversion is probably not trypsin itself but a closely related enzyme or combination of enzymes. Although this particular case may not prove to be one in which the inhibitors have a role, it seems likely that many examples will be found in which zymogens or other precursors are converted by proteolytic enzymes into their respective active products by limited proteolysis. Some of these inhibitors may play important roles in control mechanisms in such instances.

## Control of Various Proteolytic Activities

There is a long list of where protein inhibitors might possibly function. One of the most probable places is in the blood clotting mechanism. At least two inhibitors in blood serum have been demonstrated to inhibit certain of the blood clotting enzymes. The blood clotting system is a very delicately balanced system, and an additional control such as an inhibitor of one of the enzymes would be a facile way to prevent undesired clotting in the general circulation. In addition, there might possibly be small amounts of proteolytic enzymes, such as cathepsins, appearing in the blood from various tissues and the control of these enzymes by inhibitors would be advantageous to the animal. The tissues contain a variety of catheptic enzymes and esterases. These apparently are primarily present in the lysozomes where they serve a function in the degradation and catabolism necessary for tissue turnover. Control of these under certain conditions would also be advantageous.

Inhibitors have been shown to have an interesting effect when applied in another biological system. This is the inflammatory process related to the Schwartzman reaction. One of the earlier observations in this area was the demonstration that when an endotoxin (*Escherichis coli* endotoxin) and a trypsin inhibitor (soybean trypsin inhibitor) were injected intradermally, the inflammatory reaction did not occur. Figure 8-19 shows the reaction obtained when the endotoxin was injected alone and when the endotoxin plus inhibitor was injected. The subject is too complex to discuss in detail in this book which is now becoming rather large. It appears that there is a reaction involving a proteolytic enzyme or an esterase in this type of inflammatory process; and that the inhibitor stops the enzymatic action of one of these enzymes and thereby prevents the inflammatory process.

The colostrum inhibitor has been considered to have a possible function in the prevention of proteolysis of milk antibodies in the digestive tract of newborn [233], [247]. In the few mammals that have been studied, there is a much higher amount of this inhibitor found in the colostrum of those species which provide the newborn with antibodies via the colostrum. In other words, in species such as man in which the gamma globulins are provided to the fetus primarily via the placenta prior to birth, the level of colostrum inhibitor in the milk is low. In species such as pigs, in which the principal route for supplying a newborn with antibodies is via the colostrum after birth, the level of inhibitor in colostrum is high. As discussed in Chapter 4, however, our understanding of the transfer of antibody via milk is poorly understood and there has been no direct proof that the colostrum inhibitors actually protect the gamma globulin from proteolytic digestion in the newborn animal. In fact, one report with pigs disproves this hypothesis. When porcine colostrum inhibitor and porcine gamma globulin were fed together to three day old pigs, there was no effect on the absorption of gamma globulin [576]. Nevertheless, this is an attractive role for the colostrum inhibitor and this possible function deserved extensive investigation.

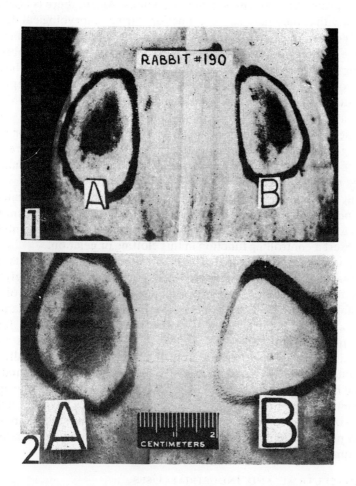

Fig. 8-19. Inhibition of Schwartzman Reaction in rabbit by soybean trypsin inhibitor. 1. Site A intradermally injected with 50 μg endotoxin alone. Site B intradermally injected with 50 μg endotoxin mixed with 50 μg trypsin inhibitor. Equal Shwartzman reactions after provocative endotoxin dose. 2. Sites A and B intradermally injected with 50 μg endotoxin. 24 hrs later (30 min prior to provocative endotoxin injection) site A intradermally injected with saline (1 ml) and site B with trypsin inhibitor (2 mg in 1 ml saline). Reaction in site B inhibited. Reproduced from [575].

## Functions in Other Systems

Unfortunately, there is no known function for the inhibitor in the biological material where an inhibitor exists in the highest amounts, the avian egg whites. None of the inhibitors in egg white have been shown to have a natural function of inhibiting any kind of enzyme! In fact, there are essentially no proteolytic enzymes in the egg whites. Again, there is always the possibility that they may serve a function during embryological development. We may know so little about the role of egg proteins in embryological development that possible functions of this inhibitor are obscured. Another possibility is an antimicrobial or antiviral role. These have not been proven. It may be necessary to invoke the idea that we repeatedly use in discussing the egg-white proteins, that some of the egg-white proteins may be merely "bulk" proteins which are present in egg white in convenient ratios in order to achieve a proper protein milieu for the developing embryo. The requirements of this milieu would be more on the basis of amino acid content and physical properties rather than on a specific biological function.

Small amounts of inhibitors are found in seminal fluids. Perhaps it is easier here to invoke a function. Seminal fluids have a mechanism which causes a gelation or clotting after the seminal fluid is ejaculated. Inhibitors might exert a controlling function in this clotting or gelation.

Plant inhibitors are usually found in the storage organs of plants. There is no evidence that they function to form complexes with proteolytic enzymes or other enzymes. Recently a change in the concentration of one of the potato inhibitors in the leaflets of young growing potato plants was observed during maturation of the plant [531, 532]. Evidence for hormonal physiological control was found (Table 8-8). There was a direct correlation between the presence of an inhibitor protein in normally growing young potato leaves and apical rhizome growth. The results indicated that the blockage of transport of materials to leaves increased the amount of the inhibitor in the leaves, suggesting that the leaves have the ability to synthesize this specific protein. Obviously, many more studies are necessary in plant biochemistry and plant physiology before an understanding of the functions of these proteins in plants will be achieved.

## Pharmaceutical and Industrial Uses

One apparent application of the inhibitors is their use in the food industry. Protein inhibitors might eventually have an application in controlling proteolytic enzymes in the processing of foods. Another possible area is the adjunctive use of inhibitors in connection with the digestion or tenderization of meats by the consumer.

Interest on the part of the pharmaceutical industry in inhibitors of proteolytical enzymes is primarily confined to the study of low molecular weight nonprotein type compounds which can be used to inhibit such enzymes as the blood clotting enzymes by direct injection into animals and humans [491]. Closely related to this is the possible function of the inhibitors present in the body as normal regulators of such systems. The pharmaceutical industry has an interest in the protein inhibitors also. A product containing the basic pancreatic inhibitor has been distributed under the name of Trasylol for clinical testing, particularly for cases of pancreatitis. The pharmaceutical aspects of this have been discussed in detail elsewhere [492], and is not discussed here.

TABLE 8-8  ACCUMULATION OF INHIBITOR I IN LEAFLETS OF YOUNG POTATO PLANTS[a] ON REMOVAL OF VARIOUS GROWTH CENTERS [531]

| Tissue Removed | Number of Leaflets Tested | Number of Positive Leaflets[b] | Percentage of Positive Leaflets |
|---|---|---|---|
| None | 76 | 5 | 7 |
| Rhizomes | 50 | 25 | 50 |
| Apex | 37 | 10 | 27 |
| Apex + rhizomes | 51 | 46 | 90 |
| Apex + lateral buds | 54 | 13 | 24 |
| Apex + rhizomes + lateral buds | 72 | 62 | 86 |

[a] Plants tested 40–60 days after planting seed.
[b] A leaflet was considered positive that gave a visible precipitin line in the double diffusion assay. Lower limit of detection of inhibitor I is 8 $\mu$g/ml.

## MORE QUESTIONS THAN ANSWERS

In the first chapter we began by discussing the intimate relationships of biochemical evolution, genetics, and comparative biochemistry. We are ending with a discussion of the very different properties of probably unrelated proteins which have one property in common, the capacity to inhibit proteolytic enzymes. We have also encountered homology and analogy between the inhibitors and between the enzymes. In one series of homologs which inhibit bovine trypsin or bovine $\alpha$-chymotrypsin, the avian ovomucoids, only one of 10 tested appears to inhibit human trypsin and only one of 15 tested is a stronger inhibitor of a bacterial proteinase than it is of the bovine enzymes. We have stated that the capacity to inhibit may be due to the presence of a fortuitous structure which binds tightly to the enzyme. Could this property be important to the species for a totally different function? The question should also be asked in a broad sense about a well-defined protein such as a cytochrome which functions in subtle coordination with other enzymes. How do some structural changes, seemingly unimportant in the test tube, affect function in the cell? The ending of this chapter and this book with more questions than answers points to the long and interesting road ahead for evolutionary biochemistry!

# References

## Chapter 1

[1] Pauling, L., H. A. Itano, S. J. Singer, and I. C. Wells, "Sickle cell anemia, a molecular disease," *Science,* **110,** 543 (1949).
[2] Hunt, J. A., and V. M. Ingram, "A terminal peptide sequence of human hemoglobin," *Nature,* **184,** 640 (1959).
[3] Zuckerkandl, E., and L. Pauling, in *Horizons in Biochemistry,* (M. Kasha and B. Pullman, Eds.), p. 189, Academic Press, New York, 1962.
[4] Bryson, V., and H. J. Vogel, Eds., *Evolving Genes and Proteins,* Academic Press, New York, 1965.
[5] Florkin, M., *A Molecular Approach to Phylogeny,* Elsevier, New York, 1966.
[6] Watson, J. D., *Molecular Biology of the Gene,* W. A. Benjamin, New York, 1965.
[7] Woese, C. R., *The Genetic Code, The Molecular Basis for Genetic Expression,* Harper & Row, New York, 1967.
[8] Dayhoff, M. O., and R. V. Eck, *Atlas of Protein Sequence and Structure 1967-68,* Natl. Biomed. Res. Found., McGregor & Werner, 1968.
[9] Anfinsen, C. B., *The Molecular Basis of Evolution,* Wiley, New York, 1959.
[10] Helinski, D. R., and C. Yanofsky, "Genetic Control of Protein Structure," in *The Proteins,* (Hans Neurath, Ed.) Vol. IV, p. 1–93, Academic Press, New York, 1966.
[11] Stebbins, G. L., *Processes of Organic Evolution,* Prentice-Hall, New Jersey, 1966.
[12] Gottschalk, A., and E. R. B. Graham, "The Basic Structure of Glycoproteins," in *The Proteins,* (Neurath, H. Ed.) Vol. IV, p. 95–151, Academic Press, New York, 1966.
[13] Partridge, S. M., "Elastin," *Advan. Protein Chem.* **17,** 227 (1962).
[14] Anfinsen, C. B., and E. Haber, "Studies on the reduction and re-formation of protein disulfide bonds," *J. Biol. Chem.,* **236,** 1361 (1961).
[15] Seifter, S., and P. M. Gallop, "The Structure Proteins," in *The Proteins* (Hans Neurath, Ed.), Vol. IV, p. 153–458, Academic Press, New York, 1966.
[16] Stevens, F. G., A. N. Glazer, and E. L. Smith, "The amino acid sequence of wheat germ cytochrome c," *J. Biol. Chem.,* **242,** 2764 (1967).
[17] Margoliash, E., and A. Schejter, "Cytochrome c," *Advan Protein Chem.,* **21,** 113 (1967).
[18] Sibley, C. G., *Discovery 3,* "Proteins: History Books of Evolution," 5–20 (1967).
[19] Margoliash, E., and E. L. Smith, in "Structural and Functional aspects of cytochrome c in relation to evolution," *Evolving genes and proteins,* (Bryson, V., and Vogel, H. J., Eds.), p. 221–243, Academic Press, New York, 1965.
[20] Neurath, H., K. A. Walsh, and W. P. Winter, "Evolution of structure and function of proteases," *Science,* **158,** 1638 (1967).
[21] Greene, F. C., and R. E. Feeney. "Physical evidence for transferrins as single polypeptide chains," *Biochemistry,* **7,** 1366 (1968).
[22] Markert, C. L., and F. Moller, "Multiple forms of enzymes: Tissue, ontogenetic and species specific patterns," *Proc. Nat'l. Acad. Sci. U.S.,* **45,** 753 (1959).

# REFERENCES

[23] Kaplan, N. O., "Symposium on multiple forms of enzymes and control mechanisms. Multiple forms of enzymes," *Bact. Revs.*, **27**, 155 (1963).
[24] Brookhaven Symposia in Biology, No. 17, Subunit structure of proteins. Biochemical and genetic aspects (1964).
[25] Shaw, C. R., "Electrophoretic variation in enzymes," *Science*, **149**, 936 (1965).
[26] Longsworth, L. G., R. K. Cannan, and D. A. MacInnes, "An electrophoretic study of the proteins of egg white," *J. Am. Chem. Soc.*, **62**, 2580 (1940).
[27] Perlmann, G. E., in *Phosphorus Metabolism*, Symposium, (W. D. McElroy and B. Glass, Eds.), Vol. 2, p. 167, Johns Hopkins, Baltimore, 1951.
[28] Colvin, J. R., D. B. Smith, and W. H. Cook, "The Microheterogeneity of proteins," *Chem. Revs.*, **54**, 687 (1954).
[29] Lumry, R., and H. Eyring, "Conformation changes of proteins," *J. Phys. Chem.*, **58**, 110 (1954).
[30] Aoki, K., and J. F. Foster, "Electrophoretic demonstration of the isomerization of bovine plasma albumin at low pH," *J. Am. Chem. Soc.*, **78**, 3538 (1956).
[31] Sogami, M., and J. F. Foster, "Microheterogeneity as the explanation for resolution of $N$ and $F$ forms of plasma albumin in electrophoresis and other experiments," *J. Biol. Chem.*, **238**, PC 2245 (1963).
[32] Stadtman, E. R., "Symposium on multiple forms of enzymes and control mechanisms. II. Enzyme multiplicity and function in the regulation of divergent metabolic pathways," *Bact. Revs.*, **27**, 170 (1963).
[33] Atkinson, D. E., "Regulation of Enzyme Activity," *Ann. Rev. of Biochem.*, **35**, 85 (1966).
[34] Hathaway, G., and R. S. Criddle, "Substrate-dependent association of lactic dehydrogenase subunits to active tetramer," *Proc. Natl. Acad. Sci.*, **56**, 680 (1966).
[35] Williams, J., "A comparison of conalbumin and transferrin in the domestic fowl," *Biochem. J.*, **83**, 355 (1962).
[36] Vaughan, M., and D. Steinberg, "The specificity of protein biosynthesis," *Adv. Protein Chem.*, **14**, 115 (1959).
[37] Itano, H. A., and S. J. Singer, "On dissociation and recombination of human adult hemoglobins A, S, and C," *Proc. Natl. Acad. Sci. U.S.*, **44**, 522 (1958).
[38] Itano, H. A., and E. Robinson, "Formation of normal and doubly abnormal hemoglobins by recombination of hemoglobin $I$ with $S$ and $C$," *Nature*, **183**, 1799 (1959).
[39] Kaplan, N. O., "Evolution of Dehydrogenases," in *Evolving Genes and Proteins* (Bryson, V., and H. J. Vogel, Eds.), p. 243–297, Academic Press, New York, 1965.
[40] Feeney, R. E., and S. K. Komatsu, "The Transferrins," *Structure and Bonding*, (C. K. Jorgensen, J. B. Neilands, R. S. Nyholm, D. Reinen, and R. J. P. Williams, Eds.) Vol. 1, p. 149–206, Springer Verlag, Berlin, 1967.
[41] Feeney, R. E., H. T. Miller, and S. K. Komatsu, "Properties of proteins from cold-adapted antarctic fish," Seventh International Biochemical Congress, Tokyo, Japan, Abst. #J-185, 1967.
[42] Somero, G. N., and A. L. DeVries, "Temperature tolerance of some Antarctic fishes," *Science*, **156**, 257 (1967).
[43] Osuga, D. T., and R. E. Feeney, "Electrophoretic studies of interactions between proteolytic enzymes and inhibitors," *Arch. Biochem. Biophys.*, **118**, 340 (1967).

## CHAPTER 2

[44] Eichholz, A., "The hydrolysis of proteins." *J. Physiol.*, **23**, 163 (1898).
[45] Osborne, T. B., and G. F. Campbell, "The protein constituents of egg white," *J. Am. Chem. Soc.*, **22**, 422 (1900).

[46]  Hektoen, L., and A. G. Cole, "The proteins of egg white. The proteins in egg white and their relationship to the blood proteins of the domestic fowl as determined by the precipitin reaction," *J. Infect. Diseases,* **42,** 1 (1928).
[47]  Sorensen, M., "The carbohydrate content of the proteins in the white of hens' eggs," *Compt. rend. trav. lab. Carlsberg,* **20,** No. 3 (1934).
[48]  Alderton, G., and H. L. Fevold, "Direct crystallization of lysozyme from egg white and some crystalline salts of lysozyme," *J. Biol. Chem.,* **164,** 1 (1946).
[49]  Fleming (Sir Alexander), "On a remarkable bacteriolytic element found in tissues and secretions," *Proc. Roy. Soc.* (London), **93B,** 306 (1922).
[50]  Boas, M. A., "LVIII. An observation on the value of egg white as the sole source of nitrogen for young growing rats." *Biochem. J.,* **18,** 422 (1924).
[51]  Parsons, H. T., and E. Kelly, "The character of the dermatitis-producing factor in dietary egg white as shown by certain chemical treatments," *J. Biol. Chem.,* **100,** 645 (1933).
[52]  Eakin, R. E., E. E. Snell, and R. J. Williams, "The concentration and assay of avidin, the injury-producing protein in raw egg white," *J. Biol. Chem,* **140,** 535 (1941).
[53]  Lineweaver, H., and C. W. Murray, "Identification of the trypsin inhibitor of egg white with ovomucoid," *J. Biol. Chem.,* **171,** 565 (1947).
[54]  Schade, A. L., and L. Caroline, "Raw Hen Egg white and the role of iron inhibition of *Shigella dysenteriae, Staphylococcus aureus, Escherichia coli* and *Saccharomyces cerevisiae,*" *Science,* **100,** 14 (1944).
[55]  Alderton, G., W. H. Ward, and H. L. Fevold, "Identification of the bacteria-inhibiting iron-binding protein of egg white as conalbumin," *Arch. Biochem. Biophys.,* **11,** 9 (1946).
[56]  Matsushima, K., "An undescribed trypsin inhibitor in egg white," *Science,* **127,** 1178 (1958).
[57]  Rhodes, M. B., N. S. Bennett, and R. E. Feeney, "The flavoprotein-apoprotein system of egg white," *J. Biol. Chem.,* **234,** 2054 (1959).
[58]  Lush, I. E., "Genetic polymorphisms in the egg albumin proteins of the domestic fowl," *Nature,* **189,** 981 (1961).
[59]  Feeney, R. E., H. Abplanalp, J. J. Clary, D. L. Edwards, and J. R. Clark, "A genetically varying minor protein constituent of chicken egg white," *J. Biol. Chem.,* **238,** 1732 (1963).
[60]  Miller, H. T., and R. E. Feeney, "Immunochemical relationships of proteins of avian egg whites," *Arch. Biochem. Biophys.,* **108,** 117 (1964).
[61]  Miller, H. T., and R. E. Feeney, "The physical and chemical properties of an immunologically cross-reacting protein from avian egg whites," *Biochemistry,* **5,** 952 (1966).
[62]  Neumann, H., and M. Sela, "Proteolytic activity in egg white," *Bull. Res. Council Israel,* **9A,** 103 (1960).
[63]  Lush, I. E., and J. Conchie, "Glycosidases in the egg albumen of the hen, the turkey, and the Japanese quail," *Biochim. Biophys. Acta,* **130,** 81 (1966).
[64]  Fossum, K., and J. R. Whitaker, "Ficin and papain inhibitor from chicken egg white," *Arch. Biochem. Biophys.,* **125,** 367 (1968).
[65]  Ketterer, B., "Ovoglycoprotein, a protein of hen's egg white," *Biochem. J.,* **96,** 372 (1965).
[66]  Lineweaver, H., H. J. Morris, L. Kline, and R. S. Bean, "Enzymes of fresh hen eggs," *Arch. Biochem.,* **16,** 443 (1948).
[67]  Almquist, H. J., and F. W. Lorenz, "Firm white of fresh and storage eggs," *Poultry Sci.,* **14,** 340 (1935).
[68]  Sugihara, T. F., L. R. MacDonnell, C. A. Knight, and R. E. Feeney, "Virus antihemagglutinin activities of avian egg components," *Biochim. Biophys. Acta,* **16,** 404 (1955).

[69] Feeney, R. E., M. B. Rhodes, and J. S. Anderson, "The distribution and role of sialic acid in chicken egg white," *J. Biol. Chem.*, **235**, 2633 (1960b).

[70] Conrad, R. M., and R. E. Phillips, "The formation of the chalazae and inner thin white in the hen's egg," *Poultry Sci.*, **17**, 143 (1938).

[71] MacDonnell, L. R., R. B. Silva, and R. E. Feeney, "The sulfhydryl groups of ovalbumin," *Arch. Biochem. Biophys.*, **32**, 288 (1951).

[72] Rhodes, M. B., P. R. Azari, and R. E. Feeney, "Analysis, fractionation, and purification of egg-white proteins with cellulose-cation exchanger," *J. Biol. Chem.*, **230**, 399 (1958).

[73] Taylor, G. L., G. S. Adair, and M. E. Adair, "Estimation of proteins by precipitation reaction," *J. Hyg.*, **32**, 340 (1932).

[74] McKenzie, H. A., M. B. Smith, and R. G. Wake, "The denaturation of proteins. I. Sedimentation, diffusion, optical rotation, viscosity and gelation in urea solutions of ovalbumin and bovine serum albumin," *Biochim. Biophys. Acta*, **69**, 222 (1963).

[75] Gagen, W. L., "The significance of the partial specific volume obtained from sedimentation data," *Biochemistry*, **5**, 2553 (1966).

[76] Foster, J. F., J. T. Yang, "On the mode of interaction of surface active cations with ovalbumin and bovine plasma albumin," *J. Am. Chem. Soc.*, **76**, 1015 (1954).

[77] Feeney, R. E., J. S. Anderson, P. R. Azari, N. Bennett, and M. B. Rhodes, "The comparative biochemistry of avian egg-white protein," *J. Biol. Chem.*, **235**, 2307 (1960a).

[78] Clark, J. R., D. T. Osuga, and R. E. Feeney, "Comparison of avian egg-white conalbumins," *J. Biol. Chem.*, **238**, 3621 (1963).

[79] Warner, R. C., and I. Weber, "Preparation of crystalline conalbumin," *J. Biol. Chem.*, **191**, 173 (1951).

[80] Fuller, R. A., and D. R. Briggs, "Some physical properties of hen's egg conalbumin," *J. Am. Chem. Soc.*, **78**, 5253 (1956).

[81] Rhodes, M. B., N. Bennett, and R. E. Feeney, "The trypsin and chymotrypsin inhibitors from avian egg whites," *J. Biol. Chem.*, **235**, 1686 (1960).

[82] Osuga, D. T., and R. E. Feeney, "Biochemistry of the egg-white protein of the ratite group," *Arch. Biochem. Biophys.*, **124**, 560 (1968).

[83] Deutsch, H. F., and J. I. Morton, "Physical-chemical studies of some modified ovomucoids," *Arch. Biochem. Biophys.*, **93**, 654 (1961).

[84] Edelhoch, H., and R. F. Steiner, "Change in physical properties accompanying the interaction of trypsin with protein inhibitors," *J. Biol. Chem.*, **240**, 2877 (1965).

[85] Alderton, G., W. H. Ward, and H. L. Fevold, "Isolation of lysozyme from egg white," *J. Biol. Chem*, **157**, 43 (1945).

[86] Canfield, R. E., "The amino acid sequence of egg white lysozyme," *J. Biol. Chem.*, **238**, 2698 (1963).

[87] Sophianopoulos, A. J., C. K. Rhodes, D. N. Holcomb, and K. E. Van Holde, "Physical studies of lysozyme. I. Characterization," *J. Biol. Chem.*, **237**, 1107 (1962).

[88] Tomimatsu, Y., J. J. Clary, and J. J. Bartulovich, "Physical characterization of ovoinhibitor, a trypsin and chymotrypsin inhibitor from chicken egg white," *Arch. Biochem. Biophys.*, **115**, 536 (1966).

[89] Fraenkel-Conrat, H., N. S. Snell, and E. D. Ducay, "Avidin 1. Isolation and characterization of the protein and nucleic acid," *Arch. Biochem. Biophys.*, **39**, 80 (1952a).

[90] Green, N. M., "The molecular weight of avidin," *Biochem. J.*, **92**, 16c (1964).

[91] Melamed, M. D., and N. M. Green, "Avidin 2. Purification and composition," *Biochem. J.*, **89**, 591 (1963).

# REFERENCES

[92] Evans, R. J., and S. L. Bandemer, "Separation of egg-white proteins by paper electrophoresis," *J. Agr. Food. Chem.*, 4, 802 (1956).

[93] Feeney, R. E., F. C. Stevens, and D. T. Osuga, "The specificities of chicken ovomucoid and ovoinhibitor," *J. Biol. Chem.*, 238, 1415 (1963a).

[94] Mandeles, S., "Use of DEAE-cellulose in the separation of proteins from egg white and other biological materials," *J. Chromatog.*, 3, 256 (1960).

[95] Stevens, F. C., and R. E. Feeney, "Chemical modification of avian ovomucoids," *Biochemistry*, 2, 1346 (1963).

[96] Fernandez-Diez, M. J., D. T. Osuga, and R. E. Feeney, "The sulfhydryls of avian ovalbumins, bovine $\beta$-lactoglobulin, and bovine serum albumin," *Arch. Biochem. Biophys.*, 107, 449 (1964).

[97] Komatsu, S. K. and R. E. Feeney, "Role of tyrosyl groups in metal-binding properties of transferrin," *Biochemistry*, 6, 1136 (1967).

[98] Simlot, M. M., F. C. Stevens and R. E. Feeney, "Reduction and reoxidation of ovomucoids," *Biochim. Biophys. Acta*, 130, 549 (1966).

[99] Feeney, R. E., D. T. Osuga, and H. Maeda, "Heterogeneity of avian ovomucoids," *Arch. Biochem. Biophys.*, 119, 124 (1967a).

[100] Forsythe, R. H., and J. F. Foster, "Egg white proteins. I. Electrophoretic studies on whole white," *J. Biol. Chem.*, 184, 377 (1950).

[101] Forsythe, R. H., and J. F. Foster, "Egg white proteins. II. An ethanol fractionation scheme," *J. Biol. Chem.*, 184, 385 (1950).

[101a] Habeeb, A. F. S. A., "The preparation and properties of guanidinated ovalbumin," *Can. J. Biochem. Physiol*, 39, 729 (1961).

[102] Lewis, J. C., N. S. Snell, D. J. Hirschmann, and H. Fraenkel-Conrat, "Amino acid composition of egg proteins," *J. Biol. Chem.*, 186, 23 (1950).

[103] Tristram, G. R., and R. H. Smith, "The amino acid composition of some purified protein," *Adv. Protein Chem.*, 18, 227 (1963).

[104] Donovan, J. W. and L. M. White, (manuscript in preparation, 1968).

[105] Clagett, C. O. "Amino acid composition of flavoprotein," (personal communication, 1967).

[106] Fevold, H. L. "Egg proteins," *Adv. Protein Chem.*, 6, 187 (1951).

[107] Warner, R. C., "Egg proteins," in *The Proteins* (Hans Neurath and K. Bailey, Eds.), Vol. II, p. 435, Academic Press, New York, 1954.

[108] Perlmann, G. E., "The electrophoretic properties of plakalbumin," *J. Am. Chem. Soc.*, 71, 1146 (1949).

[109] Boyer, P. D., "Spectrophotometric study of the reaction of protein sulfhydryl groups with organic mercurial," *J. Am. Chem. Soc.*, 76, 4331 (1954).

[110] Winzor, D. J. and J. M. Creeth, "Physiochemical studies on ovalbumin. 3. The sulfhydryl and disulphide contents of ovalbumin and an iodine-modified derivative," *Biochem. J.*, 83, 559 (1962).

[111] Smith, M. B., "Studies on ovalbumin I. Denaturation by heat, and the heterogeneity of ovalbumin," *Aust. J. Biol. Sci.*, 17, 261 (1964).

[112] Smith, M. B., and J. F. Back, "Studies on ovalbumin II. The formation and properties of S-ovalbumin, a more stable form of ovalbumin," *Aust. J. Biol. Sci.*, 18, 365 (1965).

[113] Marshall, R. D., and A. Neuberger, "Carbohydrates in protein VIII. The isolation of 2-acetamido-1-(L-B-aspartamido)-1,2-dideoxy-B-D-glucose from hen's egg albumin," *Biochemistry*, 3, 1596 (1964).

[114] Montgomery, R., Y.-C. Lee, and Y.-C. Wu, "Glycopeptides from ovalbumin. Prepa-

ration, properties, and partial hydrolysis of the asparaginyl carbohydrate," *Biochemistry*, **4**, 566 (1965).

[115] Cunningham, L., J. D. Ford, and J. M. Rainey, "Heterogeneity of β-aspartyl-oligosaccharides derived from ovalbumin," *Biochim. Biophys. Acta*, **101**, 233 (1965).

[116] György, P., and C. S. Rose, "The liberation of biotin from the avidin-biotin complex (AB)," *Proc. Soc. Exptl. Biol. Med.*, **53**, 55 (1943).

[117] Hertz, R., R. M. Fraps, and W. H. Sebrell, "Induction of avidin formation in the avian oviduct by stilbestrol plus progesterone," *Proc. Soc. Exptl. Biol. Med.*, **52**, 142 (1943).

[118] Vigneaud, V. du, K. Dittmer, K. Hofmann, and D. B. Melville, "Yeast-growth-promoting effect of diaminocarboxylic acid derived from biotin," *Proc. Soc. Exptl. Biol. Med.*, **50**, 374 (1942).

[119] Winnick, T., K. Hofmann, F. J. Pilgrim, and A. E. Axelrod, "The microbiological activity of *dl*-oxybiotin and related compounds," *J. Biol. Chem.*, **161**, 405 (1945).

[120] Wright, L. D., H. R. Skeggs, and E. L. Cresson, "Affinity of avidin for certain analogs of biotin," *Proc. Soc. Exptl. Biol. Med.*, **64** 150 (1947).

[121] Green, N. M., "Spectroscopic evidence for the participation of tryptophan residues in the binding of biotin by avidin," *Biochim. Biophys. Acta*, **59**, 244 (1962).

[122] György, P., C. S. Rose, and R. Tomarelli, "Investigations on the stability of avidin," *J. Biol. Chem.*, **144**, 169 (1942).

[123] Green, N. M., "Avidin 5. Quenching of fluorescence by dinitrophenyl groups," *Biochem. J.*, **90**, 564 (1964).

[124] Green, N. M., "Avidin 1. The use of $^{14}$C biotin for kinetic studies and for assay," *Biochem. J.*, **89**, 585 (1963).

[125] Korenman, S. G. and B. O'Malley, "Avidin assay: A new procedure for tissue fractions," *Biochem. Biophys. Acta*, **140**, 174 (1967).

[126] Wei, R. D., L. D. Wright, "Determination of avidin by use of radioactive biotin and Sephadex chromatography," *Proc. Soc. Exp. Biol. Med.*, **117**, 17 (1964).

[127] Green, N. M., "Avidin 4. Stability at extremes of pH and dissociation into sub-units by guanidine hydrochloride," *Biochem. J.*, **89**, 609 (1963).

[128] Eakin, R. E., E. E. Snell, and R. J. Williams, "A constituent of raw egg white capable of inactivating biotin *in vitro*," *J. Biol. Chem.*, **136**, 801 (1940).

[129] Wei, R., and L. D. Wright, "Heat stability of avidin and avidin-biotin complex and influence of ionic strength on affinity of avidin for biotin," *Proc. Soc. Exptl. Biol. Med.*, **117**, 341 (1964).

[130] McCormick, D. B., "Specific purification of avidin by column chromatography on biotin cellulose," *Anal. Biochem.*, **13**, 194 (1965).

[131] Norris, L. C., and J. C. Bauernfeind, "Effect of level of dietary riboflavin upon quantity stored in eggs and rate of storage," *Food Research*, **5**, 521 (1940).

[132] Winter, W. P., E. G. Buss, C. O. Clagett, and R. V. Boucher, "The nature of the biochemical lesion in avian renal riboflavinuria, II. The inherited change of a riboflavin-binding protein from blood and eggs," *Comp. Biochem. Physiol.*, **22**, 897 (1967).

[133] Deutsch, H. F., "Immunochemical analysis of egg white," *Fed. Proc.*, **12**, 729 (1953).

[134] Feeney, R. E., D. T. Osuga, S. B. Lind, and H. T. Miller, "The egg-white proteins of the Adelie penguin," *Comp. Biochem. Physiol.*, **18**, 121 (1966).

[135] Montreuil, J., B. Castiglioni, A. Adam-Chosson, F. Caner. and J. Queval, "Etudes sur les glycoprotéides VIII. L'hétérogénéité de l'ovomucoïde,"*J. Biochem.*, **57**, 514 (1965).

## REFERENCES

[136] Kanamori, M., and M. Kawabata, *J. Agr. Chem. Soc. Japan,* **38,** 367 (1964).
[137] Donovan, J. W., J. G. Davis, and L. U. Park, "Sugar nucleotides of chicken egg white," *Arch. Biochem. Biophys.,* **122,** 17 (1967).
[138] Wilcox, F. H., Jr., and R. K. Cole, "The inheritance of differences in lysozyme level in hen's egg white," *Genetics,* **42,** 264 (1957).
[139] Sibley, C. G., and P. A. Johnsgard, "An electrophoretic study of egg-white proteins in 23 breeds of the domestic fowl," *Amer. Naturalist,* **93,** 107 (1959).
[140] Baker, C. M. A., and C. Manwell, "Molecular genetics of avian proteins. I. Egg-white proteins of the domestic fowl," *British Poultry Science,* **3,** No. 3, 161 (1962).
[141] Lush, I. E., "Egg albumen polymorphisms in the fowl: The ovalbumen locus," *Genet. Res.,* **5,** 257 (1964).
[142] Lush, I. E., "The Biochemical Genetics of Vertebrates Except Man," in *Frontiers of Biology,* (Neuberger, A. and E. L. Tatum, Ed.) Vol. III, p. 118, North-Holland, Amsterdam, 1966.
[143] Kaminski, M., and Durieux, "Etude comparative des serums de poule, de coq, de poussin, d'embryon et du blanc d'oeuf," *J. Exp. Cell. Res.,* **10,** 590 (1956).
[144] Kaminski, M., "Immunochemical and electrophoretic studies of egg-white globulins. Demonstration and properties. I.," *Annales de L'Institut Pasteur,* **92,** 802 (1957).
[145] Hellhammer, D., and O. Hogl, "Paper electrophoresis of egg proteins, use of electrophoresis in the investigations of nutrients," *Mitt. Gebiete Lebensm. U. Hyg.,* **49,** 79 (1958); Chem. Abst. 53:1, 494B, 1959.
[146] Cochrane, D., and E. Annau, "Electrophoretic differences in the egg-white proteins of the hen," *Canad. J. Biochem. Physiol.,* **40,** 1335 (1962).
[147] Flavin, M., and C. B. Anfinsen, "The isolation and characterization of cysteic acid peptides in studies on ovalbumin synthesis," *J. Biol. Chem.,* **211,** 375 (1954).
[148] Canfield, R. E., and C. B. Anfinsen, "Nonuniform labeling of egg-white lysozyme," *Biochemistry,* **2,** 1073 (1963).
[149] Mandeles, S., and E. D. Ducay, "The site of egg-white protein formation," *J. Biol. Chem.,* **237,** 3196 (1962).
[150] Hendler, R. W., "Uptake of nucleic acid, protein, and carbohydrate precursors by hen oviduct lipids," *Biochim. Biophys. Acta,* **106,** 184 (1965).
[151] Carey, N. H., "Amino acid incorporation by cell fractions from the oviduct of the laying hen and the synthesis of egg-white proteins," *Biochem. J.,* **100,** 242 (1966).
[152] O'Malley, B. W., "*In vitro* hormonal induction of a specific protein (avidin) in chick oviduct," *Biochemistry,* **6,** 2546 (1967).
[153] Grau, C. R., N. E. Walker, H. I. Frita, and S. M. Peters, "Successful development of chick embryos nourished by yolk-sac perfusion with calcium-los media," *Nature,* **197,** 257 (1963).
[154] Garibaldi, J. A., and H. G. Bayne, "Iron and the bacterial spoilage of shell eggs," *J. Food. Sci.,* **27,** 57 (1962).
[155] MacDonnell, L. R., R. E. Feeney, H. L. Hanson, A. Campbell, and T. F. Sugihara, "The functional properties of the egg-white proteins," *Food Technol,* **9,** 49 (1955).
[156] Rhodes, M. B., J. L. Adams, N. Bennett, and R. E. Feeney, "Properties and food uses of duck eggs," *Poultry Sci.,* **39,** 1473 (1960).
[157] Nakamura, R., "Studies on the foaming property of the chicken egg white," *Agr. Biol. Chem.,* **28,** 403 (1964).
[158] Meehan, J. J., T. F. Sugihara, and L. Kline, "Relationships between shell egg handling factors and egg product properties," *Poultry Sci,* **41,** 892 (1962).
[159] Hoover, S. R., "A physical and chemical study of ovomucin," Ph.D. Thesis, Georgetown Univ., Washington, D.C., 1940.

[160] MacDonnell, L. R., H. Lineweaver, and R. E. Feeney, "Chemistry of shell egg deterioration: Effect of reducing agents," *Poultry Sci.,* **30,** 856 (1951).
[161] Feeney, R. E., J. M. Weaver, J. R. Jones, and M. B. Rhodes, "Study of the kinetics and mechanisms of yolk deterioration in shell eggs," *Poultry Sci.,* **35,** 1061 (1956).
[162] Feeney, R. E., R. B. Silva, and L. R. MacDonnell, "The Chemistry of shell egg deterioration: The deterioration of separated components," *Poultry Sci.,* **30,** 645 (1951).
[163] Feeney, R. E., and M. B. Rhodes, (unpublished data), 1960.
[164] Feeney, R. E., J. J. Clary, and J. R. Clark, "A reaction between glucose and egg-white proteins in incubated eggs," *Nature,* **201,** 192 (1964).
[165] Smith, M. B., and J. F. Back, "Modification of ovalbumin in stored eggs detected by heat denaturation," *Nature,* **193,** 878 (1962).
[166] Hawthorne, J. R., "The action of hen egg white lysozyme on ovomucoid and ovomucin," *Biochim. Biophys. Acta,* **6,** 28 (1950).
[167] Cotterill, O. J., and A. R. Winter, "Egg-white lysozyme. 3. The effect of pH on the lysozyme-ovomucin interaction," *Poultry Sci.,* **34,** 679 (1955).
[168] Feeney, R. E., E. D. Ducay, R. B. Silva, and L. R. MacDonnell, "Chemistry of shell egg deteriorations: The egg-white proteins," *Poultry Sci.,* **31,** 639 (1952).
[169] Rhodes, M. B., and R. E. Feeney, "Mechanisms of shell egg deterioration: Comparisons of chicken and duck eggs," *Poultry Sci.,* **36,** 891 (1957).
[170] Feeney, R. E., R. G. Allison, D. T. Osuga, J. C. Bigler, and H. T. Miller, "Biochemistry of the Adelie Penguin: Studies on egg and blood serum proteins," in *Antarctic Research Series,* Vol. 12 (Austin, O. L. Jr., Ed.), Amer. Geophys. Union, 1968.

## CHAPTER 3

[171] Welty, J. C., *The Life of Birds,* Saunders, Philadelphia, 1962.
[172] Lack, D., *Darwin's Finches,* Cambridge University Press, London, 1947.
[173] McCabe, R. A., and H. F. Deutsch, "The relationships of certain birds as indicated by their egg-white proteins," *The Auk,* **69,** 1 (1952).
[174] Sibley, C. G., "The electrophoretic patterns of avian egg-white proteins as taxonomic character," *Ibis,* **102,** 215 (1960).
[175] Romanoff, A. L., and A. J. Romanoff, *The Avian Egg,* Wiley, New York, 1949, p. 61.
[176] Canfield, R. E., and S. McMurry, "Purification and characterization of a lysozyme from goose egg white," *Biochem. Biophys. Res. Commun.,* **26,** 38 (1967).
[177] Jones, P. D., and M. H. Briggs, "The distribution of avidin," *Life Science,* **1,** 621 (1962).
[178] Hertz, R., and W. H. Sebrell, "Occurrence of avidin in the oviduct and secretions of the genital tract of several species," *Science,* **96,** 257 (1942).
[179] Landsteiner, K., L. G. Longsworth, and J. Vanderscheer, "Electrophoresis experiments with egg albumins and hemoglobins," *Science,* **88,** 83 (1938).
[180] Wetter, L. R., M. Cohn, and H. F. Deutsch, "Immunological studies of egg-white proteins V. The cross-reactions of egg-white proteins of various species, *J. Immunol.,* **70,** 507 (1953).
[181] Kaminski, M., "Study of the cross-reactions of hen and duck ovalbumins. Immunochemical relationship between native proteins and precipitating fragments obtained after proteolysis," *Immunology,* **5,** 322 (1962).

[182] Jennings, R. K., and M. A. Kaplan, "Implications of qualitative comparative serology," *Annals of Allergy*, **20**, 15 (1962).
[183] Fothergill, J. E., and W. T. Perrie, "Structural aspects of the immunological cross reaction of hen and duck ovalbumins," *Biochem. J.*, **99**, 58P (1966).
[184] Wilson, A. C., (personal communication, 1968).
[185] Arnheim, N., Jr., and A. C. Wilson, "Quantitative immunological comparison of bird lysozymes," *J. Biol. Chem.*, **242**, 3951 (1967).
[186] Wetter, L. R., M. Cohn, and H. F. Deutsch, "Immunological studies of egg-white proteins. II. Resolution of the quantitative precipitin reaction between chicken egg white and rabbit anti egg-white serum in terms of the reactions of purified egg-white proteins," *J. Immun.*, **69**, 109 (1952).
[187] Bain, J. A., and H. F. Deutsch, "An electrophoretic study of the egg-white proteins of various birds," *J. Biol. Chem.*, **171**, 531 (1947).
[188] Stratil, A., and M. Valenta, "Protein polymorphism of egg white and yolk in geese and ducks," *Folia biologica (Praha)*, **12**, 307 (1966).
[189] Baker, C. M. A., "Molecular genetics of avian proteins-VII. Chemical and genetic polymorphism of conalbumin and transferrin in a number of species," *Comp. Biochem. Physiol.*, **20**, 949 (1967).
[190] Baker, C. M. A., "Molecular genetics of avian proteins-IV. The egg-white proteins of the golden pheasant, Lady Amherst's pheasant, and their possible evolutionary significance," *Comp. Biochem. Physiol.*, **16**, 93 (1965).
[191] Baker, C. M. A., "Molecular genetics of avian proteins-III. The egg proteins of an isolated population of jungle fowl, *Gallus gallus* L.," *Comp. Biochem. Physiol.*, **12**, 389 (1964).
[192] Windle, J. J., A. K. Wiersema, J. R. Clark, and R. E. Feeney, "Investigation of the iron and copper complexes of avian conalbumins and human transferrins by electron paramagnetic resonance," *Biochemistry*, **2**, 1341 (1963).
[193] Bigler, J. C., and R. E. Feeney, "Properties of penguin ovomucoid," (manuscript in preparation, 1969).
[194] Haynes, R., D. T. Osuga, and R. E. Feeney, "Modification of amino groups in inhibitors of proteolytic enzymes," *Biochemistry*, **6**, 541 (1967).
[195] Feinstein, G. and R. E. Feeney, "Chemical modification of the methionine residues in turkey ovomucoid," *Biochim. Biophys. Acta*, **140**, 55, (1967b).
[196] Liu, W. H., G. Feinstein, D. T. Osuga, R. Haynes, and R. E. Feeney, "Modification of arginines in trypsin inhibitors by 1,2-cyclohexanedione," *Biochemistry*, **7**, 2886 (1968).
[196a] Feeney, R. E., G. E. Means, and J. C. Bigler. "Inhibition of human trypsin, plasmin, and thrombin by naturally occurring inhibitors of proteolytic enzymes," *J. Biol. Chem.*, in press (1969).
[197] Miller, H. T., H. Abplanalp, and R. E. Feeney, 1966 (unpublished).
[198] Jolles, P., "Relationship between chemical structure and biological activity of hen egg-white lysozyme and lysozymes of different species," *Proc. Roy. Soc.*, **167b**, 350 (1967).
[199] Jolles, J., G. Spotorno, and P. Jolles, "Lysozymes characterized in duck egg-white: isolation of a histidine-less lysozyme," *Nature (London)*, **208**, 1204 (1965).
[200] Bock, W. J., "The cranial evidence for ratite affinities," *Proc. Intern. Ornith. Congr.*, **13**, 39 (1963).
[201] Wilson, A. C., N. O. Kaplan, L. Levine, A. Pesce, M. Reichlin, and W. S. Allison, "Evolution of lactic dehydrogenases," *Fed. Proc.*, **23**, 1258 (1964).
[202] Simpson, G. G., "Fossil penguins," *Bull. Am. Mus. Nat. Hist.*, **87**, 1 (1946).

## Chapter 4

[203] Mellander, O., "Elektrophoretische Untersuchung von Casein," *Biochem. Z.,* **300**, 240 (1939).
[204] Webb, B. H., and A. H. Johnson, *Fundamentals of Dairy Chemistry,* AVI, Westport, Connecticut, 1965.
[205] McKenzie, H. A., "Milk proteins," *Adv. Protein Chem.,* **22**, 55 (1967).
[206] Thompson, M. P., N. P. Tarassuk, R. Jenness, H. A. Lillevik, U. S. Ashworth, and D. Rose, "Nomenclature of the proteins of cow's milk—Second revision," *J. Dairy Sci.,* **48**, 159 (1965).
[207] Waugh, D. F., and P. H. von Hippel, "$k$-Casein and the stabilization of casein micelles," *J. Am. Chem. Soc.,* **78**, 4576 (1956).
[208] Shahani, K. M., "Milk enzymes: Their role and significance," *J. Dairy Sci.,* **49**, 907, (1966).
[209] Yaguchi, M., and N. P. Tarassuk, "Gel filtration of acid casein and skimmilk on Sephadex," *J. Dairy Sci.,* **50**, 1985 (1967).
[210] Thompson, M. P., C. A. Kiddy, L. Pepper, and C. A. Zittle, "Variations in the $\alpha_s$-casein fraction of individual cow's milk," *Nature,* **195**, 1001 (1962).
[211] Aschaffenburg, R., "Inherited casein variants in cow's milk," *Nature,* **192**, 431 (1961).
[212] Kiddy, C. A., J. O. Johnston, and M. P. Thompson, "Genetic polymorphism in caseins of cow's milk. I. Genetic control of $\alpha_s$-casein variation," *J. Dairy Sci.,* **47**, 147 (1964).
[213] Gordon, W. G., J. J. Basch, and M. P. Thompson, "Genetic polymorphism in caseins of cow's milk. VI. Amino acid compositions of $\alpha_{s1}$-caseins A, B, and C," *J. Dairy Sci.,* **48**, 1010 (1965).
[214] Kalan, E. B., R. Greenberg, and M. P. Thompson, "Analysis of proteolytic digests of genetic variants of $\alpha_s$-casein," *Arch. Biochem. Biophys.,* **115**, 468 (1966).
[215] Aschaffenburg, R., *J. Dairy Res.,* "Inherited casein variants in cow milk. II. Breed differences in the occurrence of $\beta$-casein variants," **30**, 251 (1963).
[216] Mackinlay, A. G., and R. G. Wake, (private communication, 1965) quoted in [205].
[217] deKoning, T. J., P. J. van Rooijen, and A. Kok, "Location of amino acid differences in the genetic variants of $k$-casein A and B," *Biochem. Biophys. Res. Commun.,* **24**, 616 (1966).
[218] Swaisgood, H. E., J. R. Brunner, and H. A. Lillevik, "Physical parameters of k-casein from cow's milk," *Biochemistry,* **3**, 1616 (1964).
[219] Beeby, R., "Studies on the $k$-casein complex. I. The release of sialic acid-containing material by rennin," *J. Dairy Res.,* **30**, 77 (1963).
[220] Hill, R. D., and R. R. Hanson, "The effect of preparative conditions on the composition of the $k$-casein complex," *J. Dairy Res.,* **30**, 375 (1963).
[221] Yaguchi, M., N. P. Tarassuk, and N. Abe, "Distribution of lipase in milk proteins. I. DEAE cellulose column chromatography," *J. Dairy Sci.,* **47**, 1167 (1964).
[222] Patel, C. V., N. P. Tarassuk, and P. Fox, "Bovine milk lipase II. Characterization," *J. Dairy Sci.,* **51**, 1879 (1968).
[223] Brodbeck, U., W. L. Denton, N. Tanahashi, and K. E. Ebner, "The isolation and identification of the B protein of lactose synthetase as $\alpha$-lactalbumin," *J. Biol. Chem.,* **242**, 1391 (1967).
[224] Brew, K., T. C. Vanaman, and R. L. Hill, "Comparison of the amino acid sequence of bovine $\alpha$-lactalbumin and hen's egg-white lysozyme," *J. Biol. Chem.,* **242**, 3747 (1967).

# REFERENCES

[225] Brew, K., T. C. Vanaman, and R. L. Hill, "The role of α-lactalbumin and the A protein in lactose synthetase: a unique mechanism for the control of a biological reaction," *Proc. Natl. Acad. Sci.*, 59, 491 (1968).
[226] Pilson, M. E. Q., and A. L. Kelly, "Composition of the milk from *Zalophus Californianus*, the California Sea Lion," *Science*, 135, 104 (1962).
[227] Spik, G., M. Monsigny, and J. Montreuil, "Investigation of glycoprotein. Detection of a linkage between aspartic acid and the mucopolyosidic groups in human transferrin," *Compt. Rend. Acad. Sci. Paris*, 260, 4282 (1965).
[228] Blanc, B., E. Bujard, and J. Mauron, "The amino acid composition of human and bovine lactotransferrins," *Experientia*, 19, 299 (1963).
[229] Gordon, W. G., M. L. Groves, and J. J. Basch, "Bovine milk 'red protein': amino acid composition and comparison with blood transferrin," *Biochemistry*, 2, 817 (1963).
[230] Baker, E., D. C. Shaw, and E. H. Morgan, "Isolation and characterization of rabbit serum and milk transferrins. Evidence for difference in sialic acid content only," *Biochemistry*, 7, 1372 (1968).
[231] Ceppellini, R., S. Dray, G. Edelman, J. Fahey, F. Franek, E. Franklin, H. C. Goodman, P. Grabar, A. E. Gurvich, J. F. Heremans, F. Karush, E. Press, H. Isliker, and Z. Trnka, "Nomenclature for human immunoglobins," *Bull. W.H.O.*, 30, 447 (1964).
[232] Karlsson, B. W., "Immunoelectrophoretic studies in relationships between proteins of porcine colostrum, milk, and blood serum," *Acta Path. Microbiol. Scand.*, 67, 83 (1966).
[233] Laskowski, M., Jr. and M. Laskowski, "Bovine colostrum trypsin inhibitor," *J. Biol. Chem.*, 190, 563 (1951).
[234] McMeekin, T. L., "Milk Proteins," in *The Proteins*, (H. Neurath and K. Bailey, Eds.), Vol. IIA, p. 389, Academic Press, New York, 1954.
[235] Barry, J. M., "Protein Metabolism," in *Milk: the mammary gland and its secretions*, (Kon, S. K., and A. T. Cowie, Eds.) Academic Press, New York, 1961.
[236] von Muralt, G., E. Gugler, and D. L. A. Roulet, "Immuno-electrophoretic studies of the proteins of human milk and colostrum," in *Immunoelectrophoretic Analysis* (P. Grabar and P. Burtin, Eds.), p. 261, Elsevier, Amsterdam, 1964.
[237] Hanson, L. A., "Immunological analysis of bovine blood serum and milk," *Experientia*, 15, 471 (1959).
[238] Larson, B. L., and D. C. Gillespie, "Origin of the major specific proteins in milk," *J. Biol. Chem.*, 227, 565 (1957).
[239] Askonas, B. A., P. N. Campbell, J. H. Humphrey, and T. S. Work, "The source of antibody globulin in rabbit milk and goat colostrum," *Biochem. J.*, 56, 597 (1954).
[240] Juergens, W. G., F. E. Stockdale, Y. J. Topper, and J. J. Elias, "Hormone-dependent differentiation of mammary gland *in vitro*," *Proc. Natl. Acad. Sci.*, 54, 629 (1965).
[241] Lockwood, D. H., R. W. Turkington, and Y. J. Topper, "Hormone-dependent development of milk protein synthesis in mammary gland *in vitro*," *Biochim. Biophys. Acta*, 130, 493 (1966).
[242] Lockwood, D. H., A. E. Voytovich, F. E. Stockdale, and Y. J. Topper, "Insulin-dependent DNA polymerase and DNA synthesis in mammary epithelial cells *in vitro*," *Proc. Natl. Acad. Sci.*, 58, 658 (1967).
[243] Sloan, R. E., R. Jenness, A. L. Kenyon, and E. A. Regehr, "Comparative biochemical studies of milks-I. Electrophoretic analysis of milk proteins," *Comp. Biochem. Physiol.*, 4, 47 (1961).
[244] Glasnák, V., "Inter- and intraspecific differences in milk proteins of cattle and swine," *Comp. Biochem. Physiol.*, 25, 355 (1968).
[245] Bell, K., and H. A. McKenzie, "β-Lactoglobulins," *Nature*, 204, 1275 (1964).

[246] Kalan, E. B., R. R. Kraeling, and R. J. Gerrits, "Isolation and partial characterization of a polymorphic swine whey protein,"*Fed. Proc. Abstr.* **27**, No. 3129, 773 (1968).
[247] Laskowski, M., B. Kassell, and G. Hagerty, "A crystalline trypsin inhibitor from swine colostrum," *Biochem. Biophys. Acta*, **24**, 300 (1957).
[248] Chandan, R. C., K. M. Shahani, and R. G. Holly, "Lysozyme content of human milk," *Nature*, **204**, 76 (1964).
[249] Tarassuk, N. P., T. A. Nickerson, and M. Yaguchi, "Lipase action in human milk," *Nature*, **201**, 298 (1964).
[250] Stewart, R. A., E. Platou, and V. J. Kelly, "The alkaline phosphatase content of human milk," *J. Biol. Chem.*, **232**, 777 (1958).
[251] Jenness, R., and S. Patton, *Principles of Dairy Chemistry*, Wiley, New York, 1959.
[252] Campbell, B., and W. E. Petersen, "Immune milk—a historical survey," *Dairy Sci. Abstr.*, **25**, 345 (1963).
[253] Mitchell, C. A., R. V. L. Walker, and G. L. Bannister, "Studies relating to the formation of neutralizing antibody following the propagation of influenza and Newcastle disease virus in the bovine mammary gland," *Canad. J. Microbiol.*, **2**, 322 (1956).
[254] Campbell, B., M. Sarwar, and W. E. Petersen, "Diathelic immunization—a maternal-offspring relationship involving milk antibodies," *Science*, **125**, 932 (1957).
[255] Petersen, W. E., and B. Campbell, "Use of protective principles in milk and colostrum in prevention of disease in man and animals," *Journal-Lancet*, **75**, 494 (1955).
[256] Petersen, W. E., *"Rationalisierung der viehwirtschaftlichen Erzeugung."* (E. Schilling, Ed.), Max-Planck-Institut, Mariensee, **4**, 1 (1960).
[257] Lascelles, A. K., "A review of the literature on some aspects of immune milk," *Dairy Sci. Abst.*, **25**, 359 (1963).
[258] Gunther, M., E. Cheek, R. H. Matthews, and R. R. A. Coombs, "Immune responses in infants to cow's milk proteins taken by mouth," *Intern. Arch. Allergy Appl. Immunol.*, **21**, 257 (1962).
[259] Rothberg, R. M., and R. S. Farr, "Anti-bovine serum albumin and anti-alpha-lactalbumin in the serum of children and adults," *Pediatrics*, **35**, 571 (1965).
[260] Ratner, B., M. Dworetzky, S. Oguri, and L. Aschheim, "Studies on the allergenicity of cow's milk. I. The allergenic properties of alpha-casein, beta-lactoglobulin and alpla-lactalbumin," *Pediatrics*, **22**, 449 (1958).
[261] Crawford, L. V., and F. T. Grogan, "Allergenicity of cow's milk proteins IV. Relationship to goat's milk proteins as studied by serum-agar precipitation," *J. Pediat.*, **59**, 347 (1961).

# CHAPTER 5

[262] Tiselius, A., "A new apparatus for electrophoretic analysis of colloidal mixtures," *Trans. Faraday Soc.*, **33**, 524 (1937).
[263] Cohn, E. J., L. E. Strong, W. L. Hughes, Jr., D. J. Mulford, J. N. Ashworth, M. Melin, and H. L. Taylor, "Preparation and properties of serum and plasma protein IV. A system for the separation into fractions of the protein and lipoprotein components of biological tissues and fluids," *J. Am. Chem. Soc.*, **68**, 459 (1946).
[264] Sober, H. A., and E. A. Peterson, "Chromatography of proteins on cellulose ion-exchangers," *J. Am. Chem. Soc.*, **76**, 1711 (1954).
[265] Peterson, E. A., and H. A. Sober, "Chromatography of proteins. I. Cellulose ion-exchange adsorbents," *J. Am. Chem. Soc.*, **78**, 751 (1956).

[266] Flodin, P., and J. Killander, "Fractionation of human-serum proteins by gel filtration," *Biochim. Biophys. Acta*, **63**, 403 (1962).
[267] Smithies, O., "Zone electrophoresis in starch gels: Group variations in the serum proteins of normal human adults," *Biochem. J.*, **61**, 629 (1955).
[268] Putnam, F. W., "Structure and function of the plasma proteins," in *The Proteins*, Vol III, 2nd ed. p. 153 (Neurath, H., Ed.), Academic Press, New York, 1965.
[269] Wallenius, G., R. Trautman, H. G. Kunkel, and E. C. Franklin, "Ultracentrifugal studies of major nonlipide electrophoretic components of normal human serum," *J. Biol. Chem.*, **225**, 253 (1957).
[270] Pastewka, J. V., A. T. Ness, and A. C. Peacock, "Electrophoretic patterns of normal human serums by disc electrophoresis in polyacrylamide gel," *Clin. Chim. Acta*, **14**, 219 (1966).
[271] Cooper, G., "Electrophoretic and ultracentrifugal analysis of normal human sera," in *The Plasma Proteins*, (Putnam, F. W., Ed.) Vol. I, 51, Academic Press, New York, 1960.
[272] Engle, R. L., Jr., and K. R. Woods, "Comparative biochemistry and embryology," in *The Plasma Proteins*, (Putnam, F. W., Ed.) Vol. II, p. 183, Academic Press, New York, 1960.
[273] Tomasi, T. B., "Human gamma globulin," *Blood*, **25**, 382 (1965).
[274] Foster, J. F., "Plasma albumin," in *The Plasma Proteins*, (Putnam, F. W., Ed.), Vol. I, p. 179, Academic Press, New York, 1960.
[275] Harrington, W. F., P. Johnson, and H. R. Ottewill, "Bovine serum albumin and its behavior in acid solutions," *Biochem. J.*, **62**, 569 (1956).
[276] Adkins, B. J., and J. F. Foster "Subtilisin-catalyzed hydrolysis of bovine plasma albumin: Implications with respect to a postulated subunit model," *Biochemistry*, **4**, 634 (1965).
[277] Polonovski, M., and M. F. Jayle, "Peroxydases animales; Leur spécificité et leur rôle biologique," *Bull. Soc. Chim. Biol.*, **21**, 66 (1939).
[278] Holmberg, C. G., and C. B. Laurell, "Investigations in serum copper. II Isolation of the copper-containing protein, and a description of some of its properties," *Acta, Chem. Scand.*, **2**, 550 (1948).
[279] Osaki, S., "Investigation on caeruloplasmin. I. Purification and some physico-chemical properties of caeruloplasmin," *J. Biochem. (Tokyo)*, **48**, 190 (1960).
[280] Poulik, M. D., "Electrophoretic and immunological studies on structural subunits of human ceruloplasm," *Nature*, **194**, 842 (1962).
[281] Richterich, R., A. Temperli, and H. Aebi, "The heterogeneity of ceruloplasmin: isolation and characterization of two cupro proteins from human serum," *Biochim. Biophys. Acta*, **56**, 240 (1962).
[282] Kasper, C. B., and H. F. Deutsch, "Physicochemical studies of human ceruloplasmin," *J. Biol. Chem.*, **238**, 2325 (1963a).
[283] Katz, S., K. Gutfreund, S. Shulman, and J. D. Ferry, "The conversion of fibrinogen to fibrin. X. Light scattering studies of bovine fibrinogen," *J. Am. Chem. Soc.*, **74**, 5706 (1952).
[284] Shulman, S., "The size and shape of bovine fibrinogen. Studies of sedimentation, diffusion, and viscosity," *J. Am. Chem. Soc.*, **75**, 5846 (1953).
[285] Clegg, J. B., and K. Bailey, "The separation and isolation of the peptide chains of fibrin," *Biochim. Biophys. Acta*, **63**, 525 (1962).
[286] Morawitz, P., *Ergeb. Physiol.*, **4**, 307 (1905); (transl. by Hartmann, R. C., and P. F. Guenther "Chemistry of Blood Coagulation," Thomas, Springfield, 1958.)
[287] Singer, S. L., "Structure and function of antigen and antibody proteins," in *The Pro-*

*teins,* (Neurath, H., Ed.) Vol. III, 2nd ed., p. 269, Academic Press, New York, 1965.
[288] Day, E. D., *Foundation of Immunochemistry*. Williams & Wilkins, Waverly, Baltimore, 1966.
[289] Lennox, E. S., and M. Cohn, "Immunoglobins," *Ann. Rev. Biochem.*, 36, 365 (1967).
[290] Fishman, W. H., "Plasma enzymes," in *The Plasma Proteins*, (Putnam, F. W., Ed.) Vol. II, p. 59, Academic Press, New York, 1960.
[291] Allison, R. G., and R. E. Feeney, "Penguin blood serum proteins," *Arch. Biochem. Biophys.*, 124, 548 (1968).
[292] Komatsu, S. K., H. T. Miller, A. L. DeVries, D. T. Osuga, and R. E. Feeney, "Blood plasma proteins of cold adapted Antarctic fish," in press (1969).
[293] King, T. P., "On the sulfhydryl group of human plasma albumin," *J. Biol. Chem.*, 236, PC5 (1961).
[294] Witter, A., and H. Tuppy, "N-(4-Dimethylamino-3, 5-Dinitrophenyl) Maleimide: A coloured sulfhydryl reagent: Isolation and investigation of cysteine-containing peptides from human and bovine serum albumen." *Biochim. Biophys. Acta*, 45, 429 (1960).
[295] Ashton, G. C., "Serum albumin polymorphism in cattle," *Genetics*, 50, 1421 (1964).
[296] Stormont, C., and Y. Suzuki, "Genetic control of albumin phenotypes of horses," *Proc. Soc. Exptl. Biol. Med.*, 114, 673 (1963).
[297] McIndoe, W. M., "Occurrence of two plasma albumins in the domestic fowl," *Nature*, 195, 353 (1962).
[298] Quinteros, I. R., R. W. C. Stevens, C. Stormont, and V. S. Asmundson, "Albumin phenotypes in turkeys," *Genetics*, 50, 579 (1964).
[299] Haley, L. E., "Serum albumin polymorphism in quail and chicken-quail hybrids," *Genetics*, 51, 983 (1965).
[300] Giblett, E. R., and L. E. Brooks, "Haptoglobin sub-types in three racial groups," *Nature*, 197, 576 (1963).
[301] Jayle, M. F., and G. Boussier, "The biochemistry of haptoglobin, a serum $\alpha_2$-mucoprotein," *Exposes Ann. Biochem. Med.*, 17, 157 (1955).
[302] Herman-Boussier, G., J. Moretti, and M. F. Jayle, "Etude de L'haptoglobine II.- Proprietes physiques et chimiques des haptoglobines des types I et II et de leurs complexes avec l'hemoglobine," *Bull. Soc. Chim. Biol.*, 42, 837 (1960).
[303] Smithies, O., "Zone electrophoresis in starch gels and its application to studies of serum proteins," *Adv. Protein. Chem.*, 14, 65 (1959).
[304] Smithies, O., G. E. Connell, and G. H. Dixon, "Inheritance of hatoglobin subtypes," *Am. J. Human Genet.*, 14, 14 (1962).
[305] Smithies, O., G. E. Connell, and G. H. Dixon, "Chromosomal rearrangements and the evolution of haptoglobin genes," *Nature*, 196, 232 (1962).
[306] Giblett, E. R., "Variant haptoglobin phenotypes," Symposia on quantitative biology, 29, 321 (1964).
[307] Kasper, C. B., and H. F. Deutsch, "Immunochemical studies of crystalline human ceruloplasmin and derivatives," *J. Biol. Chem.*, 238, 2343 (1963).
[308] Bearn, A. G., and W. C. Parker, "Some observations on transferrin iron metabolism," An International Symposium, Aix-en Provence, France, 60, (1963).
[309] Parker, W. C., and A. G. Bearn, "Studies on the transferrins of adult serum, cord serum, and cerebrospinal fluid," *J. Exptl. Med.*, 115, 83 (1962).
[310] Robinson, J. C., and J. E. Pierce, "Studies on inherited variants of blood proteins. III. Sequential action of neuraminidase and galactose oxidase on transferrin $B_{1-2}B_2$," *Arch. Biochem. Biophys.*, 106, 348 (1964).
[311] Buettner-Janusch, J., "Transferrin differences in chimpanzee sera," *Nature*, 192, 632 (1961).

[312] Ashton, G. C., and K. A. Ferguson, "Serum transferrins in Merino sheep," *Gen. Res.*, **4**, 240 (1963).
[313] Braend, M., and C. Stormont, "Studies on hemoglobin and transferrin types of horses," *Nord. Veterinarmed.*, **16**, 31 (1964).
[314] Dessauer, H. C., W. Fox, and Q. L. Hartwig, "Comparative study of transferrins of amphibia and reptilia using starch-gel electrophoresis and autoradiography," *Comp. Biochem. Physiol.*, **5**, 17–29 (1962).
[315] Doolittle, R. F., and B. Blombäck, "Amino acid sequence investigations of fibrinopeptides from various mammals: Evolutionary implications," *Nature*, **202**, 147 (1964).
[316] Doolittle, R. F., D. Schubert, and S. A. Schwartz, "Amino acid sequence studies on artiodactyl fibrinopeptides. 1. Dromedary camel, mule deer, and cape buffalo," *Arch. Biochem. Biophys.*, **118**, 456 (1967).
[317] Mross, G. A., and R. F. Doolittle, "Amino acid sequence studies on artiodactyl fibrinopeptides. II. Vicuna, elk, muntjak, prong-horn antelope, and water buffalo," *Arch. Biochem. Biophys.*, **122**, 674 (1967).
[318] Fleischman, J. B., "Immunoglobulins," *Ann. Rev. Biochem.*, **35**, 835 (1966).
[319] Porter, R. R., "A discussion of the chemistry and biology of immunoglobulins," *Proc. Roy. Soc. (London)*, Ser. B, **166**, 113 (1966).
[320] Symposium on differentiation and growth of hemoglobin- and immunoglobulin-synthesizing cells, *J. Cell. Comp. Physiol.*, **67**, Suppl. 1 (1966).
[321] Marchalonis, J., and G. M. Edelman, "Phylogenetic origins of antibody structure II. Immunoglobulins in the primary immune response of the bullfrog., *Rana catesbiana*," *J. Exptl. Med.*, **124**, 901 (1966).
[322] Marchalonis, J., and G. M. Edelman, "Polypeptide chains of immunoglobulins from the smooth dogfish *Mustelus canis*," *Science*, **154**, 1567 (1967).
[323] Marchalonis, J., and G. M. Edelman, "Phylogenetic origins of antibody structure I. Multichain structure of immunoglobins in the smooth dogfish *Mustelus canis*," *J. Exptl. Med.*, **122**, 601 (1965).
[324] Scandalios, J. G., "Human serum leucine aminopeptidase. Variation in pregnancy and in disease states," *J. Heredity*, **58**, 153 (1967).
[324a] DeVries, A. L., and D. E. Wohlschlag, "Freezing resistance in some Antarctic fishes," in press, *Science* (1969).
[325] Good, R. A., and B. W. Papermaster, "Ontogeny and phylogeny of adaptive immunity," in *Adv. in Immunology*, (Dixon, F. J., Jr., and J. H. Humphrey, Eds.), Vol. 4, Academic Press, New York, 1964.

# CHAPTER 6

[326] Surgenor, D. M., B. A. Koechlin, and L. E. Strong, "Chemical, clinical, and immunological studies on the products of human plasma fractionation. XXXVII. The metal-combining globulin of human plasma," *J. Clin. Invest.* **28**, 73 (1949).
[327] Cohn, E. J., "Chemical, physiological, and immunological properties and clinical uses of blood derivative," *Experientia*, **3**, 125 (1947).
[328] Inman J. K., F. C. Coryell, K. B. McCall, J. T. Sgouris, and H. D. Anderson, "A large-scale method for purification of human transferrin," *Vox Sanq.*, **6**, 34 (1961).
[329] Kistler, P., H. Nitschmann, A. Wyttenbach, M. Studer, C. Hiederost, and M.

Mauerhofer, "Humanes Siderophilin: Isolierung Mittels rivanol aus Blutplasma und Plasmafraktionen, analytische Bistimmung und Kristallisation," *Vox Sanq.*, 5, 403 (1960).

[330] Boettcher, E. W., P. Kistler, and H. Nitschmann, "Method of isolating the beta$_1$-metal-combining globulin from human blood plasma," *Nature*, 181, 490 (1958).

[331] Leibman, A. J., and P. Aisen, "Preparation of single crystals of transferrin," *Arch. Biochem. Biophys.*, 121, 717 (1967).

[332] Bain, J. A., and H. F. Deutsch, "Separation and characterization of conalbumin," *J. Biol. Chem.*, 172, 547 (1948).

[333] Woodworth, R. C., and A. L. Schade, "Conalbumin: A rapid, high-yield preparation from egg white," *Arch. Biochem. Biophys.*, 82, 78 (1959).

[334] Azari, P., and R. F. Baugh, "A simple and rapid procedure for preparation of large quantities of pure ovotransferrin," *Arch. Biochem. Biophys.*, 118, 138 (1967).

[335] Montreuil, J., J. Tonnelat, and S. Müllet, "Preparation et proprietes de la lactosiderophiline (lactotransferrine) du lait de femme," *Biochem. Biophys. Acta*, 45, 413 (1960).

[336] Buttkus, H., J. R. Clark, and R. E. Feeney, "Chemical modifications of amino groups of transferrins: ovotransferrins, human serum transferrin, and human lactotransferrin," *Biochemistry*, 4, 998 (1965).

[337] Johanson, B., "Isolation of an iron-containing red protein from human milk," *Acta Chem. Scand.*, 14, 510 (1960).

[338] Blanc, B., and H. Isliker, "Isolement et caracterisation de la proteine rouge siderophile du lait maternal: la lactotransferrine," *Bull. Soc. Chim. Biol.*, 43, 929 (1961).

[339] Gordon, W. G., M. L. Groves, and J. J. Basch, "Bovine milk Red Protein: Amino acid composition and comparison with blood transferrin," *Biochemistry*, 2, 817 (1963).

[340] Laurell, C. B., "Plasma iron and the transport of iron in the organism," *Pharmacol. Reviews*, 4, 371 (1952).

[341] Schade, A. L., "The microbiological activity of siderophilin," in *Protides of the Biological Fluids*, (Peeters, H., Ed.), Proc. of the 8th Colloquium, Bruges, 1960, p. 261–263, Elsevier, Amsterdam, 1961.

[342] Fraenkel-Conrat, H., and R. E. Feeney, "The metal-binding activity of conalbumin," *Arch. Biochem.*, 29, 101 (1950).

[343] Jamieson, G. A., "Studies on glycoprotein II: Isolation of the carbohydrate, chains of human transferrin," *J. Biol. Chem.*, 240, 2914 (1965).

[344] Groves, M. L., "The isolation of a red protein from milk," *J. Am. Chem. Soc.*, 82, 3345 (1960).

[345] Montreuil, J., G. Spik, M. Monsigny, J. Descamps, G. Biserte, and M. Dautrevaux, "Etude comparee de la composition en oses et en amino-acides de la transferrine et de la lactotransferrine humaines," *Experientia*, 21, 254 (1965).

[346] Means, G., and R. E. Feeney, (unpublished data).

[347] Roberts, R. C., D. G. Makey, and U. S. Seal, "Human transferrin: Molecular weight and sedimentation properties," *J. Biol. Chem.* 241, 4907 (1966).

[348] Bezkorovainy, A., and D. Grohlich, "The behavior of native and reduced-alkylated human transferrin in urea and guanidine-HCl solutions," *Biochim. Biophys. Acta*, 147, 497 (1967).

[349] Charlwood, P. A., "Ultracentrifugal characteristics of human, monkey, and rat transferrins," *Biochem. J.*, 88, 394 (1963).

[350] Laurel, C.-B., and B. Ingelman, "The iron-binding protein of swine serum," *Acta Chem. Scand.*, 1, 770 (1947).

[351] Montreuil, J., and S. Mullet, "Isolement d'une lactosiderophiline du lait de femme," *Compt. Rend. Acad. Sci. Paris*, 250, 1736 (1960).

[352] Timasheff, S. N., and I. Tinoco, Jr., "Light scattering of isoionic conalbumin," *Arch. Biochem. Biophys.*, 66, 427 (1957).

# REFERENCES

[353] Jeppsson, J. O., "Isolation and partial characterization of three human transferrin variants," *Biochim. Biophys. Acta,* **140,** 468 (1967).
[354] Jeppsson, J. O., and J. Sjoquist, "Subunits of normal human transferrin," *Abst. Sixth Internatl. Cong. of Biochem.,* **II,** 157 (1964).
[355] Dixon, G. H., "Mechanisms of protein evolution," in *Essays in Biochemistry,* (Campbell, P. N. and G. D. Greville, Ed.), Vol. II, p. 147, Academic Press, New York, 1966.
[356] Pederson, D. M., and J. F. Foster, "Separation and characterization of subtilisin produced fragments of bovine plasma albumin," *Fed. Proc. Abst.,* **27,** 391 (1968).
[357] Phelps, R. A., and J. R. Cann, "On the modification of conalbumin by acid. II. Effect of pH and salt concentration on the sedimentation behavior, viscosity and osmotic pressure of conalbumin solutions," *Arch. Biochem. Biophys,* **61,** 51 (1956).
[358] Warner, R. C., and I. Weber, "The metal combining properties of conalbumin," *J. Am. Chem. Soc.,* **75,** 5094 (1953).
[359] Wishnia, A., I. Weber, and R. C. Warner, "The hydrogen-ion equilibria of conalbumin," *J. Am. Chem. Soc.,* **83,** 2071 (1961).
[360] Ulmer, D. D., and B. L. Vallee, "Optically active metalloprotein chromatophores III Heme and nonheme iron protein," *Biochem.,* **2,** 1335 (1963).
[361] Aasa, R., B. G. Malmstrom, P. Saltman, and T. Vanngard, "The specific binding of iron (III) and copper (II) to transferrin and conalbumin," *Biochim. Biophys. Acta,* **75,** 203 (1963).
[362] Warner, R. C., "The metal chelate compounds of proteins," *Trans. N.Y. Acad. Sci.,* **16,** 182 (1953).
[363] Aisen, P., A. Leibman, and H. A. Reich, "Studies on the binding of iron to transferrin and conalbumin," *J. Biol. Chem.,* **241,** 1666 (1966).
[364] Aisen, P., and A. Leibman, "The stability constants of the $Fe^{3+}$ conalbumin complexes," *Biochem. Biophys. Res. Commun.,* **30,** 407 (1968).
[365] Tengerdy, C., P. Azari, and R. P. Tengerdy, "Immunochemical reactions of conalbumin and its metal complexes," *Nature,* **211,** 203 (1966).
[366] Inman, J. K., "Studies on the purification and properties of human $beta_1$-metal-combining protein," Doctoral Thesis, Division of Medical Sciences, Harvard University, Cambridge, 1956.
[367] Azari, P. R., and R. E. Feeney, "The resistances of conalbumin and its iron complex to physical and chemical treatments," *Arch. Biochem. Biophys.,* **92,** 44 (1961).
[368] Michaud, R. L., and R. C. Woodworth, "The reactive tyrosines of conalbumin and siderophilin," *Am. Chem. Soc. Abst.,* C-220, (Sept. 12–16, 1966), New York.
[369] Jones, H. D. C., and D. J. Perkins, "Metal-ion binding of human transferrin," *Biochim. Biophys. Acta,* **100,** 122 (1965).
[370] Line, W. F., D. Grohlich, and A. Bezkorovainy, "The effect of chemical modification on the iron-binding properties of human transferrin," *Biochemistry,* **6,** 3393 (1967).
[371] Aisen, P., R. Aasa, B. G. Malmstrom, and T. Vanngard, "Bicarbonate and the binding of iron to transferrin," *J. Biol. Chem.,* **242,** 2484 (1967).
[372] Gaffield, W., L. Vitello, and Y. Tomimatsu, "Optical rotary dispersion of egg proteins II Environment sensitive side chain," *Biochem. Biophys. Res. Commun.,* **25,** 35
[373] Gaffield, W., L. Vitello, and Y. Tomimatsu, "Optical rotatory dispersion of egg proteins II Environment sensitive side chain chromophores in conalbumin," *Biochem. Biophys. Res. Commun.,* **25,** 35 (1966).
[374] Azari, P. R., and R. E. Feeney, "Resistance of metal complexes of conalbumin and transferrin to proteolysis and to thermal denaturation," *J. Biol. Chem.* **232,** 293 (1958).
[375] Glazer, A. N., and H. A. McKenzie, "The denaturation of proteins. IV Conalbumin and iron (III) Conalbumin in urea solution," *Biochim. Biophys. Acta,* **71,** 109 (1963).

[376] Jandl, J. H., J. K. Inman, R. L. Simmons, and D. W. Allen, "Transfer of iron from serum iron-binding protein to human reticulocytes," *J. Clin. Invest.*, **38**, 161 (1959).
[377] Morgan, E. H., and C. B. Laurell, "Studies on the exchange of iron between transferrin and reticulocytes," *Brit. J. Haematol.*, **9**, 471 (1963).
[378] Schade, A. L., "Plasma iron: Its transport and significance," *Nutrition Rev.*, **13**, 225 (1955).
[379] Kornfeld, S., "The effects of structural modifications on the biologic activity of human transferrin," *Biochemistry*, **7**, 945 (1968).
[380] Wheby, M. S., and L. G. Jones, "Role of transferrin in Fe absorption," *J. Clin. Invest.*, **42**, (7), 1007 (1963).
[381] Billups, C., L. Pape, and P. Saltman, "The kinetics and mechanism of Fe(III) exchange between chelates and transferrin," *J. Biol. Chem.*, **242**, 4284 (1967).
[382] Saltman, P., "The role of chelation in iron metabolism," *J. Chem. Educ.*, **42**, 682 (1965).
[383] Feeney, R. E., "The antagonistic activities of conalbumin and 8-hydroxy-quinoline (oxine)," *Arch. Biochem. Biophys.*, **34**, 196 (1951).
[384] Feeney, R. E., and David A. Nagy, "The antibacterial activity of the egg-white protein conalbumin," *J. Bacteriology*, **64**, 629 (1952).
[385] Schade, A. L., "Significance of serum iron for the growth, biological characteristics, and metabolism of *Staphylococcus aureus*," *Biochem. Z.*, **338**, 140 (1963).
[386] Horsfall, W. R., and O. Smithies, "Genetic control of some human serum $\beta$-globulins," *Science*, **128**, 35 (1958).
[387] Leithoff, H., and I. Leithoff, "The demonstration of blood proteins in human seminal plasma by immuno-electrophoresis," *Med. Welt.*, **21**, 1137 (1961).
[388] Ogden, A. L., J. R. Morton, D. G. Gilmour, and E. M. McDermid, "Inherited variants in the transferrins and conalbumins of the chicken," *Nature*, **195**, 1026 (1962).
[389] Szuchet-Derechin, S., and P. Johnson, "The Albumin fraction of bovine milk IV. Physico-chemical study of red protein A," *Eur. Polymer. J.*, **2**, 115 (1966).

# Chapter 7

[390] Isemura, T., T. Takagi, Y. Maeda, and K. Imai, "Recovery of enzymatic activity of reduced taka-amylase A and reduced lysozyme by air-oxidation," *Biochem. Biophys. Res. Commun.*, **5**, 373 (1961).
[391] Meyer, K., and E. Hahnel, "The estimation of lysozyme by a viscosimetric method," *J. Biol. Chem.*, **163**, 723 (1946).
[392] Meyer, K., E. Hahnel, and A. Steinberg, "Lysozyme of plant origin," *J. Biol. Chem.*, **163**, 733 (1946).
[393] Meyer, K., J. F. Prudden, W. L. Lehman, and A. Steinberg, "Lysozyme content of the stomach and its possible relationship to peptic ulcer," *Proc. Soc. Exptl. Biol. Med.*, **65**, 220 (1947).
[394] Meyer, K., A. Gellhorn, J. F. Prudden, W. L. Lehman, and A. Steinberg, "Lysozyme in chromic ulcerative colitis," *Proc. Soc. Exptl. Biol. Med.*, **65**, 221 (1947).
[395] Abraham, E. P., and R. Robinson, "Crystallization of lysozyme," *Nature*, (*Lond.*), **140**, 24 (1937).
[396] Fraenkel-Conrat, H., "The essential groups of lysozyme with particular reference to its reaction with iodine," *Arch. Biochem.*, **27**, 109 (1950).
[397] Fraenkel-Conrat, H., A. Mohammad, E. D. Ducay, and D. K. Mecham, "The mo-

lecular weight of lysozyme after reduction and alkylation of the disulfide bonds," *J. Am. Chem. Soc.*, **73**, 625 (1951).

[398] Berger, L. R., and R. S. Weiser, "The $\beta$-glucosaminidase activity of egg-white lysozyme," *Biochim. Biophys. Acta*, **26**, 517 (1957).

[399] Salton, M. R. J., and J. M. Ghuysen, "The structure of *di-* and *tetra-*saccharides released from cell walls by lysozyme and *Streptomyces* enzyme and the $\beta$ (1→4) N-acetylhexos aminidase activity of these enzymes," *Biochim. Biophys. Acta*, **36**, 552 (1959).

[400] Blake, C. C. F., L. N. Johnson, G. A. Mair, A. C. T. North, D. C. Phillips, and V. R. Sarma, "Crystallographic studies of the activity of hen egg-white lysozyme," *Proc. Roy. Soc. (Lond.)*, **167B**, 378 (1967).

[401] Jolles, J., J. Jauregui-Adell, I. Bernier, and P. Jolles, "The chemical structure of egg-white lysozyme," *Biochim. Biophys. Acta*, **78**, 668 (1963).

[402] Blake, C. C. F., D. F. Koenig, G. A. Mair, A. C. T. North, D. C. Phillips, and V. R. Sarma, "Structure of hen egg-white lysozyme A three-dimensional Fourier synthesis at 2Å resolution," *Nature*, **206**, 757 (1965).

[403] Johnson, L. N., and D. C. Phillips, "Structure of some crystalline lysozyme-inhibitor complexes determined by X-ray analysis at 6 Å resolution," *Nature*, **206**, 761 (1965).

[404] Phillips, D. C., "The three-dimensional structure of an enzyme molecule," *Scientific Amer.*, **215**, 78 (1966).

[405] Phillips, D. C., "Lysozyme and the development of protein crystal chemistry," *Proc. Int. Cong. Biochem.*, **36**, 63 (1967).

[406] Takeda, H., G. A. Strasdine, D. R. Whitaker, and C. Roy, "Lytic enzymes in the digestive juice of *Helix pomatia*. Chitinases and muramidases," *Can. J. Biochem.*, **44**, 509 (1966).

[407] Tsugita, A., M. Inouye, E. Terzaghi, G. Streisinger, "Purification of bacteriophage T4 lysozyme," *J. Biol. Chem.*, **243**, 391 (1968).

[408] Weidel, V. W., and W. Katz, "Purification and characterization of the enzyme responsible for the lipids of T-2 infected cells," *Z. Naturforsch.*, Pt.b, **16**, 156 (1961).

[409] Black, L. W., "The lysozyme of bacteriophage lambda," *Diss. Abstr.*, B **28**, 784 (1967).

[410] Ward, J. B., and H. R. Perkins, "The purification and properties of two staphylolytic enzymes from *Streptomyces griesus*," *Biochem. J.*, **106**, 69 (1968).

[411] Green, N. M., "Evidence for a genetic relationship between avidins and lysozymes," *Nature*, **217**, 254 (1968).

[412] Smith, E. L., J. R. Kimmel, D. M. Brown, and E. O. P. Thompson, "Isolation and properties of a crystalline mercury derivative of a lysozyme from papaya latex," *J. Biol. Chem.*, **215**, 67 (1955).

[413] Howard, J. B., and A. N. Glazer, "Studies of the physicochemical and enzymatic properties of papaya lysozyme," *J. Biol. Chem.*, **242**, 5715 (1967).

[414] Shockman, G. D., J. S. Thompson, and M. J. Conover, "The autolytic enzyme system of *Streptococcus facalis*. II. Partial characterization of the autolysin and its substrate," *Biochemistry*, **6**, 1054 (1967).

[415] Hawiger, J., "Purification and properties of lysozyme produced by *Staphylococcus aureus*," *J. Bacteriol.*, **95**, 376 (1968).

[416] Canfield, R. E., and C. B. Anfinsen, "Concepts and experimental approaches in the determination of the primary structure of proteins," in *The Proteins*, (H. Neurath, Ed.) Vol. I, p. 311, Academic Press, New York, 1963.

[417] Charkasov, I. A., and N. A. Kravchenko, "Enzymic lysozyme preparations," *Chem. Abst.*, **65**, 6266d (1966).

# REFERENCES

[418] Feeney, R. E., L. R. MacDonnell, E. D. Ducay, "Irreversible inactivation of lysozyme by copper," *Arch. Biochem. Biophys.*, **61**, 72 (1956).

[419] Brown, J. R., "Disulfide bridges of lysozyme," *Biochem. J.*, **92**, 13P (1964).

[420] Canfield, R. E., and A. K. Liu, "The disulfide bonds of egg-white lysozyme. (Muramidase)," *J. Biol. Chem.*, **240**, 1997 (1965).

[421] Parry, R. M., Jr., "Isolation and characterization of human milk lysozyme," *Diss. Abstr.*, **B 28**, 733 (1967).

[422] Sophianopoulos, A. J., and K. E. Van Holde, "Physical studies of muramidase (lysozyme). II. pH-dependent dimerization," *J. Biol. Chem.*, **239**, 2516 (1964).

[423] Jolles, J., and P. Jolles, "Human tear and human milk lysozymes," *Biochemistry*, **6**, 411 (1967).

[424] Manwell, C., "Molecular palæogenetics: Amino acid sequence homology in ribonuclease and lysozyme," *Comp. Biochem. Physiol.*, **23**, 383 (1967).

[425] Inouye, M., and A. Tsugita, "The amino acid sequence of T4 bacteriophage lysozyme," *J. Mol. Biol.*, **22**, 193 (1966).

[426] Koch, G., and W. J. Dreyer, "Characterization of an enzyme of phage T2 as a lysozyme," *Virology*, **6**, 291 (1958).

[427] Maass, D., and W. Weidel, "Final proof for the identity of enzymic specificities of egg-white lysozyme and phage T2 enzyme," *Biochim. Biophys. Acta*, **78**, 369 (1963).

[428] Salton, M. R. J., "The properties of lysozyme and its action on microorganisms," *Bacteriol. Revs.*, **21**, 82 (1957).

[429] Salton, M. R. J., "Chemistry and function of amino sugars and derivatives," *Ann. Rev. Biochem.*, **34**, 143 (1965).

[430] Wenzel, M., H. P. Lenk, and F. Schütte, "Preparation of tri-N-acetylchitotriose-H$^3$ and its hydrolysis by lysozyme," *Z. Physiol. Chem.*, **327**, 13 (1961).

[431] Rupley, J. A., "The hydrolysis of chitin by concentrated hydrochloric acid, and the preparation of low-molecular-weight substrates for lysozyme," *Biochim. Biophys. Acta*, **83**, 245 (1964).

[432] Sharon, N., "The chemical structure of lysozyme substrates and their cleavage by the enzyme," *Proc. Roy. Soc. (Lond.)*, **167B**, 402 (1967).

[433] Rupley, J. A., "Studies on the enzymatic activity of lysozyme," *Science*, **150**, 382 (1965).

[434] Maksimov, V. I., E. D. Kaversneva, and N. A. Kravchenko, "Character of the action of lysozyme on oligosaccharide chitin fragments," *Biochemistry*, **30**, 866 (1965).

[435] Kravchenko, N. A., "Lysozyme as a transferase," *Proc. Roy. Soc.*, **167B**, 429 (1967).

[436] Hayashi, K., T. Imoto, M. Funatsu, "The enzyme-substrate complex in a muramidase catalyzed reaction I. Difference spectrum of complex," *J. Biochem.*, **54**, 381 (1963).

[437] Hartdegen, F. J., and J. A. Rupley, "Inactivation of lysozyme by iodine oxidation of a single tryptophan," *Biochim. Biophys. Acta*, **92**, 625 (1964).

[438] Hayashi, K., T. Imoto, G. Funatsu, and M. Funatsu, "The position of the active tryptophan residue in lysozyme," *J. Biochem. (Japan)*, **58**, 227 (1965).

[439] Hartdegen, F. J., and J. A. Rupley, "The oxidation by iodine of tryptophan 108 in lysozyme," *J. Am. Chem. Soc.*, **89**, 1743 (1967).

[440] Goldberger, R. F., and C. J. Epstein, "Characterization of the active product obtained by oxidation of reduced lysozyme," *J. Biol. Chem.*, **238**, 2988 (1963).

[441] Azari, P., "Action of sulfite on lysozyme," *Arch. Biochem. Biophys.*, **115**, 230 (1966).

[442] Caputo, A., and R. Zito, *2nd Int. Symp. Lysozyme, Milan*, **1**, 29 (1961).

# REFERENCES

[443] Jolles, P., J. Jauregui-Adell, and J. Jolles, "Le lysozyme de blanc d'oeuf de Poule: disposition desponts disulfures," *Comp. Rend. Acad. Sci. Paris*, **258**, 3926 (1964).
[444] Bradshaw, R. A., L. Kanarek, and R. L. Hill, "The preparation, properties, and reactivation of the mixed disulfide derivative of egg-white lysozyme and L-cystine," *J. Biol. Chem.*, **242**, 3789 (1967).
[445] Horinishi, H., Y. Hachimori, K. Kurihara, and K. Shibota, "States of amino acid residues in proteins. III. Histidine residues in insulin, lysozyme, albumin and proteinases as determined with a new reagent of diazo-l-H-tetrazole," *Biochim. Biophys. Acta*, **86**, 477 (1964).
[446] Kravchenko, N. A., G. V. Kleopina, E. D. Kaverzneva, "Investigation of the active sites of lysozyme. Carboxymethylation of the imidazole group of histidine and of the $\epsilon$-aminogroup of lysine," *Biochim. Biophys. Acta*, **92**, 412 (1964).
[447] Yamasaki, N., K. Hayashi, M. Funatsu, "Acetylation of lysozyme. II. Mechanism of lysis by lysozyme," *Agr. Biol. Chem.*, **32**, 64 (1968).
[448] Rupley, J. A., and V. Gates, "Studies on the enzymic activity of lysozyme II. The hydrolysis and transfer reactions of N-acetylglucosamine oligosaccharides," *Proc. Natl. Acad. Sci.*, **57**, 496 (1967).
[449] Bernier, I., and P. Jolles, "Chemical evidence for involvement of tryptophans in substrate binding," *Compt. Rend. Acad. Sci. Paris*, **253**, 745 (1961).
[450] Blake, C. C. F., G. A. Mair, A. C. T. North, D. C. Phillips, and V. R. Sarma, "On the conformation of the hen egg-white lysozyme molecule," *Proc. Roy. Soc. (Lond.)*, **167B**, 365 (1967).
[451] Butler, L. G., and J. A. Rupley, "The binding of saccharide to crystalline and soluble lysozyme measured directly and through solubility studies," *J. Biol. Chem.*, **242**, 1077 (1967).
[452] Lehrer, S. S., and G. D. Fasman, "Fluorescence of lysozyme and lysozyme substrate complexes," *J. Biol. Chem.*, **242**, 4644 (1967).
[453] Chipman, D. M., V. Grisaro, and N. Sharon, "The binding of oligosaccharides containing $N$-acetylglucosamine and $N$-acetylmuramic acid to lysozyme," *J. Biol. Chem.*, **242**, 4388 (1967).
[454] Dahlquist, F. W., L. Jao, and M. Raftery, "On the binding of chitin oligosaccharides to lysozyme," *Proc. Natl. Acad. Sci.*, **56**, 26 (1966).
[455] Rupley, J. A., L. Butler, M. Gerring, F. J. Hartdegen, and R. Pecoraro, "Studies on the enzymic activity of lysozyme III the binding of saccharides," *Proc. Natl. Acad. Sci.*, **57**, 1088 (1967).
[456] Weil, L., A. R. Buchert, and J. Maher, "Photoöxidation of crystalline lysozyme in the presence of methylene blue and its relation to enzymatic activity," *Arch. Biochem. Biophys.*, **40**, 245 (1952).
[457] Galiazzo, G., G. Jori, and E. Scoffone, "Selective and quantitative photochemical conversion of the tryptophyl residues to kynurenine in lysozyme," *Biochem. Biophys. Res. Commun.*, **31**, 158 (1968).
[458] Previero, A., M. A. Coletti-Previero, and P. Jolles, "Nonenzymatic cleavage of tryptophyl peptide bonds in peptides and proteins," *Biochem. Biophys. Res. Commun.*, **22**, 17 (1966).
[459] Previero, A., M. A. Coletti-Previero, and P. Jolles, in "Relationship between chemical structure and biological activity of hen egg-white lysozyme and lysozymes of different species," [198].
[460] Churchich, J. E., "The polarization of fluorescence of reoxidized muramidase (lysozyme)," *Biochim. Biophys. Acta*, **65**, 349 (1962).

[461]   Imai, K., T. Takagi, and T. Isemura, "Recovery of the intact structure of muramidase after reduction of all disulfide linkages in 8M urea," *J. Biochem.*, **53**, 1 (1963).
[462]   Geschwind, I. I., and C. H. Li, "The guanidination of some biologically active proteins," *Biochim. Biophys. Acta*, **25**, 171 (1957).
[463]   Dickman, S. R., R. B. Kropf, and C. M. Proctor, "Reactions of xanthydrol. II. Insulin, lysozyme, and ribonuclease," *J. Biol. Chem.*, **210**, 491 (1954).
[464]   Rupley, J. A., "The binding and cleavage by lysozyme of N-acetylglucosamine oligosaccharides," *Proc. Roy. Soc.*, **167B**, 416 (1967).
[465]   Jauregui-Adell, J., and P. Jolles, "The disulfide bridges of hen's egg-white lysozyme," *Bull. Soc. Chim. Biol.*, **46**, 141 (1964).
[466]   Godfrine, P., and J. Leonis, *2nd Int. Symp. Lysozyme, Milan*, **1**, 115 (1961).
[467]   Takahashi, T., K. Hamagushi, K. Hayashi, T. Imoto, and M. Funatsu, "Structure of lysozyme. X. On the structural role of tryptophan residues," *J. Biochem.*, **58**, 385 (1965).
[468]   Schneck, A. G., L. Ledoux, and J. Leonis, *2nd Int. Symp. Lysozyme, Milan*, **1**, 103 (1961).
[469]   Jolles, J., and P. Jolles, "Isolation and purification of the biologically active deslysylvalyl phenylalanine-lysozyme," *Biochem. Biophys. Res. Commun.*, **22**, 22 (1966).
[470]   Moore, G. L., and R. A. Day, "Protein conformation in solution: cross-linking of lysozyme," *Science*, **159**, 210 (1968).
[471]   Herzig, D. J., A. W. Rees, and R. A. Day, "Bifunctional reagents and protein structure determination. The reaction of phenolic disulfonyl chlorides with lysozyme," *Biopolymers*, **2**, 349 (1964).
[472]   Charlemagne, D., and P. Jolles, "Specificity of various lysozymes toward low molecular-weight substrates from elution," *Bull. Soc. Chim. Biol.*, **49**, 1103 (1967).
[473]   Perkins, H. R., "The action of hot formamide on bacterial cell walls," *Biochem. J.*, **95**, 876 (1965).
[474]   Brumfitt, W., "The mechanism of development of resistance to lysozyme by some grampositive bacteria and its results," *Br. J. Exp. Path.*, **40**, 441 (1959).
[475]   Toennies, G., and G. D. Shockman, "Growth chemistry of *Streptococcus faecalis*," *Proc. 4th Int. Congr. Biochem.*, **13**, 365 (1958).
[476]   Streisinger, G., Y. Okada, J. Emrich, J. Newton, A. Tsugita, E. Teryaghi, and M. Inouye, "Frameshift mutations and the genetic code," *Cold Spring Harb. Symp. Quant. Biol.*, **31**, 77 (1966).
[477]   Inouye, M., E. Akaboshi, A. Tsugita, G. Streisinger, and Y. Okada, "A frame-shift mutation resulting in the deletion of two base pairs in the lysozyme gene of bacteriophage T4," *J. Mol. Biol.*, **30**, 39 (1967).
[478]   Dove, W. F., "Action of the lambda chromosome. I. Control of functions late in bacteriophage development," *J. Mol. Biol.*, **19**, 187 (1966).
[479]   Thomas, R., "Control of development in temperate bacteriophages. I. Induction of prophage genes following heteroimmune super-infection," *J. Mol. Biol.*, **22**, 79 (1966).
[480]   Dambly, C., M. Couturier, and R. Thomas, "Control of development in temperate bacteriophage II. Control of lysozyme synthesis," *J. Mol. Biol.*, **32**, 67 (1968).
[481]   Canfield, R. E., Brookhaven Symposia in Biology, No 21, June 3-5, 1968.
[482]   Canfield, R. E., (personal communication, 1968).
[483]   Nolan, C., and E. Margoliash, "Comparative aspects of primary structures of proteins," *Ann. Rev. Biochem.*, **37**, 727 (1968).
[484]   Meyer, K., "The relationship of lysozyme to avidin," *Science*, **99**, 391 (1944).

# Chapter 8

[485] Kunitz, M., and J. H. Northrop, "Isolation from beef pancreas of crystalline trypsinogen, trypsin, trypsin inhibitor and an inhibitor-trypsin compound," *J. Gen. Physiol.*, **19**, 991 (1936).
[486] Ham, W. E., and R. M. Sandstedt, "Proteolytic inhibiting substance in extract from unheated soybean meal," *J. Biol. Chem.*, **154**, 505 (1944).
[487] Bowman, D. E., "Fractions derived from soy beans and navy beans which retard tryptic digestion of casein," *Proc. Soc. Exptl. Biol. Med.*, **57**, 139 (1944).
[488] Kunitz, M., "Isolation of a crystalline protein compound of trypsin and of soybean trypsin inhibitor," *J. Gen. Physiol.* **30**, 311 (1947).
[489] Laskowski, M., and M. Laskowski, Jr., "Naturally occurring trypsin inhibitors," *Adv. Protein Chem. IX*, p. 203, Academic Press, New York, 1954.
[490] Laskowski, M., "Chymotrypsinogens and chymotrypsins," *Methods Enzym.*, **2**, 8 (1955).
[491] "Chemistry, pharmacology, and clinical applications of proteinase inhibitors," *Ann. N.Y. Acad. Sci.*, **146**, Art. 2 (1968).
[492] Vogel, R., I. Trautschold, and E. Werle, *Natürliche Proteinasen-Inhibitoren* George Thieme, Stuttgart, 1966.
[493] Bier, M., L. Terminiello, J. A. Duke, R. J. Gibbs, and F. F. Nord, "Investigations on proteins and polymers. X. Composition and fractionation of ovomucoid," *Arch. Biochem. Biophys.*, **47**, 465 (1953).
[494] Fredericq, E., and H. F. Deutsch, "Studies on ovomucoid," *J. Biol. Chem.*, **181**, 499 (1949).
[495] Kassell, B., M. Radicevic, S. Berlow, R. J. Peanasky, and M. Laskowski, Sr., "The basic trypsin inhibitor of bovine pancreas. I. An improved method of preparation and amino acid composition," *J. Biol. Chem.*, **238**, 3274 (1963).
[496] Kassell, B., and R. B. Chow, "Modification of the basic trypsin inhibitor of bovine pancreas. The $\epsilon$-amino groups of lysine and the amino-terminal sequence," *Biochemistry*, **5**, 3449 (1966).
[497] Greene, L. J., M. Rigbi, and D. S. Fackre, "Trypsin inhibitor from bovine pancreatic juice," *J. Biol. Chem.*, **241**, 5610 (1966).
[498] Cerwinsky, E. W., P. J. Burck, and E. L. Grinnan, "Acidic bovine pancreatic trypsin inhibitor I. Purification and physical characterization," *Biochemistry*, **6**, 3175 (1967).
[499] Greene, L. J., J. J. DiCarlo, A. J. Sussman, and D. C. Bartelt, "Two trypsin inhibitors from porcine pancreatic juice," *J. Biol. Chem.*, **243**, 1804 (1968).
[500] Moll, F. C., S. F. Sunden, and J. R. Brown, "Partial purification of the serum trypsin inhibitor," *J. Biol. Chem.*, **233**, 121 (1958).
[501] Bundy, H. F., and J. W. Mehl, "Trypsin inhibitors of human serum. II. Isolation of the $\alpha_1$-inhibitor and its partial characterization," *J. Biol. Chem.*, **234**, 1124 (1959).
[502] Shulman, N. R., "A proteolytic inhibitor with anticoagulant activity separated from human urine and plasma," *J. Biol. Chem.*, **213**, 655 (1955).
[503] Wu, F. C., and M. Laskowski, "Crystalline acid labile trypsin inhibitor from bovine blood plasma," *J. Biol. Chem.*, **235**, 1680 (1960).
[504] Martin, C. J., "Inhibition of trypsin, chymotrypsin, and plasmin by an inhibitor isolated from sheep serum," *J. Biol. Chem.*, **237**, 2099 (1962).
[505] Laskowski, M., Jr., P. H. Mars, and M. Laskowski, "Comparison of trypsin inhibitor from colostrum with other crystalline trypsin inhibitors," *J. Biol. Chem.*, **198**, 745 (1952).

# REFERENCES

[506] Wu, Y. V., and H. A. Scheraga, "Studies on soybean trypsin inhibitor. I. Physicochemical properties," *Biochemistry*, **1**, 698 (1962).

[507] Birk, Y., A. Gertler, and S. Khalef, "A pure trypsin inhibitor from soya beans," *Biochem. J.*, **87**, 281 (1963).

[508] Rackis, J. J., H. A. Sasame, R. K. Mann, R. L. Anderson and A. K. Smith, "Soybean trypsin inhibitors: isolation, purification, and physical properties," *Arch. Biochem. Biophys.*, **98**, 471 (1962).

[509] Yamamoto, M., and T. Ikenaka, "Studies on soybean trypsin inhibitors. I. Purification and characterization of two soybean trypsin inhibitors," *J. Biochem.*, **62**, 141 (1967).

[510] Frattali, V., and R. F. Steiner, "Soybean inhibitors. I. Separation and some properties of three inhibitors from commercial crude soybean trypsin inhibitor," *Biochemistry*, **7**, 521 (1968).

[511] Fraenkel-Conrat, H., R. C. Bean, E. D. Ducay, and H. S. Olcott, "Isolation and characterization of a trypsin inhibitor from lima beans," *Arch. Biochem. Biophys.*, **37**, 393 (1952).

[512] Jones, G., S. Moore, and W. H. Stein, "Properties of chromatographically purified trypsin inhibitors from lima bean," *Biochemistry*, **2**, 66 (1963).

[513] Haynes, R., and R. E. Feeney, "Fractionation and properties of trypsin and chymotrypsin inhibitors from lima beans," *J. Biol. Chem.*, **242**, 5378 (1967).

[514] Wagner, L. P., and J. P. Riehm, "Purification and partial characterization of a trypsin inhibitor isolated from the navy bean," *Arch. Biochem. Biophys.*, **121**, 672 (1967).

[515] Ventura, M. M., and J. X. Filho, "A trypsin and chymotrypsin inhibitor from black-eyed pea. I. Purification and partial characterization," *Acad. Brasileira de Ciencias Anais*, **38**, 553 (1966).

[516] Balls, A. K., and C. A. Ryan, "Concerning a crystalline chymotryptic inhibitor from potatoes, and its binding capacity for the enzyme," *J. Biol. Chem.*, **238**, 2976 (1963).

[517] Ryan, C. A. "Chymotrypsin inhibitor I from potatoes: reactivity with mammalian, plant, bacterial, and fungal proteinases," *Biochemistry*, **5**, 1592 (1966).

[518] Ryan, C. A., (personal communication, 1968).

[519] Kazal, L. A., D. S. Spicer, and R. A. Brahinsky, "Isolation of a crystalline trypsin inhibitor-anti-coagulant protein from pancreas," *J. Am. Chem. Soc.*, **70**, 3034 (1948).

[520] Pudles, J., F. H. Rola, and A. K. Matida, "Studies on the proteolytic inhibitors from *Ascaris lumbricoides* var. *suum* II Purification, properties, and chemical modification of the trypsin inhibitor," *Arch. Biochem. Biophys.*, **120**, 594 (1967).

[521] Kassell, B., and M. Laskowski, Sr., "The basic trypsin inhibitor of bovine pancreas. VI. Sequence studies and disulphide linkages," *Acta Biochim. Polon.*, **13**, 287 (1966).

[522] Anderer, F. A., "Strukturuntesuchungen am Kallikrein-Inaktivator aus Rinderlunge. II. Bestimmung der Aminosauresequenz," *Z. Naturforsch.*, **20b**, 462 (1965).

[523] Chauvet, J., G. Nouvel, and R. Acher, "Structure primaire d'un inhibiteur pancreatique de la trypsine (inhibiteur de Kunitz et Northrop)," *Biochim. Biophys. Acta*, **92**, 200 (1964).

[524] Dlouhá, V., D. Pospísilová, B. Meloun, and F. Sorm, "On proteins XCIV. Primary structure of basic trypsin inhibitor from beef pancreas." *Coll. Czech. Chem. Commun.*, **30**, 1311 (1965).

[525] Anderer, F. A., and S. Hörnle, "The disulfide linkages in Kallikrein Inactivator of Bovine lung," *J. Biol. Chem.* **241**, 1568 (1966).

[526] Anderer, F. A., and S. Hörnle, "Strukturuntersuchungen am Kallikrein-Inaktivator aus Rinderlunge. I. Molekulargewicht, Endgruppenanalyse und Aminosäure-Zusammensetzung," *Z. Naturforsch.*, **20b**, 457 (1965).

[527] Anderer, F. A., "Zur Identität des Kallikrein-Inaktivators aus Rinderlunge und Rinderparotis," *Z. Naturforsch.*, **20b**, 499 (1965).

# REFERENCES

[528] Birk, Y., "Purification and some properties of a highly active inhibitor of trypsin and α-chymotrypsin from soybeans," *Biochem. Biophys. Acta,* 54, 378 (1961).
[529] Frattali, V., (personal communications, 1968).
[530] Ryan, C. A., and A. K. Balls, "An inhibitor of chymotrypsin from *Solanum tuberosum* and its behaviour toward trypsin," *Proc. Natl. Acad. Sci.,* 48, 1839 (1962).
[531] Ryan, C. A., and O. C. Huisman, "Chymotrypsin inhibitor I from potatoes: a transient protein component in leaves of young potato plants," *Nature,* 214, 1047 (1967).
[532] Feeney, R. E., D. T. Osuga, and J. C. Bigler, (unpublished data, 1968).
[533] Melamed, M. D., "Ovomucoid," in *Glycoproteins,* (Gottschalk, A., Ed.), p. 317, Elsevier, New York, 1965.
[534] Feeney, R. E., and F. Greene (manuscript in preparation in *Glycoproteins,* 2nd ed., 1969).
[535] Matsushima, K., "On the naturally occurring inhibitors for *Asparagillus* Protease. III ovoinhibitor," *J. Agr. Chem. Soc. Japan,* 32, 211 (1958).
[536] Landsteiner, K., "Zur Kenntnis der antifermentativen, lytischen, and agglutinierenden Wirkungen des Blutserums und der Lymphe," *Centrbl. Bakt.,* Abt. 1, 27, 357 (1900).
[537] Jacobsson, K., "Electrophoretic demonstration of two trypsin inhibitors in human blood serum," *Scand. Clin. Lab. Invest.,* 5, 97 (1953).
[538] Sale, E. E., S. G. Priest, and H. Jensen, "Studies in the antiproteolytic activity of bovine blood," *J. Biol. Chem.,* 227, 83 (1957).
[539] Gray, J. L., S. G. Priest, W. F. Blatt, U. Westphal, and H. Jensen, "Isolation and characterization of a proteolytic inhibitor from bovine blood," *J. Biol. Chem.,* 235, 56 (1960).
[540] Laskowski, M., Jr., and M. Laskowski, "Trypsin inhibitor in colostrum" "Partial purification of the trypsin inhibitor in urine," *Fed. Proc.,* 9, 194 (1950).
[541] Astrup, T., K. Alkjaer, and F. Soardi, *Scand. J. Clin. Lab Invest.,* 11, 181 (1959).
[542] Mendel, L. B., and A. F. Blood, "Some peculiarities of the proteolytic activity of papaïn," *J. Biol. Chem.,* 8, 177 (1910).
[543] Green, N. M., "Protease inhibitors from *Ascaris lumbricoides*," *Biochem. J.,* 66, 416 (1957).
[544] Peanasky, R. J., and M. Laskowski, "Chymotrypsin inhibitor from *Ascaris*," *Biochim. Biophys. Acta,* 37, 167 (1960).
[545] Rhodes, M. B., C. L. Marsh, and G. W. Kelley, Jr., "Trypsin and chymotrypsin inhibitor from *Ascaris suum*," *Exptl. Parasit.,* 13, 266 (1963).
[546] Liener, I. E., "Toxic factors in edible legumes and their elimination," *Amer. J. Clin. Nutrition,* 11, 281 (1962).
[547] Davis, J. G., J. C. Zahnley, J. W. Donovan, "Separation and characterization of ovoinhibitors of chicken egg white," *Am. Chem. Soc. Abst.,#*C-17, (Sept. 11-25, 1967).
[548] Keller, P. J., and B. J. Allan, "The protein composition of human pancreatic juice," *J. Biol. Chem.,* 242, 281 (1967).
[549] Grossman, M. I., "Some properties of trypsin inhibitor of pancreatic juice," *Proc. Soc. Exptl. Biol. Med.,* 99, 304 (1958).
[550] Kalser, M. H., and M. I. Grossman, "Secretion of trypsin inhibitor in pancreatic juice," *Gastroenterology,* 29, 35 (1955).
[551] Kornguth, S. E., and M. A. Stahmann, "Studies on the effect of some polyelectrolytes upon the activity of trypsin," *Arch. Biochim. Biophys.,* 91, 32 (1960).
[552] Green, N. M., "Competition among trypsin inhibitors," *J. Biol. Chem.,* 205, 535 (1953).
[553] Simlot, M. M., and R. E. Feeney, "Relative reactivities of chemically modified turkey ovomucoid," *Arch. Biochem. Biophys.,* 113, 64 (1966).

# REFERENCES

[554] Gorini, L., and L. Audrain, "Influence of calcium on the stability of the trypsin-ovomucoid complex," *Biochim. Biophys. Acta*, 8, 702 (1952).

[555] Gorini, L., and L. Audrain, "Ovomucoid trypsin complex. Its proteolytic activity and the roles of certain metal ions on the stabilities of its constituents," *Biochim. Biophys. Acta*, 10, 570 (1953).

[556] Laskowski, M. and F. C. Wu, "Temporary inhibition of trypsin," *J. Biol. Chem.*, 204, 797 (1953).

[557] Sri, Ram, J., L. Terminiello, M. Bier, and F. F. Nord, "On the mechanism of enzyme action. LVII. Interaction between trypsin and ovomucoid," *Arch. Biochem. Biophys.*, 52, 451 (1954).

[558] Feinstein, G., and R. E. Feeney, "Binding of proflavine to $\alpha$-chymotrypsin and trypsin and its displacement by avian ovomucoids," *Biochemistry*, 6, 749 (1967).

[559] Lebowitz, J., and M. Laskowski, Jr., "Potentiometric measurement of protein-protein association constants. Soybean trypsin inhibitor-trypsin association," *Biochemistry*, 1, 1044 (1962).

[560] Haynes, R., and R. E. Feeney, "Transformation of active-site lysine in naturally occurring trypsin inhibitors. A basis for a general mechanism for inhibition of proteolytic enzymes," *Biochemistry*, 7, 2879 (1968).

[561] Steiner, R. F., "Reduction and re-oxidation of the disulphide bonds of soybean trypsin inhibitors," *Nature*, 204, 579 (1964).

[562] Sjoberg, L., and R. E. Feeney, "Reduction and reoxidation of turkey ovomucoid — a protein with dual and independent inhibitory activity against trypsin and $\alpha$-chymotrypsin," *Biochim. Biophys. Acta*, 168, 79 (1968).

[563] Kassell, B., "The basic trypsin inhibitor of bovine pancreas. II. Alteration of the methionine residues," *Biochemistry*, 3, 152 (1964).

[564] Finkenstadt, W. R., and M. Laskowski, Jr., "Peptide bond cleavage on trypsin-trypsin inhibitor complex formation," *J. Biol. Chem.*, 240, PC962 (1965).

[565] Ozawa, K., and M. Laskowski, Jr., "The reactive site of trypsin inhibitors," *J. Biol. Chem.*, 241, 3955 (1966).

[566] Sealock, R. W., and M. Laskowski, Jr., "Enzymatic replacement of the arginyl residue by a lysyl in the reactive site of soybean trypsin inhibitor," *Fed. Proc. Abst.*, 27, Abst. #431, 292 (1968).

[567] Rigbi, M., and L. J. Greene, "Limited proteolysis of the bovine pancreatic secretory trypsin inhibitor at acid pH," *J. Biol. Chem.*, 243, 5457 (1968).

[568] Birk, Y., A. Gertler, and S. Khalef, "Further evidence for a dual, independent activity against trypsin and -chymotrypsin of inhibitor AA from soybeans," *Biochim. Biophys. Acta*, 147, 402 (1967).

[569] Dlouhá, V., B. Keil, and F. Sorm, "A study of the complex of trypsin with its pancreatic inhibitor," *Biochem. Biophys. Res. Commun.*, 31, 66 (1968).

[570] Feinstein, G., D. T. Osuga, and R. E. Feeney, "The mechanism of inhibition of trypsin by ovomucoid," *Biochem. Biophys. Res. Commun.*, 24, 495 (1966).

[571] Haynes, R., and R. E. Feeney, "Properties of enzymatically cleaved inhibitors of trypsin," *Biochim. Biophys. Acta*, 159, 209 (1968).

[572] Feinstein, G., and R. E. Feeney, "Interaction of inactive derivatives of chymotrypsin and trypsin with protein inhibitors," *J. Biol. Chem.*, 241, 5183 (1966).

[573] Foster, R. J., and C. A. Ryan, *Federation Proc.*, 24, 473, Abst. #1905 (1965). Reactions of potato inhibitor with modified chymotrypsin.

[574] Frank, B. H., and A. J. Veros, *Federation Proc.*, 27, 392, Abst. #999 (1968). Physical studies on proinsulin — molecular weight, association behavior, and spectral studies.

[575] Chryssanthou, C., and W. Antopol, "Effect of trypsin inhibitors on Shwartzman phenomenon," *Proc. Soc. Exptl. Biol. Med.,* **108,** 587 (1961).
[576] Chamberlain, A. G., G. C. Perry, and R. E. Jones, "Effect of trypsin inhibitor isolated from sow's colostrum on the absorption of $\gamma$-globulin by piglets," *Nature,* **207,** 429 (1965).

# Author Index

Reference numbers are in parentheses.

Aasa, R., (361) 154, 158; (371) 158, 160–162
Abe, N., (221) 100
Abplanalp, H., (59) 27, 31, 33, 36, 46–48, 53, 56; (197) 86
Abraham, E. P., (395) 175
Adair, G. S., (73) 30
Adair, M. E., (73) 30
Adam-Chosson, A., (135) 47
Adams, J. L., (156) 52, 62, 66
Adkins, B. J., (276) 121
Aebi, H., (281) 122
Aisen, P., (331) 146, 149, 153; (363) 155, 156, 159; (364) 155; (371) 158, 160–162
Akaboshi, E., (477) 196
Alderton, G., (48) 25, 175; (55) 26; (85) 30
Alkjaer, K., (541) 213
Allan, B. J., (548) 215
Allen, D. W., (376) 166
Allison, R. G., (170) 57, 62, 66; (291) 127, 136
Allison, W. S., (201) 91
Almquist, H. J., (67) 28
Anderer, F. A., (522) 201; (525) 201; (526) 204; (527) 204
Anderson, H. D., (328) 146
Anderson, J. S., (69) 28, 29, 53, 57, 59, 66, 148; (77) 30, 47, 59, 65, 67, 83, 88–90
Anderson, R. L., (508) 203
Anfinsen, C. B., (9) 2, 3; (14) 4; (147) 49; (148) 49, 173; (416) 177, 180
Annau, E., (146) 48
Antopol, W., (575) 241
Aoki, K., (30) 14

Archer, R., (523) 201
Arnheim, N., (185) 68–70, 90, 177, 196
Aschaffenburg, R., (211) 95; (215) 98
Ascheim, L., (260) 116
Ashton, G. C., (295) 129; (312) 134
Ashworth, J. N., (263) 117
Ashworth, U. S., (206) 95
Askonas, B. A., (239) 110
Asmundson, V. S., (298) 129, 130
Astrup, T., (541) 213
Atkinson, D. E. (33) 15
Audrain, L., (554) 224; (555) 224, 225
Axelrod, A. E., (119) 42
Azari, P. R., (72) 30, 32, 35–37, 40, 43, 146, 175, 177; (77) 30, 47, 59, 65, 67, 83, 88–90; (334) 146; (365) 155; (367) 157, 160, 163; (374) 160, 162; (441) 187, 188

Back, J. F., (112) 41; (165) 56
Bailey, K., (285) 123
Bain, J. A., (187) 70; (332) 146
Baker, C. M. A., (140) 47, 48, 75; (189) 75, 170; (190) 75, 80; (191) 75
Baker, E., (230) 105, 107, 148–150, 165, 166, 168
Balls, A. K., (515) 203, 206, 213; (530) 207
Bandemer, S. L., (92) 31
Bannister, G. L., (253) 115
Barry, J. M., (235) 108, 109
Bartelt, D. C., (499) 202, 205, 215
Bartulovich, J. J., (88) 30, 39, 202, 209, 210
Basch, J. J., (213) 96, 97; (229) 107, 212; (339) 146
Bauernfeind, J. C., (131) 44
Baugh, R. F., (334) 146

Bayne, H. G., (154) 51
Bean, R. C., (511) 203, 207, 228
Bean, R. S., (66) 27, 46
Bearn, A. G., (308) 134; (309) 133, 170
Beeby, R., (219) 100
Bell, K., (245) 110
Bennett, N. S., (57) 26, 30, 36, 37, 39, 44, 45, 66; (77) 30, 47, 59, 65, 67, 83, 88–90; (81) 30, 47, 83, 86, 200, 202, 210, 218, 221, 225; (156) 52, 62, 66
Berger, L. R., (398) 175
Berlow, S., (495) 202, 205
Bernier, I., (401) 176, 180; (449) 188, 190
Bezkorovainy, A., (348) 148, 149, 161; (370) 157
Bier, M., (493) 202; (557) 225, 226
Bigler, J. C., (170) 57, 62, 66; (193) 83–86, 202, 209, 218; (196a) 85, 202, 203; (532) 209, 231, 242
Billups, C., (381) 168
Birk, Y., (507) 203; (528) 204; (568) 234
Biserte, G., (345) 147
Black, L. W., (409) 178, 181, 185
Blake, C. C. F., (400) 176, 188, 190–192; (402) 176, 182; (450) 190
Blanc, B., (228) 107; (338) 146
Blatt, W. F., (539) 212
Blombäck, B., (315) 135
Blood, A. F., (542) 213
Boas, M. A., (50) 25; (51) 25
Bock, W. J., (200) 91
Boettcher, E. W., (330) 146
Boucher, R. V., (132) 46, 142
Boussier, G., (301) 130; (301) 130
Bowman, D. E., (487) 200, 204
Boyer, P. D., (109) 40
Bradshaw, R. A., (444) 187
Braend, M., (313) 134
Brahinsky, R. A., (519) 201
Brew, K., (224) 101, 105, 196; (225) 103, 106
Briggs, D. R., (80) 149, 155
Briggs, M. H., (177) 67
Brodbeck, U., (223) 101
Brooks, L. E., (300) 132
Brown, D. M., (412) 177, 178, 181, 183, 195
Brown, J. R., (419) 180; (500) 202, 210
Brumfitt, W., (474) 193
Brunner, J. R., (218) 98
Bryson, V., (4) 2

Buchert, A. R., (456) 188
Buettner-Janusch, J., (311) 134
Bujard, E., (228) 107
Bundy, H. F., (501) 202, 210, 211
Burck, P. J., (498) 202
Buss, E. G., (132) 46, 142
Butler, L. G., (451) 190; (455) 191
Buttkus, H., (336) 146, 157

Campbell, A., (155) 52
Campbell, B., (252) 115; (254) 115; (255) 115
Campbell, G. F., (45) 25
Campbell, P. N., (239) 110
Caner, F., (135) 47
Canfield, R. E., (86) 30, 39, 87, 89, 176, 180, 181, 183; (148) 49, 173; (176) 63, 87–89, 178, 180, 183, 196; (416) 177, 180; (420) 181; (481) 197; (482) 197
Cann, J. R., (357) 152, 153
Cannan, R. K., (26) 14, 25, 31, 40, 152, 208
Caputo, A., (442) 187
Carey, N. H., (151) 50
Caroline, L., (54) 26, 168
Castiglioni, B., (135) 47
Ceppellini, R., (231) 105
Cerwinsky, E. W., (498) 202
Chamberlain, A. G., (576), 241
Chandan, R. C., (248) 113, 114, 178
Charkasov, I. A., (417) 177
Charlemagne, D., (472) 193
Charlwood, P. A., (349) 148, 149
Chauvet, J., (523) 201
Cheek, E., (258) 116
Chipman, D. M., (453) 191, 192
Chow, R. B., (496) 202, 204, 232
Chryssanthou, C., (575) 241
Churchich, J. E., (460) 188
Clagett, C. O., (105) 39; (132) 46, 142
Clark, J. R., (59) 27, 31, 33, 36, 46–48, 53, 56; (78) 30, 36, 65, 74, 83, 146, 170; (164) 53, 57; (192) 83, 146, 158, 160; (336) 146, 157
Clary, J. J., (59) 27, 31–33, 46, 47, 68; (88) 30, 39, 202, 209, 210; (164) 53, 57
Clegg, J. B., (285) 123
Cochrane, D., (146) 48
Cohn, E. J., (263) 117; (327) 146
Cohn, M., (180) 68, 69; (186) 68; (289) 123, 124, 137, 138, 152

## AUTHOR INDEX

Cole, A. G., (46) 25, 33, 170
Cole, R. K., (138) 47, 66
Coletti-Previero, M. A., (458) 188; (459) 188
Colvin, J. R., (28) 14
Conchie, J., (63) 27
Connell, G. E., (304) 130; (305) 130
Conover, M. J., (414) 178, 179, 193
Conrad, R. M., (70) 28
Cook, W. H., (28) 14
Coombs, R. R. A., (258) 116
Cooper, G., (271) 118, 122
Coryell, F. C., (328) 146
Cotterill, O. J., (167) 56
Couturier, M., (480) 196
Crawford, L. V., (261) 116
Creeth, J. M., (110) 41
Cresson, E. L., (120) 42
Criddle, R. S., (34) 15
Cunningham, L., (115) 41

Dahlquist, F. W., (454) 190
Dambly, C., (480) 196
Dautrevaux, M., (345), 147
Davis, J. G., (137) 47; (547) 215
Day, E. D., (288) 123
Day, R. A., (470) 188; (471) 188
Dayhoff, M. O., (8) 2, 3, 138
Denton, W. L., (223) 101
Descamps, J., (345) 147
Dessauer, H. C., (314), 135
Deutsch, H. F., (83), 30, 202; (133) 46; (173) 58, 70, 76, 170; (180) 68, 69; (186) 68; (187) 70; (282) 123; (307) 132, 133; (332) 146; (494) 202
DeVries, A. L., (42) 20; (292) 128, 140, 170; (324a) 141
DiCarlo, J. J., (499) 202, 205, 215
Dickman, S. R., (463) 188
Dittmer, K., (118) 42
Dixon, G. H., (304) 130; (305) 130; (355) 152
Dlouhá, V., (524) 201; (569) 235, 237
Donovan, J. W., (104) 39; (137) 47; (547) 215
Doolittle, R. F., (315) 135; (316) 135; (317) 135, 137
Dove, W. F., (478) 196
Dray, S., (231) 105
Dreyer, W. J., (426) 185
Ducay, E. D., (89), 30, 42; (149) 49, 165;
(168) 56; (397) 175, 185, 187; (418) 179, 188; (511) 203, 207, 228
Duke, J. A., (493) 202
Durieux, J., (143) 48, 170
Dworetzky, M., (260) 116

Eakin, R. E., (52) 25, 41, 42; (128) 43
Ebner, K. E., (223) 101
Eck, R. V., (8) 2, 3, 138
Edelhoch, H., (84) 30, 225
Edelman, G. M., (231) 105; (321) 138; (322) 138; (323) 138
Edwards, D. L., (59) 27, 31, 33, 36, 46–48, 53, 56
Eichholz, A., (44) 25
Elias, J. J., (240) 110
Emrich, J., (476) 196
Engle, R. L., (272) 118, 125, 126
Epstein, C. J., (440) 187
Evans, R. J., (92) 31
Eyring, H., (29) 14

Fackee, D. S., (497) 202, 204, 205
Fahey, J., (231) 105
Farr, R. S., (259) 116
Fasman, G. D., (452), 190, 193
Feeney, R. E., (21) 13, 20, 149–152; (40) 20, 23, 84, 145, 157; (41) 20; (43) 22, 208, 226; (57) 26, 30, 36, 37, 39, 44, 45, 66; (59) 27, 31, 33, 36, 46–48, 53, 56; (60) 27, 30, 46, 68, 71, 72, 74; (61) 27, 30, 31, 36, 39, 46, 75, 85, 87; (68) 28, 33; (69) 28, 29, 53, 57, 59, 66, 148; (71) 30, 39, 40, 56; (72) 30, 32, 35–37, 40, 43, 146, 175, 177; (77) 30, 47, 59, 65, 67, 83, 88–90; (78) 30, 36, 65, 74, 83, 146, 170; (81) 30, 47, 83, 86, 200, 202, 210, 218, 221, 225; (82) 30, 39, 63, 74, 79, 80, 83–89, 91, 170, 193, 194, 196, 209; (93) 36, 46 63, 68, 83, 208, 210, 213; (95) 36, 39, 83, 228, 231; (96) 36, 40, 63, 89, 101; (97) 36, 157, 158, 167, 208; (98) 36, 232; (99) 36, 47, 83, 90, 214; (134) 47, 62, 63, 65, 73, 79, 83, 89, 90; (155) 52; (156) 52, 62, 66; (160) 53–56; (161) 53; (162) 53; (163) 53, 55, 59, 65, 169; (164) 53, 57; (168) 56; (169) 57, 59, 61, 62; (170) 57, 62, 66; (192) 83, 146, 158, 160; (193) 83–86, 202, 209, 218; (194) 84, 105, 203, 212, 228–231; (195) 84; (196) 84, 85, 217, 230, 231; (196a) 85, 202,

203; (197) 86; (291) 127, 136; (292) 128, 140, 170; (336) 146, 157; (342) 147, 156, 157; (346) 147; (367) 157, 160, 163; (374) 160, 162; (383) 168, 169; (384) 168; (418) 179, 188; (513) 203, 206, 207, 229; (532) 209, 231, 242; (534) 208; (553) 223, 224, 228; (558) 227, 232, (560) 232, 238; (562) 232; (570) 235; (571) 235; (572) 235, 263
Feinstein, G., (195) 84; (196) 84, 85, 217, 230, 231; (558) 227, 232; (570) 235; (572) 235, 236
Ferguson, K. A., (312) 134
Fernandez-Diez, M. J., (96) 36, 40, 63, 89, 101
Ferry, J. D., (283) 123
Fevold, H. L., (48) 25, 175; (55) 26; (85) 30; (106) 38, 40
Filho, J. X., (515) 203, 206, 213
Finkenstadt, W. R., (564) 233
Fishman, W. H., (290) 125, 140
Flavin, M., (147) 49
Fleischman, J. B., (318) 137
Fleming, A., (49) 25, 172, 174
Flodin, P., (266) 117
Florkin, M., (5) 2, 10, 12, 123
Ford, J. D., (115) 41
Forsythe, R. H., (100) 36; (101) 36, 146
Fossum, K., (64) 27, 202, 210
Foster, J. F., (30) 14; (31) 15; (76) 30; (100) 36; (101) 36, 146; (274) 121, 152; (276) 121; (356) 152
Foster, R. J., (573) 235
Fothergill, J. E., (183) 68
Fox, P. F., (222) 100
Fox, W., (314) 135
Fraenkel-Conrat, H., (89) 30, 42; (102) 39, 42; (342) 147, 156, 157; (396) 175, 185, 188; (397) 175, 185, 187; (511) 203, 207, 228
Franek, F., (231) 105
Frank, B. H., (574) 239
Franklin, E., (231) 105
Franklin, E. C., (269) 120
Fraps, R. M., (117) 67
Frattali, V., (510) 203, 205, 206; (529) 206

Fredericq, E., (494) 202
Fritz, H. I., (153), 51
Fuller, R. A., (80) 30, 149, 155
Funatsu, G., (438) 187
Funatsu, M., (436) 187, 188; (438) 187; (447) 189, 190; (467) 188

Gaffield, W., (373) 158
Gagen, W. L., (75) 30
Galiazzo, G., (457) 188
Gallop, P. M., (15) 4
Garibaldi, J. A., (154) 51
Gates, V., (448) 192
Gellhorn, A., (394) 175
Gerring, M., (455) 191
Gerrits, R. J., (246) 110
Gertler, A., (507) 203; (568) 234
Geschwind, I. I., (462) 188
Ghuysen, J. M., (399) 176
Gibbs, R. J., (493) 202
Giblett, E. R., (300) 132; (306) 131
Gillespie, D. C., (238) 110
Gilmour, D. G., (388) 171
Glasnak, V., (244) 114
Glazer, A. N., (16) 7; (375) 161, 162, 165; (413) 177, 178, 181, 195
Godfrine, P., (466) 188
Goldberger, R. F., (440) 187
Good, R. A., (325) 142, 143
Goodman, H. C., (231) 105
Gordon, W. G., (213) 96, 97; (339) 146
Gorini, L., (554) 224; (555) 224, 225
Gottschalk, A., (12) 4
Grabar, P., (231) 105
Graham, E. R. B., (12) 4
Grau, C. R., (153) 51
Gray, J. L., (539) 212
Green, N. M., (90) 30, 39, 43; (91) 30, 36, 39, 43; (121) 42; (123) 43; (124) 43; (127) 43; (411) 178, 183, 197; (543) 213; (552) 218, 220, 222
Greenberg, R., (214) 98
Greene, F., (21) 13, 20, 149–152; (534) 208
Greene, L. J., (497) 202, 204, 205; (499) 202, 205, 215; (567) 234, 237

# AUTHOR INDEX

Grinnan, E. L., (498) 202
Grisaro, V., (453) 191, 192
Grogan, F. T., (261) 116
Grohlich, D., (348) 148, 149, 161; (370) 157
Grossman, M. I., (549) 215; (550) 215
Groves, M. L., (229) 107, 212; (330) 146; (344) 147, 149
Gugler, E., (236) 108
Gunther, M., (258) 116
Gurvich, A. E., (231) 105
Gutfreund, K., (283) 123
Gyorgy, P., (116) 42; (122) 43

Habeeb, A. F. S. A., (101a) 39
Haber, E., (14) 4
Hachimori, Y., (445) 189
Hagerty, G., (247) 110, 116, 203, 212, 240
Hahnel, E., (391) 175; (392) 175
Haley, L. E., (299) 130, 131
Ham, W. E., (486) 200, 204
Hamaguchi, K., (467) 188
Hansen, R. R., (220) 100
Hanson, H. L., (155) 52
Hanson, L. A., (237) 109
Harrington, W. F., (275) 121, 152
Hartdegen, F. J., (437) 187, 188, 190; (439) 187; (455) 191
Hartwig, Q. L., (314) 135
Hathaway, G., (34) 15
Hawiger, J., (415) 178, 179, 194, 195
Hawthorne, J. R., (166) 56
Hayashi, K., (436) 187, 188; (438) 187; (447) 189, 190; (467) 188
Haynes, R., (194) 84, 105, 203, 212, 228–231; (196) 84, 85, 217, 230, 231; (513) 203, 206, 207, 229; (560) 232, 238; (571) 235
Hektoen, L., (46) 25, 33, 170
Helinski, D. R., (10) 2, 3, 4, 6, 17
Hellhammer, D., (145) 48
Hendler, R. W., (150) 50
Heremans, J. F., (231) 105
Hertz, R., (117) 42; (178) 67
Herzig, D. J., (471) 188
Hiederost, C., (329) 146

Hill, R. D., (220) 100
Hill, R. L., (224) 101, 105, 196; (225) 103, 106; (444) 187
Hipple, P. H. von (207) 94
Hirschmann, D. J., (102) 39, 42
Hofmann, K., (118) 42; (119) 42
Högl, O., (145) 48
Holcomb, D. N., (87) 30, 181
Holde, K. E. van (87) 30, 181; (422) 180
Holly, R. G., (248) 113, 114, 178
Holmberg, C. G., (278) 122
Hoover, S. R., (159) 53
Horinishi, H., (445) 189
Hörnle, S., (525) 201; (526) 204
Horsfall, W. J., (386) 169
Howard, J. B., (413) 177, 178, 181, 195
Hughes, W. L., Jr., (263) 117
Huisman, O. C., (531) 207, 242, 243
Humphrey, J. H., (239) 110
Hunt, J. A., (2) 1

Ikenaka, T., (509) 203, 206
Imai, K., (390) 173, 189; (461) 188
Imoto, T., (436) 187, 188; (438) 187; (467) 188
Ingelman, B., (350) 149
Ingram, V. M., (2) 1
Inman, J. K., (328) 146; (366) 156; (376) 166
Inouye, M., (407) 178, 181–183, 196; (425) 184; (476) 196; (477) 196
Isemura, T., (390) 173, 189; (461) 188
Isliker, H., (231) 105; (338) 146
Itano, H. A., (1) 1; (37) 17; (38) 17

Jacobsson, K., (537) 210
Jamieson, G. A., (343) 147, 148
Jandl, J. H., (376) 166
Jao, L., (454), 190
Jauregui-Adell, J., (401) 176, 180; (443) 187; (465) 188
Jayle, M. F., (277) 122; (301) 130; (302) 130
Jenness, R., (206) 95; (243) 111–113; (251) 114
Jennings, R. K., (182) 68, 90

Jensen, H., (538) 212; (539) 212
Jeppsson, J. O., (353) 148; (354) 149
Johanson, B., (337) 146
Johnsgard, P. A., (139) 47
Johnson, A. H., (204) 93
Johnson, L. N., (400) 176, 188, 190–192; (403) 176
Johnson, P., (275) 121, 152; (389) 171
Johnston, J. O., (212) 95
Jollés, J., (199) 87, 89, 177, 189; (401) 176, 180; (423) 183; (443) 187; (469) 188
Jollés, P., (198) 87, 177, 178, 183, 188; (199) 87, 89, 177, 189; (401) 176, 180; (423) 183; (443) 187; (449) 188, 190; (458) 188; (459) 188; (465) 188; (469) 188; (472) 193
Jones, G., (512) 203, 206, 207
Jones, H. D. C., (369) 157
Jones, L. G., (380) 168
Jones, J. R., (161) 53
Jones, P. D., (177) 67
Jones, R. E., (576) 241
Jori, G., (457) 188
Juergens, W. G., (240) 110

Kalan, E. B., (214) 98; (246) 110
Kalser, M. H., (550) 215
Kaminski, M., (143) 48, 170; (144) 48; (181) 68
Kanamori, M., (136) 47
Kanarek, L., (444) 187
Kaplan, M. A., (182) 68, 90
Kaplan, N. O., (23) 13, 14; (39) 17, 18; (201) 91
Karlsson, B. W., (232) 105
Karush, F., (231) 105
Kasper, C. B., (282) 123; (307) 132, 133
Kassell, B., (247) 110, 116, 203, 212, 240; (495) 202, 205; (496) 202, 204, 232; (521) 201; (563) 232
Katz, S., (283) 123
Katz, W., (408) 178, 181, 185
Kaverzneva, E. D., (434) 186; (446) 188, 189
Kawabata, M., (136) 47

Kazal, L. A., (519) 201
Keil, B., (569) 235, 237
Keller, P. J., (548) 215
Kelly, A. L., (226) 103
Kelly, G. W., Jr. (545) 213
Kelly, V. J., (250) 113
Kenyon, A. L., (243) 111–113
Ketterer, B., (65) 27, 30, 47
Khalef, S., (507) 203; (568) 234
Kiddy, C. A., (210) 95; (212) 95
Killander, J., (266) 117
Kimmel, J. R., (412) 177, 178, 181, 183, 195
King, T. P., (293) 127
Kistler, P., (329) 146; (330) 146
Kleopina, G. V., (446) 188, 189
Kline, L., (66) 27, 46; (158) 52, 56
Knight, C. A., (68) 28, 33
Koch, G., (426) 185
Koechlin, B. A., (326) 146
Koenig, D. F., (402) 176, 182
Kok, A., (217) 98
Komatsu, S. K., (40) 20, 23, 84, 145, 157; (41) 20; (97) 36, 157, 158, 167, 208; (292) 128, 140, 170
Koning, P. de (217) 98
Korenman, S. G., (125) 43
Kornfeld, S., (379) 166
Kornguth, S. E., (551) 217, 219
Kraeling, R. R., (246) 110
Kravchenko, N. A., (417) 177; (434) 186; (435) 186; (446) 188, 189
Kropf, R. B., (463) 188
Kunitz, M., (485) 200, 201, 214; (488) 200, 204
Kunkel, H. G., (269) 120
Kurihara, K., (445) 189

Lack, D., (172) 58
Landsteiner, K., (179) 68; (536) 210
Larson, B. L., (238) 110
Lascelles, A. K., (257) 115
Laskowski, M., (233) 105, 240; (247) 110, 116, 203, 212, 240; (489) 201, 224; (490) 201; (495) 202, 205; (530) 202, 212; (504) 202, 212; (521) 201; (540) 212; (544) 213; (556) 224; (559) 227

# AUTHOR INDEX

Laskowski, M., Jr., (233) 105, 240; (489) 201, 224; (504) 202, 212; (540) 212; (564) 233; (565) 231, 234; (566) 234
Laurell, C. B., (278) 122; (340) 147, 166; (350) 149; (377) 166
Lebowitz, J., (559) 227
Ledoux, L., (468) 188
Lee, Y. C., (114) 41
Lehman, W. L., (393) 175; (394) 175
Lehrer, S. S., (452) 190, 193
Leibman, A. J., (331) 146, 149, 153; (363) 155, 156, 159; (364) 155
Leithoff, H., (387) 170
Leithoff, I., (387) 170
Lenk, H. P., (430) 186
Lennox, E. S., (289) 123, 124, 137, 138, 152
Leonis, J., (466) 188; (468) 188
Levine, L., (201) 91
Lewis, J. C., (102) 39, 42
Li, C. H., (462) 188
Liener, I. E., (546) 213
Lillevik, H. A., (206) 95; (218) 98
Lind, S. B., (134) 47, 62, 63, 65, 73, 79, 83, 89, 90
Line, W. F., (370) 157
Lineweaver, H., (53) 26, 86, 200; (66) 27, 46; (160) 53–56
Liu, A. K., (420) 181
Liu, W. H., (196) 84, 85, 217, 230, 231
Lockwood, D. H., (241) 110; (242) 110
Longsworth, L. G., (26) 14, 25, 31, 40, 152, 208; (179) 68
Lorenze, F. W., (67) 28
Lumry, R., (29) 14
Lush, I. E., (58) 27, 31, 32, 33, 46, 47, 68 (63) 27; (141) 48, 49; (142) 48, 75

Maass, D., (427) 185
McCabe, R. A., (173) 58, 70, 76, 170
McCall, K. B., (328) 146
McCormick, D. B., (130) 43
McDermid, E. M., (388) 171
MacDonnell, L. R., (68) 28, 33; (71) 30, 39, 40, 56; (155) 52; (160) 53–56; (162) 53; (168) 56; (418) 179, 188

McIndoe, W. M., (297) 129
MacInnes, D. A., (26) 14, 25, 31, 40, 152, 208
McKenzie, H. A., (74) 30; (205) 93–95, 98, 100, 102, 105, 108, 116, 145; (245) 110; (375) 161, 162, 165
Mackinlay, A. G., (216) 98, 100
McMeekin, T. L., (234) 108
McMurry, S., (176) 63, 87–89, 178, 180, 183, 196
Maeda, H., (99) 36, 47, 83, 90, 214
Maeda, Y., (390) 173, 189
Maher, J., (456) 188
Mair, G. A., (400) 176, 188, 190–192; (402) 176, 182; (450) 190
Makey, D. G., (347) 149, 152
Maksimov, V. I., (434) 186
Malmstrom, B. G., (361) 154, 158; (371) 158, 160–162
Mandeles, S., (94) 36, 146; (149) 49, 165
Mann, R. K., (508) 203
Manwell, C., (140) 47, 48, 75; (424) 182, 197
Marchalonis, J., (321) 138; (322) 138; (323) 138
Margoliash, E., (17) 7; (19) 7; (483) 197
Markert, C. L., (22) 13
Mars, P. H., (505) 203
Marsh, C. L., (545) 213
Marshall, R. D., (113) 41
Martin, C. J., (504) 202, 212
Matida, A. K., (520) 205, 213
Matsushima, K., (56) 26, 208; (535) 208
Matthews, R. H., (258) 116
Mauerhofer, M., (329) 146
Mauron, J., (228) 107
Means, G. E., (196a) 85, 202, 203; (346) 147
Mecham, D. K., (397) 175, 185, 187
Meehan, J. J., (158) 52, 56
Mehl, J. W., (501) 202, 210, 211
Melamed, M. D., (91) 30, 36, 39, 43; (533) 208
Melin, M., (263) 117
Mellander, O., (203) 92
Meloun, B., (524) 201

Melville, D. B., (118) 42
Mendel, L. B., (542) 213
Meyer, K., (391) 175; (392) 175; (393) 175; (394) 175; (484) 198
Michaud, R. L., (368) 157
Miller, H. T., (41) 20; (60) 27, 30, 46, 68, 71, 72, 74;.(61) 27, 30, 31, 36, 39, 46, 75, 85, 87; (134) 47, 62, 63, 65, 73, 79, 83, 89, 90; (170) 57, 62, 66; (197) 86; (292) 128, 140, 170
Mitchell, C. A., (253) 115
Mohammed, A., (397) 175, 185, 187
Moll, F. C., (500) 202, 210
Moller, F., (22) 13
Monsigny, M., (227) 107; (345) 147
Montgomery, R., (114) 41
Montreuil, J., (135) 47; (227) 43; (335) 146; (345) 147; (351) 149
Moore, G. L., (470) 188
Moore, S., (512) 203, 206, 207
Morawitz, P., (286) 123
Moretti, J., (302) 130
Morgan, E. H., (230) 105, 107, 148–150, 165, 166, 168; (377) 166
Morris, H. J., (66) 27, 46
Morton, J. I., (83) 30, 202
Morton, J. R., (388) 171
Mross, G. A., (317) 135, 137
Mulford, D. J., (263) 117
Mullet, S., (335) 146; (351) 149
Muralt, G. von, (236) 108
Murray, C. W., (53) 26, 86, 200

Nagy, D. A., (384) 168
Nakamura, R., (157) 52
Ness, A. T., (270) 121
Neuberger, A., (113) 41
Neuman, H., (62) 27
Neurath, H., (20), 13, 18, 19, 149
Newton, J., (476) 196
Nickerson, T. A., (249) 113
Nitschmann, H., (329) 146; (330) 146
Nolan, C., (483) 197
Nord, F. F., (493) 202; (557) 225, 226
Norris, L. C., (131) 44
North, A. C. T., (400) 176, 188, 190–192; (402) 176, 182; (450) 190
Northrop, J. H., (485) 200, 201, 214
Nouvel, G., (523) 201

Ogden, A. L., (388) 171
Oguri, S., (260) 116
Okada, Y., (476) 196; (477) 196
Olcott, H. S., (511) 203, 207, 228
O'Malley, B. W., (125) 43; (152) 50
Osaki, S., (279) 122
Osborne, T. B., (45),25
Osuga, D. T., (43) 22, 208, 226; (78) 30, 36, 65, 74, 83, 146, 170; (82) 30, 39, 63, 74, 79, 80, 83–89, 91, 170, 193, 194, 196, 209; (93) 36, 46, 63, 68, 83, 208, 210, 213; (96) 36, 40, 63, 89, 101; (99) 36, 47, 83, 90, 214; (134) 47, 62, 63, 65, 73, 79, 83, 89, 90; (170) 57, 62, 66; (194) 84, 105, 203, 212, 228–231; (196) 84, 85, 217, 230 231; (292) 128, 140, 170; (532) 209, 231, 242; (570) 235
Ottewill, R. H., (275) 121, 152
Ozawa, K., (565) 231, 234

Pape, L., (381) 168
Papermaster, B. W., (325) 142, 143
Park, L. U., (137) 47
Parker, W. C., (308) 134; (309) 133, 170
Parry, R. M., Jr., (421) 181
Partridge, S. M., (13) 4
Pastewka, J. V., (270) 121
Patel, C. V., (222) 100
Patton, S., (251) 114
Pauling, L., (1) 1; (3) 1
Peacock, A. C., (270) 121
Peanasky, R. J., (495) 202, 205; (544) 213
Pecoraro, R., (455) 191
Pederson, D. M., (356) 152
Pepper, L., (210) 95
Perkins, D. J.. (369) 157
Perkins, H. R., (410) 178, 183; (473) 193
Perlmann, G. E., (27) 14; (108) 40
Perrie, W. T., (183) 68
Perry, G. C., (576) 241

Pesce, A., (201) 91
Peters, S. M., (153) 51
Petersen, W. E., (252) 115; (254) 115; (255) 115; (256) 115
Peterson, E. A., (264) 117; (265) 117
Phelps, R. A., (357) 152, 153
Phillips, D. C., (400) 176, 188, 190–192; (402) 176, 182; (403) 176; (404) 176, 190; (405) 176, 193; (450) 190
Phillips, R. E., (70) 28
Pierce, J. E., (310) 133
Pilgrim, F. J., (119) 42
Pilson, M. E. Q., (226) 103
Platou, E., (250) 113
Polonovski, M., (277) 122
Porter, R. R., (319) 137
Pospisilová, D., (524) 201
Poulik, M. D., (280) 122
Press, E., (231) 105
Previero, A., (458) 188; (459) 188
Priest, S. G., (538) 212; (539) 212
Proctor, C. M., (463) 188
Prudden, J. F., (393) 175; (394) 175
Pudles, J., (520) 205, 213
Putnam, F. W., (268) 118, 119, 120, 122, 128, 130

Queval, J., (135) 47
Quinteros, I. R., (298) 129, 130

Rackis, J. J., (508) 203
Radicevic, M., (495) 202, 205
Raftery, M., (454) 190
Rainey, J. M., (115) 41
Ratner, B., (260) 116
Rees, A. W., (471) 188
Regehr, E. A., (243) 111–113
Reich, H. A., (363) 155, 156, 159
Reichlin, M., (201) 91
Rhodes, C. K., (87) 30, 181
Rhodes, M. B., (57) 26, 30, 36, 37, 39, 44, 45, 66; (69) 28, 29, 53, 57, 59, 66, 148; (72) 30, 32, 35–37, 40, 43, 146, 175, 177; (77) 30, 47, 59, 65, 67, 83, 88–90; (81) 30, 47, 83, 86, 200, 202, 210, 218, 221, 225; (156) 52, 64, 66; (161) 53; (163) 53, 55, 59, 65, 169; (168) 56; (545) 213
Richterich, R., (281) 122
Riehm, J. P., (514) 203, 206, 214
Rigbi, M., (497) 202, 204, 205; (567) 234, 237
Roberts, R. C., (347) 149, 152
Robinson, E., (38) 17
Robinson, J. C., (310) 133
Robinson, R., (395) 175
Rola, F. J., (520) 205, 213
Romanoff, A. J., (175) 62
Romanoff, A. L., (175) 62
Rooijen, P. J. van (217) 98
Rose, C. S., (116) 42; (122) 43
Rose, D., (206) 95
Rothberg, R. M., (259) 116
Roulet, D. L. A., (236) 108
Roy, C., (406) 177, 178, 181
Rupley, J. A., (431) 186; (433) 186; (437) 187, 188, 190; (448) 192; (451) 190; (455) 191; (464) 188
Ryan, C. A., (516) 203; (517) 203, 207; (518) 203, 206; (530) 207; (531) 207, 242, 243; (573) 235

Sale, E. E., (538) 212
Saltman, P., (361) 154, 158; (381) 168; (382) 168
Salton, M. R. J., (399) 176; (428) 185; (429) 186
Sandstedt, R. M., (486) 200, 204
Sarma, V. R., (400) 176, 188, 190–192; (402) 176, 182; (450) 190
Sarwar, M., (254) 115
Sasame, H. A., (508) 203
Scandalios, J. G., (324) 141
Schade, A. L., (54) 26, 168; (333) 146; (341) 147; (378) 166; (384) 168
Scheer, J. van der, (179) 68
Schejter, A., (17) 7
Scheraga, H. A., (506) 203, 206
Schneck, A. G., (468) 188
Schubert, D., (316) 135
Schütte, F., (430) 186
Schwartz, S. A., (316) 135
Scoffone, E., (457) 188

Seal, U. S., (347) 149, 152
Sealock, R. W., (566) 234
Sebrell, W. H., (117) 42; (178) 67
Seifter, S., (15) 4
Sela, M., (62) 27
Sgouris, J. T., (328) 146
Shahani, K. M., (208) 97; (248) 113, 114, 178
Sharon, N., (432) 186, 187; (453) 191, 192
Shaw, C. R., (25) 13
Shaw, D. C., (230) 105, 107, 148–150, 165, 166, 168
Shibata, K., (445) 189
Shockman, G. D., (414) 178, 179, 193; (475) 194
Shulman, N. R., (502) 202, 210, 213
Shulman, S., (283) 123; (284) 123
Sibley, C. G., (18) 7, 82; (139) 47; (174) 59, 71, 77, 78, 170
Silva, R. B., (71), 30, 39, 40, 56; (162) 53; (168) 56
Simlot, M. M., (98) 36, 232; (553) 223, 224, 228
Simmons, R. L., (376) 166
Simpson, G. G., (202) 91
Singer, S. J., (1) 1; (37) 17
Singer, S. L., (287) 123
Sjöberg, L. B., (562) 232
Sjöquist, J., (354) 149
Skeggs, H. R., (120) 42
Sloan, R. E., (243) 111–113
Smith, A. K., (508) 203
Smith, D. B., (28) 14
Smith, E. L., (16) 7, 10; (19) 7; (412) 177, 178, 181, 183, 195
Smith, M. B., (74) 30; (111) 41; (112) 41; (165) 56
Smith, R. H., (103) 39
Smithies, O., (267) 118, 130; (303) 130; (304) 130; (305) 130; (386) 169
Snell, E. E., (52) 25, 41, 42; (128) 43
Snell, N. S., (89) 30, 42; (102) 39, 42
Soardi, F., (541) 213
Sober, H. A., (264) 117; (265) 117
Sogami, M., (31) 15
Somero, G. N., (42) 20

Sophianopoulos, A. J., (87) 30, 181; (422) 180
Sorensen, M., (47) 25, 33, 146
Sorm, F., (524) 201; (560) 235, 237
Spicer, D. S., (519) 201
Spik, G., (227) 107; (345) 147
Spotorno, G., (199) 87, 89, 177, 189
Sri Ram, J., (557) 225, 226
Stadtman, E. R., (32) 15
Stahmann, M. A., (551) 217, 219
Stebbins, G. L., (11) 2
Stein, W. H., (512) 203, 206, 207
Steinberg, A., (392) 175; (393) 175; (394) 175
Steinberg, D., (36) 16
Steiner, R. F., (84) 30, 225; (510) 203, 205, 206; (561) 232
Stevens, F. C., (16) 7, 10; (93) 36, 46, 63, 68, 83, 208, 210, 213; (95) 36, 39; (98) 36, 232
Stevens, R. W. C., (298) 129, 130
Stewart, R. A., (250) 113
Stockdale, F. E., (240) 110; (242) 110
Stormont, C., (296) 129; (298) 129, 130; (313) 134
Strasdine, G. A., (406) 177, 178, 181
Stratil, A., (188) 75, 81
Streisinger, G., (407) 178, 181–183, 196; (476) 196; (477) 196
Strong, L. E., (263) 117; (326) 146
Studer, M., (329) 146
Sugihara, T. F., (68) 28, 33; (155) 52; (158) 52, 56
Sunden, S. F., (500) 202, 210
Surgenor, D. M., (326) 146
Sussman, A. J., (499) 202, 205, 215
Suzuki, Y., (296) 129
Swaisgood, H. E., (218) 98
Szuchet-Derechin, S., (389) 171

Takagi, T., (390) 173, 189; (461) 188
Takahashi, T., (467) 188
Takeda, H., (406) 177, 178, 181
Tanahashi, N., (223) 101
Tarassuk, N. P., (206) 95; (209) 96; (221) 100; (222) 100; (249) 113

# AUTHOR INDEX

Taylor, G. L., (73) 30
Taylor, H. L., (263) 117
Temperli, A., (281) 122
Tengerdy, C., (365) 155
Tengerdy, R. P., (365) 155
Terminiello, L., (557) 225, 226
Teryaghi, E., (476) 196
Terzaghi, E., (407) 178, 181–183, 196
Thomas, R., (479) 196; (480) 196
Thompson, E. O. P., (412) 177, 178, 181, 183, 195
Thompson, J. S., (414) 178, 179, 194, 195
Thompson, M. P., (206) 95; (210) 95; (212) 95; (213) 95, 97; (214) 98
Timasheff, S. N., (352) 149
Tinoco, I., (352) 149
Tiselius, A., (262) 117
Toennies, G., (475) 194
Tomarelli, R., (122) 43
Tomasi, T. B., Jr., (273) 122
Tomimatsu, Y., (88) 30, 39, 202, 209, 210; (373) 158
Tonnelat, J., (335), 146
Topper, Y. J., (240) 110; (241) 110; (242) 110
Trautman, R., (269) 120
Trautschold, I., (492) 201, 243
Tristram, G. R., (103) 39
Trnka, Z., (231) 105
Tsugita, A., (407) 178, 181, 182, 183, 196; (425) 184; (476) 196; (477) 196
Tuppy, H., (294) 127
Turkington, R. W., (241) 110

Ulmer, D. D., (360) 155; (372) 158

Valenta, M., (188) 75, 81
Vallee, B. L., (360) 155, 158, 164; (372) 158
Vanaman, T. C., (224) 101, 105, 196; (225) 103, 106
Vänngård, T., (361) 154, 158; (371) 158, 160–162
Vaughan, M., (36) 16
Ventura, M. M., (515) 203, 206, 213
Veros, A. J., (574) 239
Vigneaud, V. du (118) 42

Vitello, L., (373) 158
Vogel, H. J., (4) 2
Vogel, R., (492) 201, 243
Voytovich, A. E., (242) 110

Wagner, L. P., (514) 203, 206, 214
Wake, R. G., (74) 30; (216) 98, 100
Walker, N. E., (153) 51
Walker, R. V. L., (253) 115
Wallenius, G., (269) 120
Walsh, K. A., (20) 13, 18, 19, 149
Ward, J. B., (410) 178, 183
Ward, W. H., (55) 26; (85) 30
Warner, R. C., (79) 30, 36, 146, 154, 155; (107) 37; (358) 153, 156; (359) 153, 154; (362) 154
Watson, J. D., (6) 2, 5
Waugh, D. F., (207) 94
Weaver, J. M., (161) 53
Webb, B. H., (204) 93
Weber, I., (79) 30, 36, 146, 154, 155; (358) 153, 156; (359) 153, 154
Wei, R. D., (126) 43; (129) 43
Weidel, W., (408) 178, 181, 185; (427) 185
Weil, L., (456) 188
Weiser, R. S., (398) 175
Wells, I. C., (1) 1
Welty, J. C., (171) 58, 60, 90
Wenzel, M., (430) 186
Werle, E., (492) 201, 243
Westphal, U., (539) 212
Wetter, L. R., (180) 68, 69; (186) 68
Wheby, M. S., (380) 168
Whitaker, D. R., (406) 177, 178, 181
Whitaker, J. R., (64) 27, 202, 210
White, L. M., (104) 39
Wiersema, A. K., (192) 83, 146, 158, 160
Wilcox, F. H., Jr., (138) 47, 66
Williams, J., (35) 16, 49, 50, 142, 147, 148, 165, 166, 170
Williams, R. J., (52) 25, 41, 42; (128) 43
Wilson, A. C., (184) 68, 69; (185) 68–70, 90, 177, 196; (201) 91
Windle, J. J., (192) 83, 146, 158, 160
Winnick, T., (119) 42
Winter, A. R., (167) 56

Winter, W. P., (20) 13, 18, 19, 149; (132) 46, 142
Winzor, D. J., (110) 41
Wishnia, A., (359) 153, 154
Witter, A., (294) 127
Woese, C. R., (7) 2
Wolschag, D. E., (324a) 141
Woods, K. R., (272) 118, 125, 126
Woodworth, R. C., (333) 146; (368) 157
Work, T. S., (239) 110
Wright, L. D., (120) 42; (126) 43; (129) 43
Wu, F. C., (503) 202, 212; (556) 224
Wu, Y.-C. (114) 41

Wu, Y. V., (506) 203, 206
Wyttenbach, A., (329) 146

Yaguchi, M., (209) 96; (221) 100; (249) 113
Yamamoto, M., (509) 203, 206
Yamasaki, N., (447) 189, 190
Yang, J. R., (76) 30
Yanofsky, C., (10) 2, 3, 4, 6, 17

Zahnley, J. C., (547) 215
Zito, R., (442) 187
Zittle, C. A., (210) 95
Zuckerkandl, E., (3) 1

# Subject Index

Albumin, bovine milk and plasma, 142
  chicken serum and egg white, 142
  plasma, amino acid compositions, 128
  in milk, 95, 118
  model, 121
  polymorphisms, 129
  sulfhydryls, 127
Allergy to milk proteins, 116
Analogous proteins, 10, 13
  detection and quantitation, 20
Antarctic fish, 20
Antihemagglutination, 29
Apoprotein, *see* Flavoprotein
Avidin, 41
  assay, 43
  avian egg white, 65
  biological activity of, 25
  history, 25
  and lysozyme, 197
  physical properties, 30
  preparation from egg white, 43
  quantitation, 67
  role of tryptophan, 42
  synthesis, 50
  biotin complex, 25

Bacteriophage, lysozymes, 182
Biotin, binding factors in egg white, 42
Black-eyed pea inhibitor, amino acid composition, 206
Blood clotting, 123
  control of, 240

Blood plasma, composition, human, 122
  protein composition, 117, 118, 122

Casein, 93
  $a_s$-casein, 93, 94, 95
  amino acid composition, 97
  $\beta$-casein, 94, 98
  $\gamma$-casein, 94
  $\kappa$-casein, 94, 98
  origin of, 109
Cassowary, egg, physical structure, 59
Catalase, in chicken egg white, 46
Ceruloplasmin, 131
  history, 122
Chalazae, 27, 59
  formation of, 28
Chinese dried-egg-white, injury-factor, 25
$a$-Chymotrypsin, homologous proteins, 18
  inhibitors of, 202, 203
Colostrum, 108
  colostrum inhibitor, 105, 116
  and antibodies, 240
  human, lysozyme in, 114
  serum proteins of, 108
  trypsin inhibitors, 110
Conalbumin, *see* Ovotransferrin
Cytochrome c, 7

Deterioration of eggs, 52
  chicken, 52
  duck, 57
  glucose-protein reaction, 53

## SUBJECT INDEX

goose, 57
penguin, 57
theories, 56
turkey, 57
Doublet protein, 90
DNA, and protein subunits, 151
in protein synthesis, 2

Eggs, chicken, as a food, 62
deterioration, 53, 59
duck, as a food, 62
food use, species comparison, 61
incubation, 52; *see also* Deterioration of eggs
physical characteristics, species comparisons, 59
Egg yolk, index, 59
Egg white, changes on incubation, 53, 59
chicken, composition of, 27, 29
electrophoresis, 31
index, 59
physical structure, 27
protein composition, 29, 30
species protein composition, 64, 65
thick egg white, 27, 53
thin egg white, 27
Egg white proteins, antihemagglutination activity, 29
avidin, 41
biosynthesis, 49
catalase, 46
chemical composition of chicken, 38, 39
chicken, 24
electrophoretic studies, gel, 70, 72
paper, 70, 71
enzymatic activity, 27
food uses, 52
fractionation of, 31
functions, 50, 51
genetics, 47
genetic globulins, 27, 46
genetic information, 58
history, 24
immunological comparisons, 68
inhibitor of ficin, 27
inhibitor of papain, 27

ion-exchange, 36
lysozyme, 25
molecular filtration, 36
ovalbumin, 25
ovoglycoprotein, 27
ovoinhibitor, 26
ovomacroglobulin, 27
ovomucin, 25
ovomucoid, 25
ovotransferrin, 26
peptidase, 27
physical properties, 30
preparation of, 83
quantitation, 63
species comparison, 63
riboflavin-binding protein, 26
salt fractionation, 33
sialic acid, 28
taxonomic relationships, 90
Elastase, 18
Enzymatic activity, in egg white, 27
in plasma, 124, 140
Enzymes, in blood plasma, 124
in milk, 97, 116

Fibrinogen, 123, 135
properties of, 123
Ficin, protein inhibitor of, 27
Flavoprotein, 26, 44, 142
in avian egg white, 65
and diet, 44
physical properties, 30
quantitation, 66
riboflavin binding, 45
Fungal proteinase, inhibitor of, 214

Gamma globulins, 123
and homologs, 13
in milk, blood and eggs, 142
properties of, 123
Genetic code, 2, 3
Genetic globulins, $A_1$ and $A_2$, 27, 46
Globulins, $a_1$ and $a_2$, in blood plasma, 118

Haptoglobins, 121, 130
phenotypes, 103

# SUBJECT INDEX

Hemagglutination, viral, 28
Hemoglobin, 7, 8, 17
Homologous proteins, 8, 13
  detection and quantitation, 20
  origin, 15

Immunoglobulins, 109, 123, 135
  in colostrum, 105
  in milk, 115
  nomenclature, 123
  and placental structure, 142
  structure, 124
Inhibitors of proteolytic enzymes, amino
    acids of animal, 205
  amino acids of plant, 206
  association, physical properties, 224
  association rates, 220
  association with inactive enzymes, 235
  from avian egg white, 208, 209
    blood, 210
    colostrum, 56
    milk, 212
    potato, 207
    soybean, 204
  basic pancreatic, 201
  biological function, 239
  chemical modification, 216, 227
  comparative biochemistry of, 214
  complex study, 218
  different forms, 214
  enzymatic assay, 218
  enzymatic modification of, 233
  essential amino acid, 228, 231
  fluorescence quenching, 225
  function in egg white, 242
  general properties, 201
  history, 200
  homologous inhibitors, 215
  and human trypsin, 243
  Kallikrein, 201, 204
  Kazal, 201
  Kunitz, 201
  mechanism of action, 217
    theories, 237
  naturally occurring, 199, 202, 203
  pancreatic, 201

  pharmaceutical uses, 242
  sources of, 202, 203, 212
  structure and function, 216
  and synthetic compounds, 217
  temporary inhibition, 224
  and zymogen activation, 239
Ion-exchange, fractionation of egg white
    protein, 36
Isologous compounds, 10
Isozymes, of proteins, 13

Kallikrein, inhibitor of, 204
Kiwi, taxonomic relationships, 90
Kazal inhibitor, 201, 204
Kunitz inhibitor, 201

$\alpha$-Lactalbumin, 94, 95, 101
  biosynthesis, 109
  and lactose synthetase, 103
  and lysozyme, 102
Lactic dehydrogenase, 17, 18
$\beta$-Lactoglobulin, 94, 100
  amino acid composition, 102
  biosynthesis, 109
  comparative biochemistry of, 110
Lactose synthetase, 103
Lactotransferrin, 103
  amino acid composition of, 107
  bovine, 171
  human, 171
  molecular weights, 149
  preparation, 146
  rabbit, 171
  amino acids, 166
Leucine aminopeptidase, blood plasma, 140
Lima bean inhibitors, 207
Lysozyme, 50, 172
  active site studies, 185
  and $\alpha$-lactalbumin, 101
  avian, amino acid compositions, 89
  biosynthesis, 49
  chemical modification, 187, 192
  chemistry, 175
  chicken egg white, 180
    conformation, 180
  comparative biochemistry in milk, 112

crystallization, 175
deterioration of egg white, 57
distribution, 12, 13, 175, 176, 178
in duck eggs, 57
egg white turbidity, 61
enzymatic activities, 193
enzymatic characteristics, 185
evolution, 195
Fleming, Sir Alexander, 172
genetics, 196
history, 25, 172
immunological comparisons, 69
in human colostrum, 114
in human milk, 114
index of dissimilarity, 70
inhibitors, 189
isolation, 177
physical and chemical properties, 30, 179
physical constants, 181
properties of, 87
related proteins, 196
structure, 176
substrates, 176
quantitation, species comparison, 65, 68
transglycoslyation, 186
$\alpha$-Lytic, protease, 18

$\alpha$-Mannosidase, in chicken egg white, 27
Milk, alkaline phosphatase in, 113
  allergy, 116
  biosynthesis of, 108
  bio-utilization of, 114
  crop milk, 108
  enzymes in, 97
  human, lysozyme in, 112
    serum proteins of, 108
  immunoglobulins, 105
  immunology, 115
  protein composition, 93, 95
  serum albumin in, 118, 119
Milk proteins, caseins, 94
  classification, 95
  comparative biochemistry of, 110
  immunoglobulins, 105
  lactotransferrin, 103
  properties of, 94

rennin action on, 115
Molecular filtration, of egg white proteins, 36
Multiple molecular forms, and biological control mechanism, 15
  of proteins, 13, 14
Muramidase, see Lysozyme

$\beta$-N-acetylglucosaminadase, in chicken egg white, 27
N-acetylneuramic acid, see Sialic acid

Ovalbumin, 31, 40
  $A_1$, $A_2$ and $A_3$, 25, 40
  biosynthesis, 49
  carbohydrate, 38, 39, 41
  deterioration of egg white, 56
  electrophoresis, 74
  history, 25
  immunological comparisons, 69
  index of dissimilarity, 69
  phosphate, 40
  physical properties, 30
  properties of, 88
  quantitation, 63
  in salt fractionation, 35
  and serum albumin, 118
  S-ovalbumin, 41
  sulfhydryl, and food, 62
  sulfhydryl groups, 40
  whipping ability, 52
Ovoflavoprotein, see Flavoprotein
Oviduct, in protein synthesis, 50
Ovoglycoprotein, 27, 47
  physical properties, 30
Ovoinhibitor, 26, 65
  amino acid composition, 209
  physical properties, 30
Ovomacroglobulin, 46, 65
  amino acid compositions, 87
  identification, 27
  immunological comparisons, 68
  physical properties, 30
  properties of, 85
Ovomucin, 28, 65
  and chalazae, 28

## SUBJECT INDEX

chemical compositions of, 88
deterioration of egg white, 56
history, 25
in salt fractionation, 35
Ovomucoids, 208
chemical compositions of, 86, 209
history, 25
properties of, 30, 83, 85
quantitation, 63
in salt fractionation, 36
Ovotransferrin, 64
antimicrobial substance, 51
bacterial inhibitory capacities, 26
biosynthesis, 50
carbohydrate, 147
chemical composition of, 84
chemical modification of, 157
chicken, amino acids, 166
comparative, 170
deterioration of egg white, 53
electrophoresis, 74
iron-binding, 26
molecular weights, 149
physical properties, 30
preparation, 146
properties of, 83
quantitation, 66
in salt fractionation, 35
and serum transferrin, 170

Pancreatic inhibitor, 201
Papain, protein inhibitor, 27
Penguin, taxonomic relationships, 91
Peptidase, activity in chicken egg white, 27
Plasma albumin, see Albumin
Plasma proteins, 117
comparative biochemistry, 125
electrophoresis, 117
enzymes, 124
gamma globulins, 142
history, 117
immunoglobulins, 135
leucine aminopeptidase, 140
properties of, 118
transferrins, 132
Potato inhibitor, 207

amino acid composition, 206
biological function, 242
Progesterone, and avidin biosynthesis, 50
Protein structure, 3
and evolution, 7
Proteolytic enzyme, control of activation, 239
Proteolytic enzyme inhibitors, 199

Ratites, taxomic relationships, 90
Red protein, in milk, 103, 107
Rennin, 98
action on milk, 115
Riboflavin, binding by flavoprotein, 44
egg white color, 59
RNA, and protein subunits, 151
in protein synthesis, 2, 3

Schwartzman reaction, inflammation inhibition, 240
Serum protein, in human milk, 108
Sialic acid, in avian egg white, 63
species comparison, 65
in $\kappa$-casein, 100
in duck eggs, 57
in egg deterioration, 57
virus antihemagglutination, 28, 29
Sickle-cell anemia, 17
Siderophilin, see Transferrin
S-ovalbumin, 41
Subtilisin, inhibitors of, 214
Sulfhydryl groups, in avian egg white, 65
in $\kappa$-casein, 98
in chicken ovalbumin, 40, 41
in $\beta$-lactoglobulin, 101
masked, 40
in ovalbumins, 89
in plasma albumin, 127
Soybean inhibitors, 204
amino acid composition, 206

Thrombin, action on fibrinogen, 123
Tinamou, taxonomic relationships, 90
Transferrin, 144
absorption spectra, 153
amino acid compositions, 166

of amphibians, 135
antimicrobial activity, 168
biological function, 166
biosynthesis, 165
in blood, 118
bovine, amino acid composition, 107
carbohydrate, 147
chemical modification, 157
chicken serum, amino acids, 166
chicken ovotransferrin and serum, 142
circular dichroism, 158
cotton effect, 158
determination, 144
distribution, 12
electron spin resonance, 158
iron transport, 166
isoelectric points, 152
metal binding, 144, 155
  pH dependence, 154
  site, 157
metal complex, structure, 157
metal-free properties, 153
molecular weights, 149
nomenclature, 145
physical properties, 148
plasma, 132
preparation, 145
rabbit milk and serum, 142
rabbit serum, 171
  amino acid composition, 107, 166
of reptiles, 135
shape and size, 155
stability, 160
subunits, 148
Trypsin, and colostrum inhibitor, 105
  homologous proteins, 18
  inhibitor distribution, 12
  inhibitors of, 200, 202, 203
  in colostrum, 110, 212
Tryptophan synthetase, 4

Whey proteins, 93

Yolk membrane, deterioration, 28, 53
Yolk sac, 143